FROM DEATH TO IMMORTALITY

Can Science take us there?

Learn the Art of Living Long

From Death to Immortality

Can Science take us there?

Learn the Art of Living Long

Ananda Murti Vemuri, Ph.D.

INFINITY
PUBLISHING.COM

ISBN 978-0-7414-4525-4

Published by:

INFINITY PUBLISHING.COM

1094 New DeHaven Street, Suite 100
West Conshohocken, PA 19428-2713
Info@buybooksontheweb.com
www.buybooksontheweb.com
Toll-free (877) BUY BOOK
Local Phone (610) 941-9999
Fax (610) 941-9959

Printed in the United States of America

Published March 2013

In memory of my mother,
Sītamma Vemūri
And my father,
Someswara Rāo Vemūri
From whom I learned the Value of
Austerity

From the unreal, lead me to the Real
From darkness, lead me to Light
From death, lead me to Immortality

---Brihad-aranyaka Upanishad, (1.3.28)

ACKNOWLEDGMENTS

I am deeply indebted to Dr. K. R. Seshan, formerly senior scientist in the Medical College of Ohio, for his library research assistance, for reading the manuscript of this book, for offering suggestions for improvement, and for his constant encouragement and unstinted cooperation.

I thank the following copyrights holders for permissions to use extracts from their published works. Full citations are furnished at appropriate places, where their contributions are used.

Dr. Richard C. Allsopp, professor, University of Hawaii, Honolulu, HA

Dr. Robert B. Setlow, Professor, Brookhaven National Laboratory, Upton, NY

Dr. Robert G. Jahn, professor, Princeton University, Princeton. NJ

Permissions administrators of:

Science magazine,

Journal of Nutrition,

Journal of Applied Physiology,

Preventive medicine of Elsivier publishing company, and

American Medical Association.

Note on Transliteration and Pronunciation

In this book, many ideas from Sanskrit literature are employed as they bring into bold relief the problems of life and death. Use of some Sanskrit words was inevitable, as they have no proper English equivalents. People who are not familiar with oriental languages may not be able to pronounce them correctly due to the vagaries of English spelling.

To overcome the inconsistencies in pronunciation associated with spelling, diacritical marks are employed following the convention of International Congress of Orientalists held in 1912 in Athens. According to this convention, sixteen marks (One tilde, one accent, four macrons and ten dots) are adequate to represent correctly all the Sanskrit letters. Using this scheme, a pronunciation key with some examples is provided below.

KEY TO PRONOUNCIATION

VOWELS

Letter	Sounds like
a	o, in s*o*n
ā	a, in m*a*ster
i	i, in *i*f
ī	ee, in f*ee*l
u	u in f*u*ll
ū	oo, in b*oo*t
r	somewhat between *r* and *ru*
e	a, in ev*a*de
ai	y, in m*y*
o	o, in *o*ver
au	ow, in n*ow*

CONSONANTS

Letter	Sounds like
k	k, in *k*een
kh	ckh, in blo*ckh*ead
g	gh, in lo*g-h*ut
n	ng (somewhat)
c	ch, in *ch*alk
ch	tch, in ma*tch*
j	j, in judge
jh	dgeh in he*dgeh*og
n	n (somewhat)

CONSONANTS (contd)

th	th, in an*th*ill
d	th, in *th*en
dh	dh, in go*dh*ood
n	*n*, in u*n*der
t	t, in *ti*
n	
th	th in *th*umb
d	th in *th*en
dh	thh, in wi*th-h*old
n	n in *n*ot
p	p in *p*en
ph	ph in *loop-h*ole
b	b in *b*ag
bh	bh in a*bh*or
m	m in *m*other
y	y in *ya*rd
r	r in *r*un
l	l in *l*uck
v	v in a*v*ert
ś	s, in *s*ample
ṣ	sh, in *sh*ow
s	s in *s*un
h	h in *h*ot
m	m in su*m*
h	half h in hu*h*!

CONTENTS

LISTS OF FIGURES AND TABLES

Figures

Tables

Tables in Appendix

PREFACE

What question has been asked more often, what question is nearer to the human heart, what question is more inseparably connected with our existence, what question has led more men to search the world for an answer, than this one, the question of Immortality? While the question is the same, the emphasis has changed from time to time and people to people. While people in one part of the world and at one time struggled and dedicated their lives to find out the truly immortal and to attain it, others have sought to make immortal what is generally considered mortal. It was the theme in the lives of saints and sages, prophets and priests, and it was the topic of discussions in the councils of the learned and the courts of kings. The interest in this theme has not died yet nor will it ever die. Thousands before us, who grappled with this problem left to us whatever solutions they could offer but the search is not over as the search for better solutions continues.

In spite of striking achievements in science and technology, the problems of mortality and aging continue to haunt the human minds. The rich and the poor, the mighty and the commoner, everyone is concerned about these problems. It is impossible to be alive and not be concerned about the processes of living and their end. Clinging to life is instinctive to all creatures. Every living being is concerned about the duration of its life, known as life span. Human beings measure the length of their life span on a time scale and use that knowledge for various purposes. The distressing thing they have learned by this measurement is that this span is short and limited.

We recognize this fact when we see others who existed before our eyes, cease to exist after some time. We also see that some people cease to exist earlier than some others. Thinking men naturally try to understand why they cease to exist after some time. Having recognized that fact, they estimate the length of the average lifetime and maximum lifetime. Merely by observing the longest-lived individual in a species, one recognizes maximum life span. This information has helped the human beings to understand their place and significance in the world, to plan their lives and to pursue their future aspirations.

Normal expected life spans are reflected in the adages like 'The Biblical Span of three-score and ten and the' Vedic benediction '*Satamaanam bhavati'* (May the measure of your life span be hundred!) but they are not the maximum attainable limits. Similar is the greeting 'May you live in interesting times', implying that interesting time is always yonder.

We cannot bring to the study of these problems virgin minds that are untouched by the views of many generations of thinkers who have gone before us. They may influence us directly or indirectly. In the distant past, efforts were made to achieve immortality by performing penance and pleasing gods or by drinking elixirs and potions.

In the recent past, penance has given way to science and scientists got involved in this project seriously. As we enter the new millennium, these efforts have been intensified. In spite of the discouraging warnings of the evolutionary biologists, hectic efforts have been going on in the biology laboratories around the world to capture the culprit that robs us of our lives. Of late, we have begun hearing the cries of 'Eureka' from the boardrooms of some interested corporations. Murmurs of dissent are also heard from the academic halls of universities. This is not an easy subject in which quick and easy fixes can be found. It remains to be seen who will win in this debate, the evolutionary theorists with their precise arguments or the molecular biologists who come up now and then with some encouraging evidence gathered from experiments on lower forms of life.

If the molecular biologists win the debate, what is the scope of the technology emerging from it? If it has a wide scope, what will be the social consequences? The consequences of success in prolonging the life of mankind in general are drastic. However there is hope if we make the quest for longevity a subjective affair instead of viewing it as an object that can be exploited on a commercial scale.

This book is meant for lay readers. It takes them through a conducted tour of the myriad scenes of human efforts to conquer old age and death. Every effort is made to take the lay reader along, by using reader friendly language, by devoting an entire chapter for background theories and explanations, and by providing glossary. Some contextual anecdotes and diversionary material are included to relieve the tension of running into the intricacies of modern biology. I hope they serve also as a didactic tool.

My motivation for writing this book started as a curiosity to examine some surprising claims in the popular press about the possibility of extending human lifespan far beyond the traditional century mark. As I began examining these claims, I was inevitably led into the labyrinths of biology and its branch known as gerontology, which deals with the study of the process of aging. As my study progressed, I came across many conflicting opinions. Though I compromised that it is a sign of endeavors in any emerging field of knowledge, I was dismayed that some people are exploiting the unsettled nature of this field from a personal angle by issuing books with deceptive titles and by selling longevity potions to lure the gullible public susceptible to exploitation. Gerontology is indeed a topic in which the public has a stake and therefore they are entitled to know the reasonableness of the claims and counterclaims about longevity found in the popular press and scientific journals. As the scientific investigations in this area are in a state of disarray while each investigator clings to his own experimental result as gospel truth without a common underlying theory, the laity is left to look upon this field with suspicion. The public should know the potential of this research because they have a stake in it. Its success has far-reaching socio-economic consequences but a gap is left between the scientists and the public. As I began feeling the imminent need to bridge this gap, my curiosity grew into strong motivation to write this book.

One of the tools used by the gerontologists is genetic engineering. No doubt, genetic engineering is very useful but it is a double-edged sword and can be highly disastrous if misused. As human nature is unpredictable, we should be wary about the ugly side of this technology. Seldom has the world focus zeroed in on a technology long before it set foot on the market place. This distinction belongs to biotechnology though the world is not new to biotechnology per se. For hundreds of years humans used microbial enzymes to produce bread, beer, yogurt, wine, vinegar, and cheese; selectively bred plants and animals, through ingenious use of biotechnology methods; all for their ordered survival. What sets apart the modern biotechnology is its portentous grave dimension, its ability to manipulate the heart of biological life, the cell nucleus and its ability to induce radical and irreversible changes in the genetic material sheltered by in the recess of nucleus for thousands of years.

Humans can now destroy old forms of life create new forms of life while sustaining old forms or destroying them by using genetic engineering techniques. Taken to the extremes, this genetic meddling can lead to eugenic racism or it may spawn floral and faunal monsters among which humans may find themselves prisoners of their own making. The only protection the public has against this contingency is the honesty and diligence of the scientists who agreed in the Asilomar conference of 1974 to regulate themselves in safe pursuit of this technology.

Therefore, it is imperative that society should have well informed citizens to participate in deciding how best to use this technology. This book is aimed at informing and educating lay people about such problems, the stakes involved, and possible solutions to those problems. I hope this work enables the lay readers to get a good idea of what is going on in biology in general and gerontology in particular, helps them from falling prey to the longevity peddlers and provides them with the necessary tools for living healthy and long lives.

To achieve this objective, relevant scientific ideas have to be popularized. Scientific works known by the generic label 'Popular Science' have to be lucid expositions of difficult subjects in non-technical language. Science has to be brought to the level of lay people without sacrificing the essence and precision of ideas. Keeping this norm in view, effort is made to avoid technical jargon as much as possible without obscuring the ideas. Where it has become unavoidable due to the complexity of the subject, the jargon is explained either parenthetically or in chapter notes. When certain theories required lengthy explanations, those explanations are relegated to a special chapter. Although this book is meant for the layperson, a little scientific orientation will be helpful.

Biology and Computer science are two subjects that have

been rapidly changing in recent years. In the field of computer science, though today's theories and products become obsolete tomorrow, better ones replace them. Biology is not enjoying such luxury. Today's theories are contradicted tomorrow without presenting a better theory or more convincing experimental result. For the present, it looks like a random walk but the biologists are saying the puzzle is slowly but steadily getting resolved. It is difficult in this scenario to come up with specific and fool proof solutions to problems addressed in the book. The struggle to mellow the travails of old age and to postpone death has not ended and final solutions cannot be expected in the near future. For that reason no book in this field can claim to have the final say at present.

In a book like this that explores the perennial problem of humanity, one must put forward not only the aspirations of contemporary scientists but also give due place to the views of the galaxy of eminent thinkers who grappled with this problem before us, and tried to find solutions. I have therefore not hesitated to use some of their views in spite of the risk of opening myself to the charge of being unscientific, as I believe that science has no monopoly of Truth. Even as I have been trying to popularize the admirable work done by the modern scientists, I could not dismiss all the views of our forefathers as frivolous.

By the nature of the subject I have undertaken, I had to refer now and then to the group of ideas associated with the word 'God', but that does not mean that I support the idea of a personal God who distributes only favors to human beings. I used the word as a shorthand notation to indicate the superhuman agency who orders the complex universe and who dispenses both rewards and retributions to people according as they deserve.

If readers discern some ideas expressed in this work go against the current trend, it is not because I do not appreciate the achievements of science. Far from it, I know it is the most valuable thing we have, however small it is, in an ocean of unknown things in the universe. I am only bothered that we have not been able achieve some of our goals by tenaciously confining ourselves to a scientific paradigm that is necessary but not sufficient. If biologists, philosophers of science, professionals in life sciences, and especially public find some helpful ideas in this book, it will have achieved its purpose.

The world is now beset with the monumental problems of overpopulation and destruction of our ecosystem by the pernicious goal of unlimited economic growth to feed the unlimited wants fostered artificially. I am afraid that technological advances cannot solve these problems in the long run unless the ameliorative measures we plan to take are based on understanding of the biological character of these problems.

Fortuitously, the quest for immortality leads us to the roots of our biological origins, where you find the truth about yourself, that 'Thou art That', the Ultimate Reality.

Writing a book on a controversial and unsettled topic that simmers on the frontier of science is not easy. In order to bring the essence of the current knowledge to the lay readers, the writer has to keep himself abreast of the latest happenings in the field continuously and weigh the claims of various ideas and conflicting results of experiments. I could not have ventured into this effort without the ungrudging help and attention given to me by my friend Dr. K.R.Seshan, formerly a molecular biologist in the Medical College of Ohio, at Toledo. He laid me under debt of gratitude by continuously watching the progress of my writing and encouraging me through the tedium of this work.

I have the pleasure to acknowledge the assistance given by my daughter Kiranmayi, a student of biology, in reading some proofs of the manuscript. As usual, all errors and any shortcomings found in the book are mine alone.

CHAPTER ONE

Introduction

> It must be so, Plato,thou reasonest well
> Else whence this pleasing hope,this fond desire,
> This longing after Immortality?
> Or whence this secret dread and inward horror
> Of falling naught? Why shrinks the soul
> Back on herself, and startles at destruction ?
> 'Tis the divinity that stirs within us;
> 'Tis heaven itself that points out an hereafter
> And intimates eternity to man
> Eternity! thou pleasing, dreadful thought!

---Joseph Addison (Cato, Act V. Sc.1)

The famous essayist, Joseph Addison, who was also the acclaimed author of the tragic play 'Cato', tells us through this soliloquy of the hero of his play, that the dread and aspiration of humanity have a common source. Whence, this dread of death and the aspiration for immortality have come. He commends Plato for his reasoning that both these feelings arise from a divinity within us that transcends our corporeal existence. He asserts that it is the divinity within us that points out to us the life after this.

We have also had daydreamers who reveled in their utopian world of perennial life and happiness with a single body. Dreaming up utopias has always been a favorite amusement of humankind since times immemorial. In the late eighteenth century, William Godwin in England and Marquis de Condorcet in France voiced their visions of utopia in which the world of humans would be free from disease and death. In his 'Political Justice', Godwin expressed the view that he believed a time would come when we would be so full of living that we need not die and the need for marriage would be gradually lost by emphasis on developing intellect. He rhapsodized:

> Other improvements may be expected to keep pace with those of health and longevity. ... Besides this, there will be neither disease, anguish, melancholy, nor resentment. Every man will seek with ineffable ardor the good of all.

In short, all humans would be as angels. Of course, these beautiful bubbles invited pricking and the needle was furnished by the young economist-cleric Thomas Robert Malthus, whose theory was later to become not only one of the corner stones of economics but also a landmark in the development of biology. Similarly, generations of explorers tantalized by the ideas of "fountain of youth" and "conquest of death"made valiant efforts, but were unable to realize their utopian ideals. Now, scientists fascinated by the same ideas have been producing a bewildering array of theories but are unable to arrest the progress of degeneration and decline leading to the inevitable end.

The Greatest Wonder

In the great Sanskrit epic Mahabharata, a story is told how king Yudhishtira, the hero, when asked by Yama, the god of death, to tell what was the most wonderful thing in the world, replied, that it was the persistent belief of mankind in their own deathlessness, in spite of witnessing death everywhere almost every moment of their lives. This is the greatest wonder of life.

THE LACUNA:

In spite of the arguments to the contrary urged in different times by different thinkers, in spite of the inability of reason to penetrate the veil of mystery, which hangs between life and death, yet humans cannot face the idea of their extinction. Behind this struggle lies the irrefutable fact that our minds are unable to conceive of our own extinction. Percolating through generations, this incorrigible optimism has infected the present day biologists, who are under the spell of mechanistic view of the universe.

They argue that our physical bodies can somehow be made permanent by the power of science, but there is a lacuna in this argument. Any explanation claiming to be scientific cannot omit one part of fact in order to explain the rest

The whole universe is an ever-changing mass irrevocably

bound by the law of causation. Nothing bound by the law of causation can be permanent. Yet, an impression is produced in human mind about the existence of some permanent entity. Ordinarily, when I speak of myself, I cannot think of anything else other than my body. When I think of my permanence, I think of myself as a body, but obviously this body, being part of the changing nature cannot be permanent. If so, is there something that is permanent behind the changing universe or is the human belief about permanence merely a delusion? If not a delusion, there should be something free and permanent common to us and to the universe. What is that? Can our present science answer this question? Anyone who asserts the validity of logical conclusions relating to external world should be willing to investigate the internal also with the same tool and be ready to go wherever it leads us unmindful of our hopes and dogmas.

Without bridging this lacuna, how can science deliver us from the throes of old age and death? Can it lead us from death to immortality? Everyone should ponder about that question. Answering that question is the central idea of this work.

The ancient problems of aging and death have remained unsolved and they continue to bewilder us. The glitter of everlasting youth and immortality has not faded since times immemorial. Off and on, since ancient times, individuals have attempted to retain youth, to defy death and extend life span. Instances of success in such attempts have however been limited to anecdotes.

Janus faced world

Like the faces of Janus, life and death are the two inseparable faces of the world. We don't see anywhere in the world one without the other. Generally, life and death are considered antithetical phenomena but there is a viewpoint that considers life and death as a single phenomenon. The two being the different faces of the same phenomenon, they are regarded as different modes of the same consciousness that pervades the space-time continuum. When it is limited by locations of space and time, this consciousness manifests as individual life. Otherwise, the individual vanishes.

In Sanskrit language, the words denoting life and death are

derived from the same root '*Mru*', meaning to kill or destroy. The Sanskrit word '*mrutyu*' signifies mortification. '*Mriti*' means death. 'Murti' signifies embodied matter, and '*martya*' means human being. Sanskrit grammar defines death as, '*maarayati iti mrutyuh*', 'destroys, hence death'. Life and death are equally sacred to the ancient Hindu. Life is seen as the energy component of matter wherein the energy is hidden in potential form. Death is therefore regarded as existence in non-physical state.

It is the concept of continuity of soul and it's reintegration with matter in another form about which science and philosophy seem to differ most. The idea of Hindus that the process of death is an integral part of life is worth serious consideration in this context. It is the basis of their famed doctrine of transmigration of souls, purveyed persistently through their numerous philosophical treatises.

Even then, the desire for prolonging life in the same body, as long as possible was entertained, ostensibly, with the purpose of continuing efforts for attaining salvation, as a short life is not adequate to finish such a great task. This was evidenced by the idea of immortality being mentioned in the Vedas, the earliest literary monuments of humankind. The Vedas describe an elixir called 'Soma' which is said to confer everlasting youth and immortality on the recipients.

Do we know today anything more of life and death than our ancestors knew thousands of years ago? Has science made our knowledge of either life or death radically different from what earlier thinkers recorded? If so, can we justifiably reject all those early beliefs as sheer superstitions? This book helps the reader to ponder these questions.

Normative Aspects of Life Extension

Assuming the availability of such an elixir to any individual who wants it, does the extension of lifespan caused by its consumption lead to increase of his (her) total welfare? Does it lead to the welfare of society in general? Trying to answer these questions will land us in problems involving semantics, ethics and logic. What is meant by welfare of an individual? It is generally understood that pleasure or happiness as perceived by a particular individual is his welfare. Pain is the opposite.

Jeremy Bentham, who popularized the utilitarian philosophy, used the words 'pleasure' and 'pain' as synonyms of good and bad. Therefore, one state of affairs is better than another, if it yields a greater balance of pleasure over pain or, at least a smaller balance of pain over the alternative. The other basic principle of utilitarian philosophy is that each individual follows what he believes to be his own happiness, which has been the bedrock of social sciences. The lacuna in this principle is that it cannot satisfy the criterion of collective happiness. The governors of society cannot imagine the multifarious pleasures of the governed and come up with a suitable governing policy that will result in general happiness.

The ethical part of the utilitarian philosophy says: those desires and actions are good which, in fact, promote general happiness. According to them, this general happiness need not be the intention of an action but only its effect. That is, even if an individual undertakes an action with a purely selfish motive that is desirable if it results in some general happiness. Utilitarian philosophy was built up on this definition of desirability. That is the foundation on which Adam Smith; the priest of capitalist societies, built up his theory of economics.

What is meant by the word 'desirable'? A thing becomes 'desirable' when it **ought** to be desired. Thus "desirable" is a word presupposing ethical judgment; we cannot infer what is "desirable" from what is desired.

As emphasized in *Katha Upanishad, "Sreyas"* is different from *"Preyas." "Sreyas"* is that which yields enduring good, and *"Preyas"* is temporary sensual pleasure.

There are two distinct questions involved in the premises of the utilitarian doctrine. First, does each man really pursue only his own happiness or seek something else? Are there no exceptions? Second, is the general happiness or group welfare the true end of human action? Can it not be something personal?

Are there any objective measures for individual and group welfare? Various measures like daily intake of calories, access to comforts of life, leisure time, possession of good health and even longevity were suggested to indicate individual welfare; but none of them enjoys universal acceptability like the standard measures of physical sciences. A glutton enjoys eating more than anything else. A tyrant revels in the display of his tyranny. A saint has no utility for any material things and wants to be left alone. Even the purpose of a desire varies from a person to person; for example a tyrant wants longevity to derive pleasure by satisfying his ego by exerting power over people for longer time. A saint values longevity as it provides more time for his efforts to do good things and to attain salvation. Amid such diversity of desires and valuations, how can an objective measure of individual utility be chosen? To handle this intractable situation economists have come up with a working rule, though it is not quite satisfactory. Their suggestion is to replace the amorphous utility map of the individual by a proxy called the choice map because choice is observable.

The meaning of this choice map is simply this. Given the opportunity of choice, if an individual chooses situation 'A', over situation 'B', it implies that the individual has higher utility in 'A'. One objection to this approach is that it speaks of anticipated or conjectured choice rather than choice of experienced states. Defining welfare, as *ex-post* situation is not any better, as many untenable situations would arise in that case. For example, can we ask an individual to tell whether he was better off when he lived for 100 years or 150 years! On the other hand you can tell an individual that means of extending life span up to 150 years are now available and ask him to choose among several available alternative life spans with their associated consequences. As neither of these definitions is completely satisfactory, we settle down for *ex-ante* definition and take present tastes as the basis rather than past experiences.

Group Welfare: Defining group or social welfare is more difficult. The difficulty arises because of the heterogeneity of the groups. Can it be said that group welfare is the sum of the welfare of each individual? This can't be made unless we are able to quantify the welfare concept and make interpersonal comparison of satisfactions. Many thinkers tried to get ahead of this impediment by trying to obviate its necessity. Wilfred Pareto is the most famous among them. The celebrated criterion formulated by him and embraced by the economists is this: Suppose a policy action improves the lot of at least one person without making anyone else worse off, then the society's welfare is said to have increased. When the policy action is no longer able to improve the lot of at least one person without making somebody worse off, then that action produces maximum welfare. That state is said to be Pareto-Optimal state. The attentive reader must have already guessed the shortcoming of this definition too, but this definition has been quite handy to the economists as a mental construct from which further thought can proceed.

If not, some paternalistic authority has to make the decision based on it's own ethical judgment that has nothing to do with the individual preferences. That is to say, that the state may issue a fiat stating that it is desirable for everyone to live up to 150 years, and order research to achieve that goal. In the absence of such an authority, groups seldom make unanimous decisions. They may make majority decisions but they do not lead to the welfare of everyone. Majority decisions have become the rule of the day under the presumption that the decision-making bodies are paternalistic and they have nothing but the welfare of the group in their hearts. Few people, in these days, agree with this presumption as honesty of the governing classes has all but disappeared in many lands. That leaves the idea of group welfare as murky as ever.

Study of normative topics like "welfare" differs from the positive subjects in a distinct way. Study of positive subjects involves making some assumptions and testing the results of the model. If the results are verified to be true, the theory is accepted; else, it is abandoned along with its assumptions. That is the procedure adopted in science. In welfare topics, this procedure doesn't work, as the conclusions deduced from the premises are not easily testable. Propositions of welfare theories are logical deductions from a set of definitions and assumptions which may or may not be realistic, and which may or may not be ethical. Once it is done, the theory helps us to deduce conditions under which welfare will increase or decrease. There is a snag in this argument. If the conditions are satisfied and if we can find out whether welfare has increased, there is no problem, but unfortunately, there are no reliable tests to verify the changes in welfare. We have to depend on the validity and reasonableness of the assumptions themselves. Therefore, we have to be very careful in making definitions and formulating our assumptions in normative subjects. Similar difficulties in formulation of hypotheses will be encountered in life sciences.

One way of doing this is to lay down the currently conventional conclusions, find out the assumptions necessary for such conclusions, and test the validity of those assumptions. Another is to treat the matter historically trying to rationalize the tacit acceptance of assumptions in various periods. Following strictly either method is again very difficult, as the first becomes very formal and the second very voluminous.

Therefore, a combination of these methods is used in this book to lay down the assumptions and their validity is examined from time to time. When normative aspects are discussed, they refer until chapter eight to the individual. However, as the supposed success in extending life spans inevitably impinges on the society, the normative aspect is extended to problems of group welfare from chapter eight.

It is extremely difficult to classify any topic involving human beings as being of completely positive or of completely normative in nature. What we can do at best is to identify which areas belong to which. Scientific investigations are positive in nature but the previous advances in science, especially in atomic physics, have taught us that we cannot pursue even science independent of its normative aspects. It is more so in the field of biology, which has made rapid strides in recent times and threatens to metamorphose human society. The desirability aspect, which falls into the normative compartment, is discussed first, and then its practicability. The present methods of achieving longevity, being scientific, the related discussion proceeds on positive and empirical grounds and finally the limitations of empirical methods are pointed out.

Are We Trying To Play The Role Of God?

The great advances in science, if properly understood, reveal not the omnipotence of matter but the superiority of human mind to the world of matter. The law of conservation of matter tells us that matter can neither be created nor destroyed. What the scientist has been doing is exploitation of the created matter into various other forms of matter and energy. No human being has ever created matter or energy. So also, with 'life', which is neither matter nor energy. When some scientists claim that life can be created, what they really mean is, that they can create life from some inorganic chemicals, not out of nothing. However, no one has ever succeeded in creating life even from inorganic matter.

Evolutionary theorists argue that death is a necessary condition for evolutionary change. This idea can be intuitively understood by observing the evolution of a tree from its seed. As the sapling grows into a tree the seed perishes. Gerontologists do not lay any claim for creating life but dream of extending somehow its upper limit indefinitely. In both cases, there is denial of providence or superhuman intelligence. The difference between the two is this. Evolutionists are sure that life started spontaneously and individuals perish, so the lifeline may continue. Gerontologists hope that they can do something to save the individual from perishing. That is the idea behind their struggle to decelerate and stop the march of aging. The question whether such an event is possible at all does not arise in the minds of those who think that life can be created from chemicals.

What is the basis of the optimism of those who think that human life can be extended without limit? Although the oft-repeated maxim of the evolutionary theorists that death is an evolutionary necessity still prevails, the evidence of certain long living animal species and the ageless redwood trees of California stimulate the interest of biologists to investigate whether the maxim of their theory could be set at naught.

Biology has come a long way shedding it's false stereotypes that it was only study of 'natural history' and a science of classification in contrast to physics and chemistry which produced things of value to human beings. With the isolation of DNA, understanding its structure and the invention of electron microscope, biology has come out of its backwoods and started making substantial contributions to society. We are now at the threshold of not only a new century but also a new era in which biology will play a prominent part in human lives for better or worse. Biologists are now poised to make startling contributions and usher in a new era of biological revolution.

Gerontology has now acquired the status of a science, drawing ideas from molecular biology, genetics and medicine. Taking cues from the processes of aging and death, biologists have been investigating whether causes of death could be removed and the processes of aging could be neutralized. Several theories that have been in vogue are discussed briefly and the subsequent research that flowed from the discovery of the structure of DNA and the emergence of Hayflick paradigm is discussed more elaborately. More importantly, earnest endeavor is made to bring the ideas about recent advances in genetics, into the easy reach of the public.

In order to measure the success of life span extension methods, one should understand the patterns of aging and longevity of groups of people. Demography helps us to do that. Demography is used to formulate our basic premises about aging and longevity based on the law of large numbers and to project how the average and maximum life spans would change if our wishes were fulfilled. Actuarial predictions based on demographic trends are some of the most accurate predictions and they are very useful in the study of longevity patterns.

Analysis of Aging

Change is the norm of life. Everything, plants, animals and human beings change just by the drift of time. The changes we observe in objects through passage of time are loosely called aging. Subjective estimations of ages do not quite often correlate with the chronological age. These measures denote functional age as distinguished from the chronological age.

It appears that not only different people age at different rates but also different parts of the body age at different rates. It was observed that life spans shrink drastically for smaller bodied mammals, but there is an anomaly in the argument connecting body size with life span. The life clock in each one of us differs, and it is profoundly affected by the individual's heredity, life style, life experiences, and mental attitudes.

Aging is expressed as a process of inevitable and cumulative loss of functions of the body. One of the major factors of aging is the sometimes gradual and sometimes unpredictable breakdown of the body's defenses and body's reaction to stress. Throughout our lives, we are subjected to a variety of attacks of life-threatening forces, mental stresses, physical traumas, and invasions by bacteria and viruses. As body ages its reserve capacity to withstand stress becomes less and less. From the biological point of view, aging is merely a process of change, a continuous change that takes place throughout our lives. Some parts of body may age faster than other parts, and one's abilities may fade at different rates. The differences in life spans and aging of different species give us some clues. The onset of old age is not sudden; it sneaks in gradually and imperceptibly as years are counted. Some biologists hope to unravel this mystery and claim to be on the verge of success but we have no reason to be elated because of the stupendous problems that will arise in the wake of such success.

Theories of Aging

Several theories like those based on the idea of evolution, those based on molecular biology like damage theory, cellular process theory, program theory, immunology theory, and some other miscellaneous theories have been in vogue. None of them except evolutionary theory has so far been able to come up with a coherent theory to explain the phenomenon of aging. Other theories based on molecular biology have not been able either to identify the exact causes of aging or, if identified, to offer a reliable method of neutralizing the causes. Of late, some of them have become more prominent than others.

Is Aging a Disease?

Aging and diseases are so much mixed up that it is difficult to disentangle and tell what is disease and what is aging. A better understanding of the process can be gained by viewing aging as a two-part process, viz. Primary aging and Secondary aging. Primary aging is dependent on genetic factors while diseases cause secondary aging.

Damage theory appears superficial as it simply says that damage occurring to the body and the genetic material called DNA leads to old age and death but it is an important idea worthy of serious consideration. It tells us to prevent damage to DNA.

Program theory goes deeper into the problem by trying to find out why these damages occur in the first place. It tries to identify the mechanisms, which develop in the body for sometime and then run it down systematically with clockwork precision into old age and death. Why should the body behave as if it had been following some pre-drawn program? After vain search of the hormonal system for some sort of clock analogue running the program, scientists have now settled down on the cells in which the clock controlling the aging program is supposed to be hidden. They have now veered round the view that aging is intrinsic to the body, and that it is a fundamental cellular process.

Cellular Process Theory

For a long time, biologists were under the spell of the misguided theory of Alexis Carrel, which postulated wrongly that all cells are immortal. They were not able to understand why people should grow old and die if their cells are self-renewing and never die. This magic spell continued until Leonard Hayflick shattered this illusion with his landmark experiments and proved that cell division has a limit in vitro and in vivo (in laboratory and in real life). The emergence of Hayflick paradigm set them on the correct path and eventually led to what is called "The Telomere Hypothesis".

Telomeres are the end portions of chromosomes. It was observed that they have a tendency to shorten due to end replication problem of DNA. Statistical investigation revealed that there is association between telomere shortening and aging and created flutter in the scientific community. Then they were quick to realize that somatic cells reach Hayflick limit because telomerase (an enzyme), the cell elixir, is not active in them. From then on, stopping the cellular clock has meant finding suitable methods of bringing telomerase enzyme to act upon the shortening telomeres and stop their shortening. It is however not known how to do it as it still bristles with several problems like discovery of proper quality and quantity of the enzyme and developing precise clinical methods of administering it. Any deviation from the optimum dose may either be useless or encourage cancer as shortening of telomeres, is recognized as Nature's method of protecting the body from cancerous growths.

Even if a perfect telomerase enzyme is found and precise clinical methods are developed, it is a moot question whether telomere therapy is panacea. Obviously, it is not a cure-all therapy and not a panacea for all the ills of human body as there are definitely some problems with longer telomeres and in some cases, telomere therapy is not relevant and hence it is not indicated.

Rate of living theories propound the concept of "Endowment", which says that all living beings in a species are given, at the outset, certain amount of equal endowment and it is up to the individual to use it at the rate he can. The logical deduction from this premise is that the metabolic rate determines the life span. This sequence of logic leads us to look for relations between metabolic rate and core temperature on the one hand and core temperature and hibernation related physiological states on the other. Hibernation as a physiological state can be observed in animal kingdom but not in humans. Vain searches were made to find out what triggers hibernation state. Many bizarre experiments like suspending animation by deep freezing and even transposition of heads were attempted. After a long and vain search, scientists seem to have realized that life in the slow lane is the best alternative to hibernation and started experimenting with model animals using this concept. The implication of this idea is reduction of food intake that stokes the metabolic fire.

Some Suggested Solutions

Caloric Restriction and Aging

Food as a regulator of health and longevity is in fact an ancient idea. Some ancient ideas startle us with their strikingly modern character. One such ancient concept picked up by the modern researchers is the effect of food on the vital functions and longevity of human body. Research on this aspect called 'Caloric Under-Nutrition research' has been progressing almost parallel with the telomere research. The fancy modern phrase 'caloric under-nutrition' corresponds to the ancient concept of moderation in eating.

Experiments conducted on animals with relatively small life spans yielded encouraging results and the technique called 'Caloric Under-nutrition' came into being. Well-designed experiments could not yet be extended to humans due to obvious practical reasons, though it is strongly believed that similar encouraging results will ensue.

Thus, scientists have not so far found any viable method of extending life span at will. They realized that it is not as easy as liberating a gas in the laboratory. They can now liberate awesome atomic energy, they can send a man to the moon and bring him back, and they can make wonderful devices like computers but, alas! They are not able to extend the life span of human beings.

Even if they succeed in extending the life span either by telomere therapy or by some other method, which can be used on a mass scale, the consequences of that so called success will be far from favorable. A host of stupendous social and economic problems will follow and it will not be easy to solve them. If there is a made-easy method to longevity, the first and foremost problem we have to face will be long-living societies. Current social institutions are designed to cater to people of limited life not exceeding hundred years. If communities consisting of people with limitless desires begin living longer, the burden of human population on earth will be so severe that the ecological system of our small planet will not be able to carry. If it is said that the successful longevity medicines will be used only selectively on a limited scale, the question of choice arises. Who chooses the candidates for receiving longevity therapy and on what criteria? It is an intractable problem.

This kind of problems are inevitable as long as longevity is treated as an objective problem i.e., expecting somebody to provide drugs and elixirs to give you as many years as you want.

A better solution would be to make the concept of longevity a subjective affair. If there is a method, which succeeds or fails depending on one's own choices and deservedness, many of the foregoing problems can be tackled. In the first place, it is for the individual to choose whether he (she) wants to live for overlong period. The burden of choice should be left on the individual, as he (she) alone knows how to trade off between longer life with its attendant problems and a normal life. Once life longer than usual is chosen, the method involved should require higher price in the shape of superior qualities of head and heart. One such longevity scheme is presented in this book. The four-fold scheme presented here emphasizes *right food*, *right exercise*, *right state of mind*, and *right choices*.

Right food implies the appropriateness of both quality and

quantity of food ingested. Though scientifically designed experiments on the effects of caloric under nutrition on human beings are not yet completed, we can say it is a commendable method, as we know from studies in nutrition theory that we eat generally wrong quality of food and in quantities more than necessary and invite all sorts of ailments. This topic is dealt with elaborately in a chapter on nutrition.

In modern medical practice, mind is considered a separate entity unrelated to the rest of the body. It has been well known for a long time that mind can act on the tiniest parts like the muscles of eyelids. Charles Darwin demonstrated how mental states produce physical gestures and postures in animals including human beings. Results of recent studies on mind-body relation have started to change that absurd position. Right state of mind is a necessary ingredient in the longevity recipe.

Exercise as an unavoidable routine in health maintenance is well established but it is widely misunderstood. Health promoting exercises are not the same as muscle building exercises. By discriminating between the two and following a regime of Right exercise along with Right food, health and longevity can be promoted.

Finally, we are the arbiters of our lives. If we make wrong choices, we lose. The choices we make every hour of the day affect us. If you choose to jaywalk on the road and you are hit by a speeding vehicle, no longevity elixir can save your life. You will have to face hundreds of such choices that will have impact on your longevity. Right from your routine household chores to choosing and pursuing a career, you have to exercise your common sense and act properly to protect your life. With every wrong choice, the probability of long life diminishes. They are all common sense choices in making which you trade off longevity with some other desired objective. The choice I make may not be good enough for you. You have to make your choices with longevity in view.

The final chapter discusses the limitations of using an inadequate scientific paradigm to achieve our goals. It provides insights into the idea of immortality and leaves the reader pondering whether science can take him there.

CHAPTER TWO

Quest for Immortality

> When death comes breaking into the circle of our friends, words fail us, our mental machinery ceases to operate, all our little stores of wit and wisdom, our maxims, our mottoes accumulated from daily experience , evaporate and are of no avail. These things do not seem to touch or illuminate in any effective way the strange vast Presence whose wings darken the world for us.

---Edward Carpenter (The drama of love and death).

Edward Carpenter describes the state of our minds when death comes breaking into our "Circle of our friends", but it would be worse if death breaks into the inner circle. What would be our state of mind if it approaches us? One is virtually paralyzed by the very thought of it. Death holds out such terror to humanity. It is this fear of death and the anxiety to evade its terror that prompted adventurers to set out in quest of immortality.

Aging and death problems are as old as creation. Life is sweet and death is a terrifying prospect to the vast majority of people. Whoever has a foot in the grave tries to hold the other as firmly as possible. Even the oldest entertain hopes of living one more day and look upon death as the greatest evil from which they should retreat. To some sentimental people, mere mention of the word "death" or any related thought is a taboo. As Shakespeare said even "*that most loathed worldly life*" is considered better than death.

Ancient Aspirations

All humans desire to be immortal. The one ambition of humans since the Vedic times has been to be immortal like gods. The Vedic Aryans recognized the great divide that separated them from gods is immortality. No wonder they aspired for it. Heraclitus expressed this gulf between humans and gods in an epigrammatic style when he said: "*What are men? Mortal gods. What are gods? Immortal* men."

The ambition of Vedic Aryans led them into the quest for immortality in various ways, one of them being ingestion of the famous Soma juice. Whitney observes:

> ...The simple minded Aryan people, whose whole religion was worship of the wonderful powers and phenomena of nature, had no sooner perceived that the liquid had the power to elevate the spirits and produce temporary frenzy under the influence of which the individual was prompted to and capable of deeds beyond his natural power, than they found in it something divine, it was to their apprehension a god, endowing those into whom it entered with godlike powers; the plant which afforded it became to them the king of plants, the process of preparing was a holy sacrifice, the instruments used therefore were sacred.[1]

One worshipper who had the opportunity to drink Soma Juice, exclaimed:

> We have drunk the Soma, we have become immortal, we entered into light, and we have known the gods.[2]

It is not known whether it was physical intoxication or spiritual ecstasy, but William James tells us that drunken consciousness is a bit of mystic consciousness. Whatever it might be they were so much impressed by that, they personified it into a god. The god of inspiration was then given the name 'Soma.'

Hymns addressed to Soma can be found in ninth mandala of Rg Veda. See the ecstasy in the following hymn addressed to Soma:

> Where there is eternal light, in the world,
> Where the sun is placed, in that immortal, imperishable world,
> Place me O, Soma.
>
> Where the son of Vivaswat reigns as King,
> Where the secret place of heaven is,
> Where these mighty waters are,

There make me immortal.
Where life is free,
in the third heaven of heavens,
Where the worlds are radiant,
There make me immortal.

Where wishes and desires are,
Where the bowl of the bright Soma is,
Where there is food and rejoicing,
There make me immortal.[3]

Annals of the Pioneers

This great divide between the gods and others is illustrated in the episode of churning the ocean and retrieving the pot of nectar, which was consumed by gods only, while the others were tricked out.

Hiranyakasipu, a demonic king belonging to the progeny of one of those tricked out of nectar distribution, started the rebellion against this exclusive possession of immortality by gods. He made valiant efforts to attain immortality but did not succeed because gods were averse to give that privilege to anybody outside their clan.

Then comes the story of Gilgamesh, the Sumerian king, who raised the banner of revolt against death and went on his adventure to acquire immortality, He found the 'Tree of Life' but having failed to take the fruit of the tree, he became one with us, the ordinary mortals.

The story of Adam and Eve, the biblical progenitors of man-kind, is similar. They were expelled from the Garden of Eden for plucking the fruits from the 'Tree of Life' and hence their progeny are condemned to disease and death.

In these stories, a couple of which are narrated below, the message is clear, that gods kept immortality to themselves and gave others death. We have mythologies, which chronicled the vain efforts of ambitious persons to wrest the boon of immortality from God. As if to confirm the inevitability of death, God was always wary of granting the boon of immortality even to his ardent devotees. Yet, there are anecdotes of God relenting to satisfy the wishes of his devotees to some extent by granting conditional immunity from death. Some ambitious devotees, usually tyrants, succeeded in obtaining such immunity under a variety of conditions, which they thought were exhaustive enough to spare them from the clutches of death forever. When these tyrants misused this newly acquired power against death, to harass people, the stories tell us, that the almighty God outwitted those villains by devising methods to finish them without violating his promised boons. In the annals of immortality, the story of demon-king Hiranyakasiipu stands preeminently.

The Anecdote of Hiranyakasipu:

The novel procedure adopted by God Vishnu to kill Hiranyakasipu without violating any of the boons granted earlier to the villain is very interesting and reminds us of the adage, "Inscrutable are the ways of God."

Hiranyakasipu, the demon-king wanted to attain lordship over all worlds. He cannot achieve an ambition of that magnitude without a very long life and invincibility in warfare. Therefore, he performed penance, pleased Brahma and prayed for the boon of immortality. Reluctant to grant the boon of immortality, Brahma gave him the option to ask for an alternative. As Hiranyakasipu was insistent on his chosen boon, Brahma relented to some extent, and agreed to grant him conditional immortality and asked him to set his own conditions in which he would not like to die. Seeing no way out, Hiranyakasipu tried to bend this concession to his best advantage and said:

"My Lord, these are my conditions in which death should not approach me: Make me immune from death either by humans or animals, either on the land or in the water, or in the air; either during day or night, either within buildings or without, and immune from death by any weapon."

Brahma understood the scheme of the villain and replied, "So be it Hiranyakasipu, I grant you thy wish, but never misuse my boon."

So saying he vanished. Hiranyakasipu sought protection under a well thought out set of conditions that are normally sufficient to escape death. The range of protections he obtained is a formidable challenge to anyone who intended to oppose him. Emboldened by this newly acquired shield against death, Hiranyakasipu began harassing the worlds even more, ignoring the warnings of God. The king did not spare even his own young son Prahlada from harassment because he worshipped Vishnu, whom he considered his enemy. Puffed up with the pride of newly acquired strength, he worked up his hatred of Vishnu and increased the tempo of his harassment. Unable to bear the sufferings the victims sought the protection of lord Vishnu. The young Prahlada tried to reform and save his father by convincing him about the omnipotence and omnipresence of Lord Vishnu, but the arrogant king ridiculed the ideas of omnipotence and omnipresence of Vishnu and challenged his young son one day saying,

"You, impudent fellow, if you are so confident that your God is so omnipresent, show him in this pillar or I'll make you regret your intransigence."

Throwing the gauntlet, the challenger lifted his mace and gave a strong blow to a pillar supporting his mansion, confident of showing nothing but the empty space in the broken pillar. Taking the blow, the pillar crackled and gave way. Out of it emerged a strange biological form with its upper half as a lion and the lower half as a man, to the consternation of everyone. Hiranyakasipu, who was awe-struck and terrified at the sight of that unusual and fierce biological form, was quickly overpowered by it. Fiercely roaring, the man-lion form (which became known as the Narasimha incarnation of Lord Vishnu) seized the tyrant, dragged him to the threshold of his palace, laid him on his lap and sheared him to death with its claws, roaring and panting with rage. The incarnation of Vishnu took due care to ensure that the tyrant breathed his last on his lap just when the daylight was merging into darkness. Thus, God fulfilled all his promises and yet circumvented all of them to slay the evil fellow who became almost immortal and misused that privilege.

A tyrant's selfish efforts to achieve immortality with the purpose of harassing the world thus ended ignominiously.

This story reminds me of Macbeth, who confiding in the prophecy of the witches,

"No man that is born of woman,

Shall ever have power upon you", boasted with Macduff in the battlefield:

"Let thy blade fall upon vulnerable crests

I have a charmed life, which must not yield

To one born of woman"

Macduff taunts him with the repartee,

"Let the angel whom thou hast still served

Tell thee, Macduff was from his mother's womb,

Untimely ripped"– and slays him.

What is the moral of these anecdotes? Whatever be the ploys one employs to evade death; all of them can be foiled by death.

These stories reveal the fact that the desire for immortality is not new and that persistent efforts made through ages to attain individual immortality have failed, and that mankind's revolt against death never succeeded in the past. Death is accepted as the final status of all living beings, a status supposed to be in conformity with "Natural Order". Yet, mankind's dream of making that "Natural Order" a little more subservient to itself never ceased. As man's desire to tame the forces of Nature increased, daydreaming and fantasizing yielded place to pragmatic efforts that resulted in the ushering of industrial revolution. With the advent of industrial revolution, many dreams of mankind began to materialize. The subsequent progress in science and technology has brought before us undreamt of vistas, but nothing has happened to the status of old age and death. Does this all mean that everyone wishes to live forever?

Do you wish to become immortal assuming it were possible? Probably your first impulse is to say 'Yes', but after careful reflection about the implications of immortality, you may say 'No.' If immortality means simply prevention of physical death without prevention of the decline of bodily and mental powers, probably you may not entertain such a desire.

Simply prevention of death without restraining the decline of bodily and mental vigor is not an attractive prospect. Who wants to hang on to a wrinkled, decrepit and tattering body? Why should one live on as a graybeard forever while one's senses are fading away?

However, justification is sought for our feeling that our present average life of about sixty-to-seventy years is inadequate because almost one third of it passes unknown while we are asleep and we spend the rest of it in our struggles for survival and in enduring old age. We use a very small part of it for enjoyment for which every one craves. That is perhaps why the idea of conquering death and achieving immortality exercised the minds of adventurous people from time to time. Judged from the evidence accumulated since the beginnings of creation, we know that death of every created being is certain. Nobody has ever succeeded in vanquishing death. Scientists who are proud to have conquered Nature are baffled by aging and death. How often the shortness of life has been described in poetic terms! You are not alone in bemoaning the shortness of life. William Shakespeare wrote, "Out, out, brief candle. Life is but a walking shadow."

Is death a biological or Evolutionary necessity?

The idea of immortality could not be dismissed easily as it has been a lingering wish of mankind from ancient times. Setting aside our wishes let us see what science says about immortality. Opinion is divided among scientists. Evolutionary biologists say it is impossible. Modern molecular biologists believe that they can eventually control death. The idea of evolutionary necessity as the cause of aging in vertebrates was first introduced by the great biologists J.B.S. Haldane and Peter Medawar, and further developed by W. D. Hamilton, and George C. Williams. They argued that animals are not designed to live forever. Most of the animals do not live into ripe old age because they are killed either by predators or by accidents. So Nature prefers using its resources for breeding prolifically early in life, to using them for extending barren life. They believe that this happens in the case of sexually reproduced organisms as this kind of reproduction imposes a trade off between fertility and longevity.

Evolutionary biologists say that there is neither single reason that is responsible for aging nor is it derived from any genetic program. It does not occur because of a single universal defect in all cell types. If some single flaw caused every cell to fail, no animal can escape aging, but some animals do escape. Some primitive creatures are virtually immortal. Sea anemones kept in aquariums live on for decades looking hale and healthy. The jellyfish are produced from a stub known as schyphistoma larva. Though each animal produced from the stub has a limited lifetime, the stub appears to carry on indefinitely. Therefore, we consider jellyfish as immortal among the more complex organisms.

The argument of the evolutionists is that senescence is the byproduct of natural selection, which afflicts human beings and other vertebrates but not asexual creatures like anemones. When it comes to that, death has become part of Nature's plan of evolution that favors creatures that die after reproduction in order to enable it to modify the species by natural selection to meet changing conditions. Creatures, which lived forever without reproduction, become extinct. A new species, for better or worse, can start only with new life. Therefore, their conclusion is that death is an evolutionary necessity rather than a biological necessity.

Molecular biologists have not been able to counter these arguments but hope to find the key to aging at the molecular level that somehow compromises with evolutionary process.

The giant redwood trees and the bristle cone pines of Nevada and California seem to live on forever. Using radiocarbon dating scientists have estimated the age of redwood trees to be around four thousand-six hundred years and that of bristle cone pines to be five thousand years. Some desert creosote bushes in Riverside have even better record with about ten thousand years. Bushes growing in clusters are believed to be the offshoots of clones of a single individual plant that thrived thousands of years ago.[4] Such instances prompt us to believe that cumulative cell death is not inevitable. In that way, the human family can be said to be immortal as human beings have been there generation after generation, for millions of years though individuals have been dying but that is not what we are looking for. The issue is about individual life and not perpetuation of a species.

Perhaps the most famous account, in recorded history, of a man who grappled with the idea of conquering old age and death is that of the Buddha. Unlike Hiranyakasipu, the Buddha, in his quest for immortality, was motivated by his profound concern for the welfare of humanity. The story of this great human being is evergreen. It was retold innumerable number of times, as it illuminates the hearers and it enables them to ponder deeply about the questions of old age and immortality.

The search of Buddha for finding a solution to the mystery of old age and death eventually culminated in the founding of a religion whose central theme is that birth, death and the intervening old age are but a single phenomenon separated only by instants of time.

We will see elsewhere in this book how the Buddha's conclusion that growing old and dying are built into the phenomenon of life, is born out by modern biology. Having seen how people through ages desired immortality, let us now see whether it is really desirable.

Is Immortality Desirable?

The old age issue divides our original problem of immortality into two. The first is the desirability of attaining immortality. The second is the possibility of immortality. Paradoxically, the sting of facing death is softened by the burden of old age. The pains and infirmities of old age relent the desire to avoid death. If the misery of senile life cannot be avoided, there would probably be an overwhelming vote in favor of compromising with death. We know from history and experience that old age and death are the inevitable final status of all living beings. People who are obsessed with science and technology think that belief in the inevitability of death need not be axiomatic and this almost axiomatic belief may be changed with further progress in science.

Even so, there has always been a large body of opinion against the desirability of immortality. Great philosophers, influential writers, men of medicine and ordinary people have been among those who did not look upon physical immortality as a worthy goal. Most of the thinkers on this subject felt it is not worthwhile worrying about prevention of death for one or both of the following reasons.

- Life is not worth living when the infirmities of old age intervene.

- Living overlong life is boredom as it involves repetition of the same kind of experiences.

Some people advance the first reason, not because they have achieved satiety of pleasures, but because their instruments of pleasure, namely the senses, fade away and do not report pleasure as efficiently as they used to do. It is a feeling of resignation about the inevitable. The other group thinks that we must realize that sensual pleasures are of the same type and that experiencing pleasures for some time should satisfy us. They assert that we must give up endless craving for sensual pleasures. They are of opinion that it is better to mentally prepare oneself for deliverance from the boredom and travails of old age. This suggests a practical point of view.

The Greek philosopher Epicurus whose name is associated with pleasure mongering did not in fact teach endless craving for pleasures. He said it is better to live a full life and to accept the inevitable end quietly instead of desiring to live an overlong life craving for the selfsame round of sensual pleasures. Among the Indian philosophers, Cārvāka expressed a similar view while the others sanctioned only limited enjoyment of pleasures that are consistent with right conduct and the reality. Socrates and Plato sided with the dominant opinion of the Indian philosophers in denying the presumption that death terminates the identity of an individual and in consoling the individual that death need not either be feared or lamented.

As a satire on immortality concept, Jonathan Swift, in his masterpiece Gulliver's Travels, gives a graphic description of the old age sufferings of "*Struldbrugs*", the mutant human beings of the island of *Luggnagg*, who keep on aging but never die:

> When they came to Fourscore Years, which is reckoned the Extremity of living in this Country, they had not only all the Follies and Infirmities of other old Men, but many more which arose from the dreadful Prospect of never dying. They were not only opinionative, peevish, covetous, morose, vain, talkative; but unacceptable of friendship, and dead to all natural Affection, which never descended below their Grand-children. Envy and impotent Desires, are their prevailing Passions. But those Objects against which their Envy seems

> principally directed, are the Vices of the younger sort, and the Deaths of the old. By reflecting on the former, they find themselves cut off from all Possibility of Pleasure; and whenever they see a Funeral, they lament and repine that others are gone to a Harbour of Rest, to which they themselves never can hope to arrive[5].

Though these "*Struldbrugs*" are born exempt from the constant apprehension of death suffered by others, they are not exempt from the ravages of old age. They suffer more, as they, being immortal, have no release from life with its intrinsic quality of suffering.

There is a similar story in Greek mythology.

> Usas or Eos as the Greeks called her fell in love with a mortal. Then she went up to the gods and said: 'Pray, confer immortality on my lover.' They granted her prayer saying 'He will be immortal.' Later on, the man grew old and senile, became weary of life and longed to die. Eos said 'When I asked the gods to confer the boon of immortality on you, I forgot to ask them perpetual youth as a condition of immortality. So, the man continued to be decrepit and languished in oldness. Immortal body is useless without perpetual youth.

What is the moral you draw from these anecdotes?

Contemporary Opinion

Let us leave aside for a while the anecdotes and opinions of ancient philosophers and try to find out whether our contemporary common man really desires to choose immortality given other options. Suppose you tell a smoker that smoking is injurious to health and advise him to give up smoking in the interests of better health and longer life. Does he readily give up smoking in favor of the other choice? It is hard to come across a smoker who makes such a voluntary choice. He may say that he values better health and longer life but unable to resist the temptation to smoke. Observe that, excuses apart, he continues to smoke revealing his preference to smoking over the other choice. The choice is indicated not by his words but by his action. The individual's choice that is revealed by the preference shown by his action is his real choice. Similar is the case with drug addicts. It is said that the pain of withdrawal from drugs is so terrible that the addict prefers death to the pain of withdrawal. Similar is the case of people suffering with incurable diseases. So, can we say unequivocally that everyone prefers immortality in every situation?

As far as longevity is concerned, the present wisdom is that it is desirable to raise the life expectation of the community because it is generally agreed that a government should concern itself with improving the community health, but no government can accept the obligation of extending maximum lifespan, because maximum life span of individuals is not a public good. Trying to raise the maximum limit should be understood purely as a private endeavor of individuals. How then does an individual make his decisions about longevity?

As an individual tries to maximize his utility of an action, he tries to weigh the pros and cons of an action and looks at the net outcome on which to base his decision. He knows that extra longevity is, verily, not an unmixed blessing because he knows that quality of life declines with age and old age is very burdensome with all its infirmities. All are not alike in this respect. Some people may want to extend it and they may be willing to trade off some quality of life for the extra years. This idea is illustrated in the following hypothetical tradeoff diagram.

Fig.2-1: Trade off between length and quality of life (Not to scale).

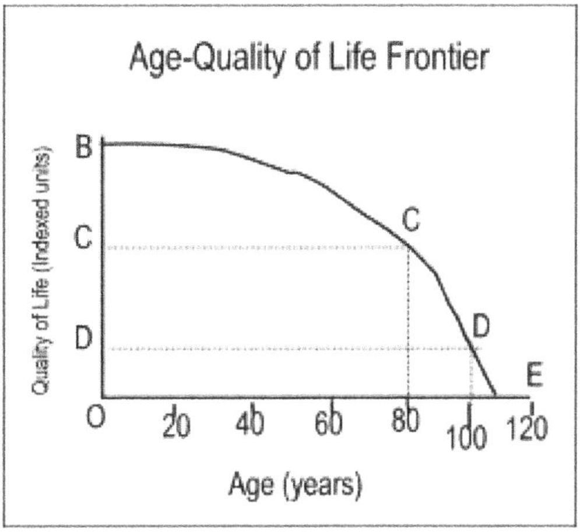

The curve BCDE represents all maximum possible combinations of lengths of life with their associated qualities. If an individual at point C on the frontier wants to extend his life, say by twenty years (assume it is possible), he will have to sacrifice CD units of the quality of life. If an individual at point D on the frontier wants to increase the quality of life, say by DC units, he will have to give up about 20 years.

Fear of Death

> The weariest and most loathed worldly life
> That age, ache, penury and imprisonment
> Can lay on nature, is a paradise
> To what we fear of death.

---Shakespeare (Measure for Measure)

Why then the notion, that immortality is a widespread desire, prevails? Perhaps this is a misunderstood notion; perhaps, it is the *fear of death* that is misinterpreted as desire for immortality. The mighty Hiranyakasipu desired to avoid death in order to achieve his ambition of over-lordship of the worlds, but why do ordinary people with no claims to such grand plans desire to avoid death? Our general attitude towards death is fear, fear of the unknown and fear of leaving the familiar and slipping into the unknown. Our intense attachment to our experiences makes us desire avoidance of death that robs us of that attachment. That desire creates the feeling that anything is better than death.

That is why we try to postpone death, but is it really to be feared? Do we fear to leave behind our experiences as a toddler as we grow up into youth? The Bhagavad-Gita says that:

The soul, just as it takes on the stages of boyhood, youth and old age on its gross body, so also it acquires an alternate gross body and the wise are not deluded about this'[6].

Taking on a new body is an inevitable stage in the evolution of the soul. The Bhagavad-Gita drives home this point with a remarkable simile.

Just as a man discards the worn and torn garments and wears new ones, so also the soul discards worn out body and wears a new one[7]

Anyway, does the fear of death make living anymore enjoyable? Very doubtful, since the alternative is old age with all its infirmities. Fear of death makes it impossible to enjoy the available living moments. Human body is largely self-renewing and lasts a remarkably long time but bodies do wear out and everyone must come to an end. As Shakespeare put it through the mouth of Julius Caesar, death seems to be the necessary end.

There is no point in being afraid of the inevitable. Instead, isn't it better to search the ways and means to extend the healthy span of life and live the latter part of life as healthily as the former?

That being the background, a question arises in our minds. If these great minds had accepted the idea of immortality, would it have made any difference in the present status of longevity? In other words, is the science of gerontology languishing due to lack of enthusiastic support by the earlier thinkers? Perhaps it would not have mattered because they knew that such a condition as immortality, if possible, could be attained only by a few rare individuals and not by all and sundry. We can imagine the consequences if our intellectual forebears spent their time and energy on developing methods of achieving immortality for all and succeeded. With everyone already born on the earth staying alive and the birth rate growing or remaining constant, this small planet would have been plunged in an unthinkable catastrophe. People would have been choked to death due to congestion, let alone the ravages of famine, pestilence and disease decimating them. If not, people suffocating under unbearable congestion and lack of adequate resources for living, would have killed each other anyway, and thus neutralizing the effect of contrived immortality. Therefore, even if a few individuals succeed in attaining their ambition of immortality, it may not be a desirable goal for the society taken as a whole. This is glaring example of the fallacy of composition.[8]

Is Immortality Possible?

Life has been there on this planet for billions of years and it will continue to be there long after you and I are gone. Lifeline is immortal. Your inheritance can be traced back to the origin of life on this planet. The process of life is immortal but the individual is not. We may think that we are immortal only in the sense that we may look backward or forward into our genealogy indefinitely, but that is all. We cannot expect to stay alive in the same bodies and maintain the same egoism associated with these bodies forever. Death remains inevitable for the individual no matter what we do to extend our life spans. We may be able to contrive some methods of living longer, say a few more decades but death will still knock at our doors.

Why is life immortal while the individual is not? Life is propagated by germ cells. Germ cells are carried from generation to generation for billions of years. They have no death. Our partial identity is carried through our genes in our progeny and their progeny and theirs. That partial identity is immortal but not our bodies, because our bodies are made up of somatic cells that age and die. As the somatic cells age and die, so we do. So, what consolation is there that our genes are immortal? You are dissimilar from your genes because what you call 'I' is not your body or genes. You have a consciousness and self-awareness, an apparatus for feeling and enjoyment. Therefore you treasure conscious life however short it may be and so your concern is with prolonging that 'I conscious life' as long as possible but not with the thread of life which may be immortal. That is what molecular biologists are trying to do now.

Two Sides of the Same Coin

As the Buddha declared two thousand years ago, the strange property of individual life is that both aging & death are built into it. Aging is the process through which final dissolution is accomplished. We are constantly at war with forces that break us apart--free radicals, infections, and external attacks on the gross body. If we lose any of these battles at any level-biochemical, cellular or systemic, old age and death will not be far off. Similarly, attacks on the gross body may result in death.

However, why do all these changes take place at all leading us to death? Even if we are carefully protected from external attacks by an efficient security system, the friendly foes built into the body system will bring us down. What is it that triggers the aging process leading to death? There is a wealth of biological and clinical data surrounding these questions although it is difficult to discern the exact cause. Exploration for the root cause has become impossible (at least till now) because the proximate causes, distant causes, and their effects are all intermingled that it is difficult to isolate causes from their effects or to trace the thread of causation. Is it the increased activity of free radicals, is it decline of immune function, or is it slower protein turnover or decline in ability to repair the damaged DNA? It may be one of these, some of these, may be all of these and more; but one can be certain that aging and death are built-in like the other side of the same coin. Life processes have death built into them. There is no wonder that we cannot live forever. However, we need not live decrepit lives weighed down by the infirmities of old age. Non-aging is a different matter. Aging is not death; perhaps it can be avoided.

Agelessness and Immortality

On the other hand non-aging is not the same as immortality. One may be ageless and yet die. When scientists talk about 'immortal cells', they mean ageless cells and not cells that never die. Even ageless cells die under hostile conditions. Just for the same reason the organism, which is composed of mortal cells cannot be immortal.

Even if you have ageless cells, that is not a guarantee that you won't age and die because you are not just a random collection of cells. You are a highly organized system of purposeful groups of cells, of course, presided over by consciousness. In fact these ageless cells, themselves may turn against you as in the case of cancer. What is important is that the organization of cells and their individual and group behavior also must be conducive to life even if your cells are immortal. It does not matter even if a few of them die as long as the arrangement and social behavior of the rest is intact because the dead cells can be replaced. In fact, many cells die as the body constantly renews itself. Cell is not relevant to human life. Cancer cells are immortal but can we make a body out of these immortal cells? Cells become cancerous when they defy the organizational discipline for any reason. For you to be immortal you need more than ageless cells. They are necessary but not sufficient. You need immortal cells that cooperate with each other and maintain the social decorum required by the constituents of the body to preserve and maintain the integrity of body. No cell is an island unto itself. Each cell depends on others in the system.

I wondered often whether a scientific explanation could be found for the unusual biological feats attributed to the legendary heroes Hanumān of Rāmāyana epic and the Greek hero Hercules. The story tells us that Hanuman was able to expand his body to great dimensions and reduce it to tiny dimensions at will. What we are concerned with here is not whether the episodes are all true in their entirety. We cannot think of them as any more ridiculous than the current efforts of biologists to vanquish death. We should take such events as biological conundrums and think whether they can be interpreted in terms of modern biology. Patanjali, the author of Yoga Sutras, listed these feats with the names "Garima" and "Aṇima" in his famous Yoga Aphorisms. It looks like gene expression was at the beck and call of Hanumān who could multiply all groups of cells or dissolve all groups of cells in perfect proportion in a controlled fashion so as to expand or contract all his limbs resulting in integral changes in the stature of his body. He was able to do this by some power, perhaps the power of his will. Perhaps we should explore this idea to know whether we can improve the stature of human species. Hanuman is exceptional; he is both ageless and immortal and cannot be emulated. So let us leave it at that.

Then there is the legend of Bhīṣma, the grandsire of Mahābhārata, who was granted the boon of immortality by his doting father to satisfy whose lustful desire for a ferry maiden, the young prince abdicated his right to succeed to the throne and remained a lifelong celibate. This venerable man was able to choose both the mode and time of his death. His decision to die was voluntary for he knew the futility of maintaining the physical body too long. He however used the powers granted him for a short-range objective of prolonging life until a desired time with a specific motive. Bhīṣma lived a healthy and meritorious life with an ageless and powerful body and invited death when he wanted it. He even chose the mode and time of his death.

He not only set up an example of virtuous and unselfish life but also remained a role model for longevity seekers who wanted to emulate his healthy and vibrant life until the end.

Let alone all other aspects of his life, he demonstrated that subduing old age is more important than clinging to life at any cost. Our general attitude towards death is fear, that anything is better than death. We cling to the body. That is why we try to postpone death though such postponement means sinking into old age. Of what avail is the extended life if you cannot discern the sight of the beautiful, cannot hear the melodies of music, cannot feel the taste of a dainty dish and you are deprived of the faculties of the brain which inheres all pleasures?

However, if you harbor a feeling of regret that death may knock you down too soon before you achieve your modest ambitions, senescence overwhelms before you have spent a reasonable time with your beloved ones, you are not alone.

John Milton lamented that 'time, the subtle thief of youth' had stolen his 'three and twenty years'.

Therefore, a more rational attitude would be aiming at postponing death to a reasonably distant time and endeavoring to live a healthy and vigorous life until the end. Our motto should therefore be '*Stay youngish till the end.*'

We will now examine some facts and figures about people who lived long lives, about their relative proportions in the total population, of people who lived to different ages and about the idea of life expectation. A few instructive diagrams will help to understand the dimensions of the problem of aging and a few others to project the future trends. To get some critical analysis of these ideas, we will have to navigate into a territory known as demography.

CHAPTER THREE

How to Think about Aging with Facts and Figures

Demography of Aging

Average & Maximum Life Span Concepts

When we talk of extension of life span, we should know the answer to the question, extension from where. What is the present limit that we want to extend? It is not a clearly identifiable point. From our experience we know that different species have different life spans varying from as short as a single day of Mayfly to as long as one hundred and fifty years of a large tortoise.

Observed maximum life spans of various animals vary widely prompting us to ponder whether there is some relation between their longevity and characteristics of body weight. These life spans are listed in Table (3-1)

What is the allotted life span of human beings?

The proverbial life span of human being is one hundred years but this limit has been broken now and then. No other subject has been more obscured by deceit, fraud and vanity, than human longevity. Extreme claims have been made on behalf of the very old, some of them mythical and we are unable to verify some of them though the claims have come from honorable sources. According to Bible, Adam lived 930 years, Noah 950 and Methuselah 969 years.

Table 3- 1: Maximum life spans of Selected animals

S #	Animal	Life span	S #	Animal	Life span
1	Mayfly	1 day	15	American Buffalo	26 years
2	Shrew	2 years	16	Domestic Cat	30 years
3	Mouse	3½ years	17	Grizzly Bear	30 years
4	Rat	4 years	18	Horse	35 years
5	Skunk	8 years	19	Whale	40 years
6	Vampire bat	13 years	20	Golden Eagle	50 years
7	Rabbit	15 years	21	Chimpanzee	50 years
8	Kangaroo	16 years	22	Gorilla	55 years
9	Moose	17 years	23	Asian Elephant	80 years
10	Domestic goat	20 years	24	Human	113 years
11	Dolphin	23 years	25	Tortoise	200 years
12	Dog	20 years	**Maximum Life Spans of some animals**		
13	Rhesus Monkey	24 years			
14	Lion	25 years			

In recent times, Benjamin Franklin informed the readers of his "Poor Richard's Almanac" that a Shropshire farmer lived to be about 150 years and Henry Jenkins of Yorkshire lived 169 years but later these claims were disputed.

Some claims were made of people living into their second century but the present situation is that people living beyond 110 are an extreme rarity. Since Britain introduced compulsory registration of births and deaths in 1834, no case of any one living beyond 109 years was recorded there. In Sweden where alone a proper and thorough official investigation follows the death of every allegedly very old citizen, none has been found to surpass 110 years. List of age record holders in economically developed countries compiled from various sources is furnished in Table (A3-1). Table (A3-2) depicts top ten oldest living men in economically developed countries.

In India there used to be a saint named Trilinga Swami who was reported to have lived 280 years. He lived in Varanasi for a long time. Many of his contemporaries claimed that even their forefathers knew him and were familiar with him. Sri Ramakrishna, Vivekananda's master met this yogi and testified about his extraordinary powers. Paramahamsa Yogananda of 'Self Realization Fellowship' of Los Angeles claimed that his guru Lahiri Mahasaya was the contemporary of this great saint and knew him well[9]. Paul Brunton paid rich tributes to this extraordinary yogi in his book '*A Search in Secret India*.'

More centenarians might have lived in various parts of India but as record keeping had not been vogue in this country, many long-lived people disappeared unknown to us. However, a small sample of long-lived yogis in Telugu speaking area is compiled and appended as Table (A3-3).

From a perusal of these life spans, we may be tempted to draw the inference that the proverbial centenary-mark is not inviolable if only one knows how to do it. However, we cannot expect the ordinary mortals to possess the extraordinary powers of yogis.

That leads us to the quest for other methods that can be used by ordinary people.

Benchmarks

Any effort to extend life span has to contend with two different but related measures. Average life span and maximum life span. For the sake of convenience, we can set up three different benchmarks viz., the ages 100, 110 and 120 years. The proverbial centenary mark has been there like a rock since ancient times as the resting point of human life. While attaining 110 is difficult, reaching 120 is very rare in this age. To make it less rare is our motivation for this study and remains our proximate goal and we will examine whether higher limits may be possible.

Patterns of Survival Curves

We are aware that most of the people do not reach even the lowest of these benchmarks. Their lives are ended either by disease or by accident before they reach the centenary mark. By observing the evidence of a few people, however few, living up to a benchmark, we conceive the idea of a biologically possible maximum age. By observing people dying at varying ages below the maximum mark, we conceive the idea of Average life span. In other words, average life span is the age at which 50% of people in any given population group die. It is easier to visualize the survival patterns of populations and to measure their changes if we use a graphical device involving these two measures. In this device the percentage survival is measured on the vertical axis against various ages on the horizontal axis. The resulting plot is known as Survival Curve. In Fig (3-1), different types of survival curves are shown.

Fig.3- 1: Patterns of Survival Curves

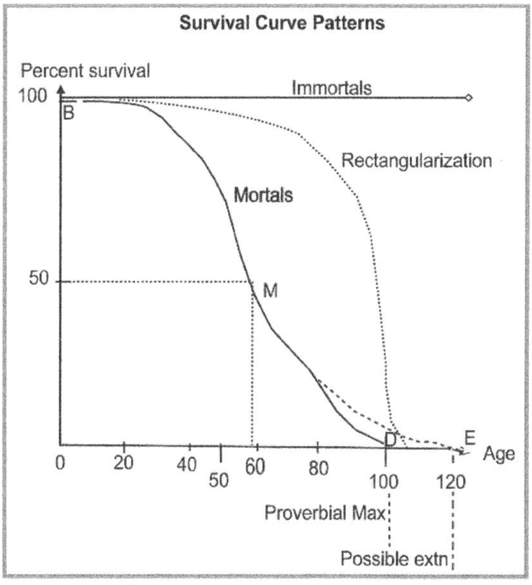

The horizontal line at the top shows a community in which hundred percent of people survive forever. They may be angels or Gods. It shows that everyone is eternally alive like demigods who drank nectar. The solid line BMD shows a more realistic representation of the survival curve. It is the current pattern. It shows the percent survival at younger ages is very high and the survival rate declines gradually so that very small percent of people live up to hundred years. Point M on the curve denotes the age up to which 50% of the population survives. This is a very important concept and is known as 50% survival age in demography. This point tells us that 50% of population attains the age a little less than sixty. Curve BME shows the possibility of extension up to one hundred and twenty years.

The dotted line just below the horizontal line may be viewed as the bulging of BMD towards North East representing what is known as rectangularization of survival curve. It means more percentage of people attain higher maxima. That is exactly the aim of current gerontology research.

From these graphs, we can pick up two yardsticks of measurement, viz., 50% survival and maximum survival. We have agreed that maximum survival age limit has been there as if a rock at 120 for thousands of years but the survival curve itself has been bulging upwards with both ends anchored. What does this bulging mean to us? It is easy to guess by looking at Fig (3-2).

Fig.3- 2 Historical Survival Curves

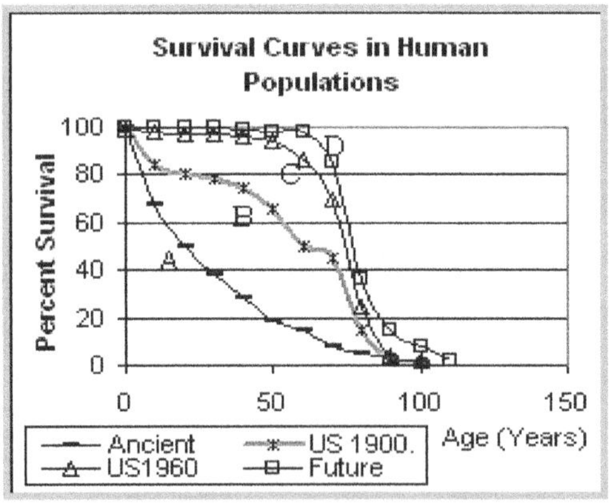

The curves in Fig. (3-2) shown above represent historical populations, but let us suppose that curves A, B, and C represent three different clones of populations. We see that people on the curves A and B do not enjoy the squared survival as those on the curves C and D. No doubt, curves A and B can be stretched outwards with some effort to approach C, enhancing their average life spans. That dos not necessarily imply extending the maximum life span ordained for that clone species. In terms of the diagram, our question now is how to move on from a curve on the left to the one on the right? Translated into plain English, how to enhance the average life spans

A bulging to the North East direction means that for any given age more percent of people are living or 50% of the community are now living for a longer period than before subject to, of course, the upper limit. A shrinking towards the SW has the opposite meaning.

As you can see from Fig. (3-2), 50% survival age has increased from about 25 years in ancient Europe to 40 years in the nineteenth- century. In USA, it has risen from 50 years at the beginning of this century to 67 years in 1946, to 72 in 1960 and to 75 years now. The maximum survival age has not budged much. Why? What makes these two measures differ in their response to human intervention?

Fifty percent survival age reflects our genetic susceptibility to diseases and interactions with our environment. These susceptibilities are extremely individualistic in the sense that they vary from individual to individual, race to race and location to location. On the other hand, maximum survival age, while it varies from species to species, remains stable for a given species.

Maximum survival age seems to be determined by some sort of innate biological clock built into each species. Even the most guarded member of a species cannot escape extinction when the count down of its clock is completed.

The rate at which this clock ticks is a species characteristic rather than an individual characteristic. The upper limit of our life span has not budged for thousands of years in spite of our advances in science and social reorganizations but there have been remarkable changes in average life span.

The age of pandemics like cholera and plague, which marched across the earth wiping out large populations in their stride, has receded. This has been made possible largely due to several sociological changes like increased awareness of personal and family hygiene and community sanitary measures like improved drainage and protected water supply systems. The dreaded contagion of small pox was all but eradicated because of splendid international cooperation in preventive medicine. Deaths from tuberculosis, diphtheria, measles and mumps have been drastically curtailed.

Change in Patterns of mortality

These beneficent changes in hygiene, sanitation and preventive medicine have led to the increase in 50% survival age as reflected by the straightening out of the survival curve as two sides of a rectangle. If we ever stumble into that fortuitous circumstance of eliminating all diseases and if all people live (excepting accidents) up to the age of 100 years, how would the survival curve look like? Guess! Looks like a rectangle with a horizontal side passing through 100% mark and a vertical side passing through the age mark hundred. It implies that all people live up to the age of hundred years and fall off like autumn leaves. Therefore, even if we perfect the so-called 'rectangularizing technologies' (the technologies that help to transforming the survival curve to an approximate rectangle shape), it does not necessarily mean automatic extension of maximum survival age. In fact, we have not succeeded much even in rectangularizing technologies. The age of epidemics and pandemics has yielded place to the age of degenerative diseases. Look at the progressive ascendancy of degenerative diseases into the top ten killer diseases from (Table3-2)

Table 3- 2:Cause-Specific Deaths

Year 1900	Year 2002
1. Pneumonia and Flu	1. Heart Disease
2. Tuberculosis	2. Cancer
3. Diarrhea and Intestinal ills	3. Stroke
4. Heart Disease	4. Lower respiratory diseases
5. Stroke and brain lesions	----------(Medical errors)
6. Kidney inflammation	5. Accidents
7. Accidents	6. Diabetes
8. Cancer	7. Alzheimer's disease
9. Senility	8. Influenza
10. Diphtheria	9. Nephritis
	10. Septicemia
(Source: Centers for Disease Control & prevention). Medical errors were not ranked by CDC, but their rank comes above accidents according to a study by NAS	

Does the elimination of these diseases lead to extension of life span and prevention of the onset of senescence? The changing pattern of age distribution in several geographical areas suggests that to some extent that is possible

IS MAXIMUM-LIFE-SPAN SHOWING AN UPWARD TREND?

In 1980, the Census bureau counted fourteen thousand seven hundred Americans who reached centenary mark and this figure was expected to reach 100,000 by the turn of the century.

According to census bureau projections, the number of people hundred years or older will approach one million in about a hundred years from now. In 1979, Social Security Administration listed 11,891 persons who are over hundred years or older. See the progressive increase of people sixty-five years and older in Fig (3-3) represented by vertical hatched area.

Fig.3- 3:Changing patterns of Age distribution in USA

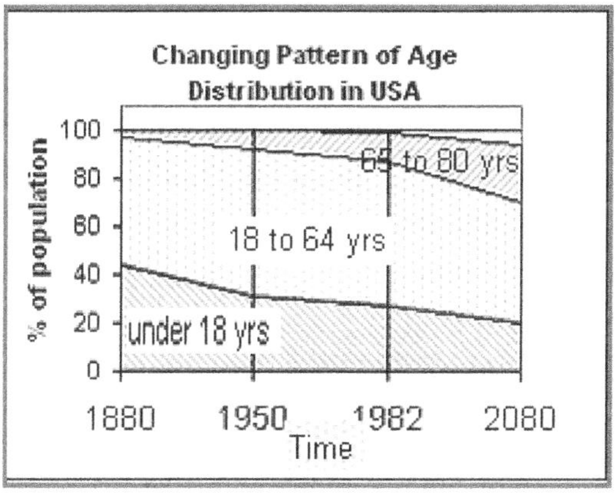

The demographic determinants of a population's size and age structure are fertility, mortality and migration. It has long been recognized that the first factor is the most important of these. Children, if they survive to maturity, will become parents of children who may themselves reproduce. Thus, each birth represents not merely a single addition to the current generation but an exponentially increasing contribution to the future generations.

The bar charts shown in Figures 3-4 and 3-5 show the projected future increases in sizes of older cohorts by the year 2020 in selected countries.

Fig.3- 4: Population changes in cohorts 85 & older

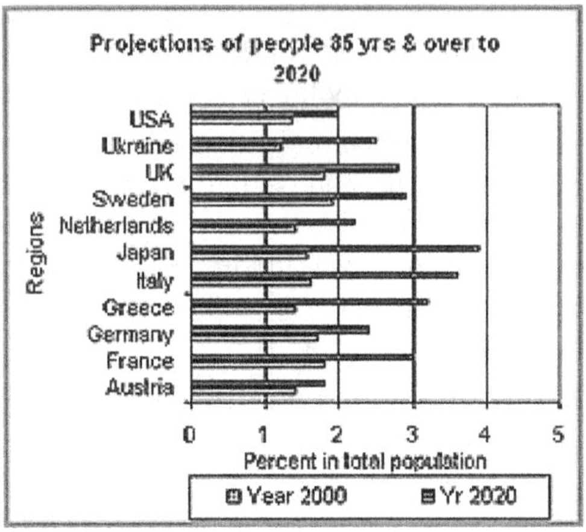

Fig.3- 5: Population changes in cohorts aged 65 & over

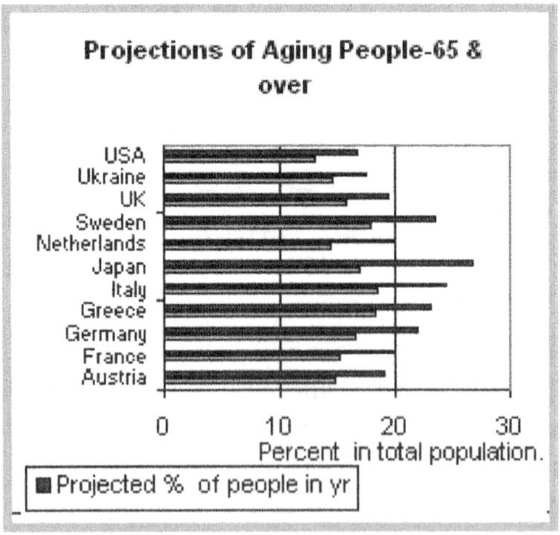

<u>Population Aging</u> – A shift towards older age structure- is primarily a consequence of long-term downward trends in fertility. During 19th and the early 20th centuries, this shift towards older

populations occurred in most of the developed countries. Declines in death rates at older ages and further shift to very low fertility rates in some regions have resulted in further concentration of the aged in many populations. Both the charts given above show increases in life expectations of older cohorts from 2000 to 2020. However, it is debatable whether it is a temporary phenomenon or a permanent trend. This has become a contentious issue among demographers and much heat has been generated over it.

We will consider first what life-tables suggest about this matter, as they are the basis of the success of insurance companies. Actuarial calculations of insurance companies for collection of premiums and for setting terms for covering risk depend upon the fundamental premise that man is mortal and a finite limit to maximum life exits.

The question being asked now is whether this limit is immutable. Picking up the cue from the world record of 122½ years of life from Jeanne Calment (see page 394), a group of demographers[10] investigated recently whether there is any secular movement of the upper limit of life span. If there has been an upward change recently, what accounts for the change? One obvious explanation is improvement in an individual's own ability to survive longer due to any reason. The other is probabilistic. The larger size of contemporary populations increases the probability that at least one individual will survive to an extreme old age. In either case, does it suggest a trend? The research group says that their sophisticated statistical analysis of life-tables of Sweden indicated that there was marked acceleration of maximum age at death around the year 1969. They report that maximum life span increased on average of 0.44 years per decade from 1861 to 1969 and at the rate of 1.1 years per decade from 1969 to 1999.

Another interesting question that has been bothering the researchers is this. If all the diseases were eliminated, which of course is doubtful, can we make large additions to maximum life span? If we assume that deaths occur only due to accidents and suicides, some argue that life span can be extended, at least theoretically, by leaps and bounds.

An interesting study was made to see how the survival curve would look like if accidents alone were the cause of all deaths. The proposition is not verifiable on the human beings as there is a plethora of diseases still bothering us. As such, the hypothetical situation was simulated on the lives of porcelain coffee cups at home. Coffee cups do not age but surely, they do break. The resulting percentage of surviving cups is plotted against time tracing the curve shown in Fig. (3-6).

Fig 3-6: Figure showing simulation of Accidental loss of life

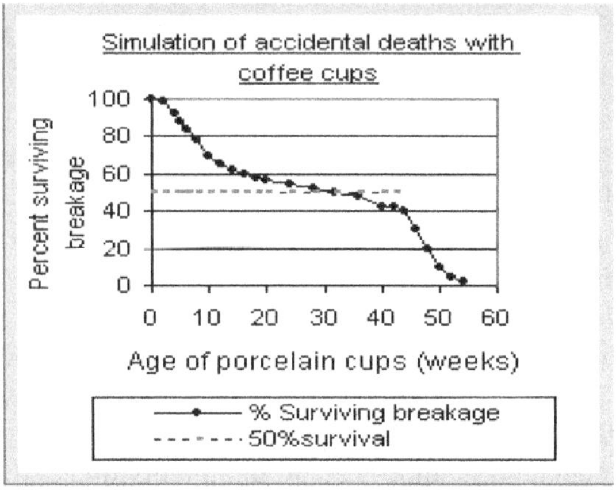

This study suggests that even if all diseases are eliminated and aging is prevented death is inevitable.

In such a simplified and hypothetical case, maximum life span can be computed as function of accidental death rate alone and that life can be extended by hundreds of years[11]. This calculation of course depends on the naïve assumptions that primary aging is absent and a mathematical model can represent longevity. Therefore, it is only of theoretical interest.

Depending on which death rate we have used, the potential life-spans range from a few hundred years to a few thousand years. The reader should be careful to note that these calculations are based on the observed death rates which implies death itself is not ruled out even at distant time. All this exercise does not necessarily imply automatic prevention of the onset of senescence. Even if the killer diseases are eliminated and serious ailments controlled, yet degradation of youthful traits will set in.

Various accounts put the maximum life span of human beings

between 100 and 120 years. Indian astrologers maintain that the maximum life span of ordinary human beings should tally, in ideal conditions, with the sum of the major periods of all the twelve planets in the zodiac, which equals 120 years. Ayurveda recognized long ago the possibility of living 120 years if one lived a disciplined life right from the beginning according to its principles. Stories of people who lived super long lives approaching the maximum allotted life-span come to light off and on. The Hunza tribe in India, the natives of Vilcabamba in Ecuador and a group from the Caucasus Mountains in Russia are some examples who claim extreme longevity though the claims could not be verified. Greek writers reported that during Greek invasion of India they found people, aged 150 and over were found living there.[12]

Controversy Among Demographers

It is not clear whether the perceived increases in life span of certain communities is the secular trend of the life expectations of populations or whether it is a temporary phenomenon reflecting the consequences of eliminating infant mortality and improvements in sanitary conditions and availability of modern medical aids in those communities. Demographers are divided into two groups on this issue. One group claims that there are finite limits to reduction in mortality rates while the other claims that actual experience shows that there are no such limits. Both groups use time series data but interpret them differently. The cleavage between these two groups needs to be reviewed to have a better understanding of the status of this topic.

Most sexually reproducing species show signs of senescence with advancing age. Evolutionary biologists developed a theory, which contends that the post reproductive span of life should be short because natural selection acts against the mutations occurring after the cessation of reproductive stage. For several decades, this theory of evolutionists remained un-assailed due to the absence of any compelling counter theory. Now, doubts are being raised about the infallibility of this theory based on recent evidence of observed decline in old age mortality.

Opponents of evolutionary theory say that they[13] tested the key hypothesis that mortality rate accelerates as reproduction declines with the age of the subject group, using nematodes, yeast, wasps, fruit flies, human beings and even automobiles and claim to have found declining trends of mortality in all these groups. However, they are unable to reconcile this finding with the established evolutionary theory that post reproductive mortality must accelerate. What are we to conclude from this controversy? It is not easy to dismiss an established theory without compelling arguments. Then how do we reconcile the experimental findings with the theory, which is at variance with the recent findings? The answer to this enigma lies, perhaps, in the interpretation of the experimental findings. These findings are extrapolated from the past using regression analysis on the assumption that the future will behave like the past. What guarantee is there that future will behave like the past? If the range of data is changed different results can be obtained and from them different inferenccs have to be drawn. The short-term decline in mortality rates may be the outcome of public health improvements and medical advances made in the previous century. The analysts are not able to show any other specific endogenous factors that will sustain this trend forever. It is unrealistic to conclude that the trend of declining death rates will continue forever until they become insignificant and the life expectation at birth will grow to infinity. We will revert to this topic at a different place in this book and assess its plausibility.

Population growth curves

A population grows, if more individuals enter it than leave it by death or immigration during a given period. In such a case, a net increase in number is said to have occurred. All populations have a potential for explosive growth in numbers. By the nature of reproduction, populations grow in geometric proportion rather than in arithmetic proportion. After some time this growth takes the form exponential growth and rises steeply. When plotted against time it looks roughly like the 'J' shape but fortunately, populations do not increase indefinitely as Malthus predicted. There seems to be an upper limit to the number of individuals in each species in a habitat. This limit is called the carrying capacity of that habitat. It is the maximum number of members, which can be supported over a long time by the available resources. When population exceeds the resource limits of the life supporting habitats, this 'J' shape may crash sharply downwards bringing the population back into equilibrium with resources.

If many natural populations are observed over a long period, we can observe a gradual rise in the beginning, next steep growth and then finally gradual tapering off. If these values are plotted against time, this gives us a curve resembling roughly a leaning and elongated 'S' shape. All populations have the potential to engulf the world with their progeny but none have done so yet. This happens because the growth of most species is held in check by environmental resistance.

There are several limiting factors, which work against rapid increases in populations; these are mostly environmental factors, which by their deficiency or excess prevent explosion of populations. The factor that is most responsible for preventing such explosion is called primary limiting factor. It is the weakest link in the population's life support system. There are two laws that describe these limiting factors. One is called Liebig's Law of minimum and the other is called Shelford's Law of the Maximum. Liebig's law says that limit to population is imposed by some scarce nutrient or some least available ecological factor. The other law says that excess of factor imposes the limit. For example, heat is necessary for survival but excess of it can also prove disastrous.

Two more categories of factors control populations. They are called density dependent and density-independent factors. There are several density-dependent factors like predation, disease and reaction to overcrowding but the most important and all embracing factor is competition. Competition among the increasing number of members diminishes the available food causing starvation to some.

CHAPTER FOUR

Mystery of Aging

> Thou madest man, he knows not why,
> He thinks he was not made to die.

—Tennyson (Memoriam)

As Tennyson said man knows that the foot of god, his maker is always on his head and yet he always thinks that he is not made to die. So he keeps on dreaming of staying alive forever. That is the mystery of life. As we said earlier, the greatest wonder of the world is that man behaves as if he is permanent resident of this planet. In spite of all arguments to the contrary urged by different schools of thought in different times, in spite of the inability of reason to penetrate the veil of mystery, man is somehow persuaded that he cannot die.

Theoretical Framework

As observed earlier, we may investigate all our lives but we cannot bring the problem of life and death, to the plane of rational demonstration. All that we have been doing is that we are trying to cling to some empirical results and believe them to be eternal truths in this ever-changing world.

Change is the characteristic of time. Everything around us, animate and inanimate, changes. The elemental forces change the landscapes and physical structures. The force of time changes plant life, animals and human beings. The changes we observe in objects through passage of time are loosely called aging. Perhaps we use the word 'aging' in case of objects whose origin is known at least roughly. We do not say that 'this mountain is aging' or 'this river is aging,' much less 'this world is aging', as we do not know their origin in time. We do not also use the word 'aging' to connote every case of change. We do not generally say 'this baby is aging,' rather we would say 'this baby is growing.' In that sense aging denotes some sort of deterioration rather than mere changes that may be developmental. We say a building is aging when it show signs of wear and tear.

Does it mean that those people who show signs of wear and tear are to be considered to be aging and vice versa? Not necessarily, because we are aware of numerous cases of people who exhibit signs of wear and tear are not as old chronologically as those who do not exhibit such symptoms. When we make sometimes a ball park guess of somebody's age by his/her appearance, what we are trying to do is to guess the functional or the biological age of the subject, but not doing the arithmetic of subtracting the subject's date of birth from the current date. Subjective estimations of ages do not quite often correlate with the chronological age. What we need therefore is a biological marker that can be measured as a function of passage of time. This measure denotes functional age as distinguished from the chronological age.

People of almost same chronological age who outwit our guesses by their varying appearances suggest that we age at different rates. It appears that not only different people age at different rates but also different parts of the body age at different rates. Changes in cell tissues and different organs may begin at different times and proceed at different rates defying any precise definition of biological aging.

Biological Clock

The flame of life incorporates the continuous turnover of the panoramic spectrum of life units, from the smallest organelles in the cells to the largest arrays of plant species. Examination of the range, role and mechanisms of the processes of death led the biologists to three suggestions.

- Death of organisms is the central component of evolution; it is a basic component of evolutionary selection. It is inconceivable that the evolutionary processes could give rise to an organism that could evade this requirement.
- The death of organisms need not be passive. There is evidence that higher plant species have utilized programmed lethal stages for development of a wide variety of ecological, physiological and adaptive advantages.

- Given the essentiality of the death of organisms in the evolutionary process, one may presume that senescence imposes death with associated advantages whereas predation and stress impose death in a more passive manner.

According to Strechler, age related changes must meet the following four conditions.

- They must be deleterious, i.e., they must reduce function
- They must be progressive, i.e., they must take place gradually.
- They must be intrinsic, i.e. they must not be the result of modifiable external agents.
- They must be universal, i.e., all members of a species should show such changes.

There is an important idea that has been intimately embedded in everybody's conscience that aging means passage of time, thus assigning time a causal role in the physiological changes. Does time itself play a causal role in the progression of an organism from birth to death?

The fundamental units of a biological clock are transcription units. The fundamental units of a sidereal clock are day and night, which are established on the relationship between the Sun and the Earth. Involvement of sunlight connects these two otherwise disparate clocks thus connecting time to aging.

For example, the biological clock in fruit flies, Drosophilae and some other organisms depends upon the cyclic interaction between two specific gene products that usually repress their own transcription on exposure to light. This phenomenon provides the rhythmic circadian output* that is characteristic of a biological clock.

* In human beings, the regular metabolic, glandular and sleep rhythms may persist through dislocation of day and night caused by high-speed travel involving crossing of time zones. These physiological rhythms associated with the 24-hour cycles of earth's rotation, are known as Circadian rhythms.

Definition of Aging

Even though the precise definition of aging defies our imagination, scientists provided a couple of working definitions based on these characteristics. Strechler defined aging like this:

- Aging is time independent series of cumulative, progressive, intrinsic and deleterious functional and structural changes.

Another definition is:

- Aging is genetically determined, environmentally modulated and event dependent process.

It appears that the phrase 'time-independent' is introduced into the first definition because of the phenomenon of circadian output. However, we feel that this definition should be restricted to 'Functional aging' and not 'Chronological-aging', because no observable event can be extricated from time.

Chronological vs. Functional Aging

To die youthful does not mean dying at the age of thirty, forty, or even fifty. To think like that is an acquired mental habit. The terms 'Young' and 'old' are species relative and should be understood in the context of total length of life span. The labeling of periods in human life as 'young' and 'old' should be understood in the context of total length of life span and human experience of the decline of bodily and mental powers. For example, if we agree that the word 'old' should be used after two-thirds mark of life span; the age of 13 years at which oldness sets in for a dog or a goat is still adolescence for a horse (which has a maximum life span of 35) and childhood for a human being (with a life span of 100).

Therefore, the terms old age and young age should be viewed as relative and not absolute. We have given these labels to various segments of life depending upon productive efficiency. So, we may change the labels and say a person at sixty is 'young' if she (he) can maintain the youthful vigor. In that sense, it is better to die youngish at age 80 rather than die old at 120. That is, we should aim at cutting short the unproductive and senile period and extending the youthful and vigorous period instead of aiming at prolonging the burdensome old age in the name of longevity. If maximum life span is extended by some means we can suppose, other things remaining constant, that the youthful part of the life span also can be extended and vice versa. Think of a rubber string. When it is stretched, the elongation occurs in every part of it and not merely at one end. The problem of overcoming old age is not formidable if we take our mind out of the old age rut and think afresh. There are numerous instances of people living healthy and vigorous lives until the end. Now there is little doubt that we can curtail the unproductive and senile period of life and extend the youthful and productive part. How are the life spans of different species determined by the nature?

WHAT INFLUENCES AGING? METABOLISM OR BRAIN SIZE

Basic principles of thermodynamics tell us that there is association between mass of an animal and the rate of its heat production. The first law of thermodynamics is essentially a statement of the law of conservation of energy. Living animals are as much subject to its sway as calorimeters are. This law tells us that any object, whether it is a building, a human being or a worm, if it has to maintain a steady internal temperature, it has to give up as much heat from its surface as it is produced within its volume.

Metabolism of an animal is closely related to its body size. To be more precise, surface area-to-volume ratio is the criterion for measuring heat production per unit volume.[14]

It is generally believed that body size is related to longevity via metabolism. As seen from the Fig. (4-1), metabolism comes down, as the size of body increases. Elephants have very low metabolism and live up to 60 years under favorable conditions.

Life spans shrink drastically for smaller bodied mammals, but there is an anomaly in this argument. All birds do not die as fast as a shrew in spite of their higher metabolism. Somehow, birds like Eagles, owls and parrots, which are smaller than some mammals can live as long as 50 or more years. A human being who is much smaller than the elephant definitely lives longer than the elephant. How do we reconcile these contradictions then? Does it depend on the facility of evolution? In most creatures, protection of a genetic material called DNA was an essential feature of evolution

Fig.4- 1: Association of metabolism with body weight of selected species

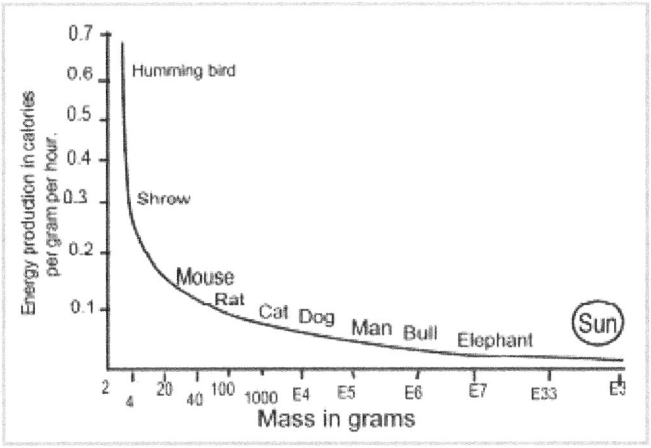

The graph is supposed to be extended to the far right to include Sun. The graph shows the relationship between mass and rate of heat production in animals.

Furs on animals, shell on tortoise and feathers of birds are supposed to be some examples of evolution's effort to protect the DNA. The degree of protection of DNA is supposed to be the determining factor of life spans of different species.

On this criterion human being is certainly not a good candidate for longevity with his/her body structure vulnerable to all outside influences. Unlike many animals, human being emerges from mother's womb utterly helpless. A newborn human baby cannot do anything except suck, cry, swallow and sleep. Even after being risen by parents for several years, a human being can neither run fast enough and escape when threatened by wild animals nor can he/she stand and fight the threat with his/her bare hands, nails and teeth. Then it makes no sense that human who is physically more vulnerable than many other creatures should live longer than almost any other warm-blooded animal. There must necessarily be something in Nature's plan that makes the human being so eminently suitable a candidate for longevity. No one has solved that puzzle yet. With physical strength no match to most of the animals, human being subjugated all of them with his mind. A more plausible explanation therefore is that longevity of species is perhaps related to the size of the brain rather than the size of the body.

Characteristics of Old-Age

Is growing old natural? Old age is like hell of a long sentence with no time off for good behavior. The body seems filled with marks of self-protection gathering itself over those vital parts, folding shoulders and arms like a fading flower. To age is to learn the feeling of no longer growing, and of struggling to do old routine chores. The body seems to give up slowly, stopping now and starting again as if to tease you with the memory of your strength of the bygone days. To walk to the corner, to climb the stairs, or to ride in a car, each action seems to take away the precious energy needed to stay alive. Everything is limited by the scarce energy you try to hoard dearly. Your needs decrease too, your food intake diminishes, and you sleep less. Maintenance of even these minimum human needs becomes increasingly difficult. You dread the eventual prospect of being reduced to a heap of flesh and bones somehow breathing and ingesting some nourishment and depending on somebody for everything you need to survive.

There is nothing to prepare you for the experience of growing old. Nothing can simulate it. Yes, it is natural. Life is an irreversible process starting at birth and progressing towards old age and death but the progression differs from person to person. The life clock in each one of us differs and it is profoundly affected by the individual's heredity, life style, life experiences and mental attitudes.

The Biological Expression of Aging:

Aging is expressed as a process of inevitable and cumulative loss of functions of the body. The pace may differ from organism to organism and individual to individual. Some parts of body may age faster than others and ones own different abilities may fade at different rates. The structure and functions of the body are gradually impaired; motivation, perceptions and emotional stability change, every function of body worsens increasing the potential for disease and death as time marches on.

Between the ages of thirty and ninety, weight of our muscles, and the power we can exert fall by 30% to 40%. The number of nerve fibers in a nerve trunk falls by a quarter. Average weight of brain falls from three pounds to about two and one-fourth pounds as brain cells die and are not replaced. The nephrons in kidneys fall in number almost by half. Our senses of sight and sound decline and our capacity to taste declines as the taste buds in the tongue dwindle with age.

One of the major factors of aging is the sometimes gradual and sometimes unpredictable breakdown of the body's defenses and body's reaction to stress. Throughout our lives, we are subjected to a variety of attacks of life threatening forces – mental stresses, physical traumas and invasions by bacteria and viruses. Fortunately, normal body is capable of repelling the attacks of a vast majority of bacteria and viruses and only a tiny fraction of these turns out to be life-threatening enemies. The ductless glands of our body rally themselves to marshal the body's resources and protect the integrity of body in times of trauma and stress. When this protective system breaks down or weakens, as it happens with aging, the body is subjected to debilitating and degenerating diseases.

As body ages, its reserve capacity to withstand stress becomes less and less. The doctor may give it some name like influenza, pneumonia or simply call it some syndrome when he himself does not know the nature of the illness, but the body knows that the once failsafe mechanism has now failed and that the failure exists within. The body that was once invulnerable to an array of bacterial organisms now becomes an easy prey to many of these invaders. It may be noticed that all undesirable qualities have a positive correlation while the desirable qualities have a negative correlation.

About a quarter of our lives is spent in growing up healthy and the remainder is spent in degenerating. Growth and development are orderly while aging is disorderly and chaotic. Why do all these things happen as they do? What is the scientific explanation?

Speculations on the Basis Of Aging

Entropy and Homeostasis

A branch of physics known as thermodynamics, studies the relation between heat processes and life phenomena. The second law of thermodynamics is stated in various ways but the substance is the same.

There is a constant tendency in Nature towards disorder and disruption that cannot be overcome automatically without suitable expenditure of energy. This phenomenon is called entropy.

If so, how is organic matter highly organized? Living beings are in a high state of automatic internal organization without which life and growth are impossible. Does it mean that organic matter defies this law? On closer examination, this apparent impression is found to be erroneous because the neat organization of plant life going through the process of photosynthesis is not confined to nutrients and carbon dioxide alone. It needs the sun's energy to complete the process of transformation and to stabilize it. This energy that has a temperature of about 6000^0K when it emanated from the sun comes down to room temperature, while retaining the same wavelength when it reaches the plants on earth. This dilution of radiant energy, known as negative entropy' or entropy deficiency, enables the plants to organize the nutrients, water and carbon dioxide into plant life. The rest of the living organisms including the human beings, are simply parasites of plants, at least from the point of view of plants. Thus, even the living organisms cannot violate the universal law of entropy.

Life is a balancing process between the forces of preservation and destruction. Once created, it must engage in a constant war that must always be won in order to survive. At every moment, we are engaged in a struggle for existence, a struggle to repel attacks from other species, struggle against starvation, struggle among ourselves and struggle at the molecular level to maintain the integrity of body. If we lose the battle at any level, the result will be catastrophic. The forces of anarchy are always on the prowl to swallow you outright or dissolve you slowly. Scientists call this perpetual tendency to disorder in nature, by the name 'Entropy.' Entropy has many faces. It can attack you from any side. At the cellular level, it can break down your molecules, degrade proteins or cross-link them, let loose free radicals, or slow down the turnover rate of molecules. At the gross level, it can appear as injuries, accidents, attacks by animals or duels with fellow humans. Expenditure of energy is necessary to overcome this disorder either in the physical world or in the biological world.

Entropy is constantly at work to wear you down and dissolve you eventually. You are however able to live and conduct yourself in this world because homeostatic mechanism of your body comes to your rescue against the onslaught of entropy. Homeostasis is a mechanism that maintains internal stability of an organism by automatic production of compensatory responses against external changes. When it is hot outside you sweat and evaporation on your body cools you. When it is cool, you shiver to produce warmth and crouch to preserve it and maintain body temperature. When your body is dehydrated due to excessive heat you are forced to drink, when your blood sugar level falls you are prompted to eat and so on. When a cell detects a shortage of molecules it produces and when there are excess molecules the excesses are pumped out. The homeostatic forces pervading the body are in general well balanced against entropy, in your youth, but eventually they fail as you age and entropy takes you over, disrupts your body and kills you.

However, different species have different maximum life spans and people die at different ages. Why? The only plausible explanation is that some timing mechanism in every organism tips the balance in favor of entropy in the fight between entropy and homeostasis. As the final time approaches you are abandoned and entropy takes over and destroys.

Like the Hindu Triumvirate Brahma, Vishnu and Rudra; the creative force, homeostasis and entropy are constantly at work maintaining the balance in this world. When once Brahma creates, Vishnu takes over to sustain the world of Beings with homeostatic defenses and Rudra intervenes to limit their lives by letting loose the forces of entropy. The procession of functions performed by this Triumvirate to perpetuate the universe is what we metaphorically call 'evolution'.

The Biomarkers of Aging and degeneration

Before we embark on our study of exploring the methods of life extension, it is necessary to recognize the biological markers of aging. When we are in perfect state of health, life flows so smoothly that we are not even aware of the various parts of our bodies. Somewhere in the middle of life, we begin to notice the many changes that have been imperceptibly taking place in our bodies. We become aware of the weakening of our senses of sight and sound, general loss of muscle power and flexibility of joints.

Skin shrivels and mental ability is blunted.

The visible biomarkers, viz., graying of hair, visual accommodation, hearing ability, and the skin elasticity are of course well known to everybody. In addition to these visible changes, other vital organs are affected.

The most important among the invisible biomarkers are:

- Lung capacity
- Basal metabolic rate
- Pumping efficiency of heart
- Kidney filtration rate
- Velocity of nerve signal transmission

Among the general effects, which are not considered diseases, are loss of elasticity of tissues and loss of respiratory efficiency. As aging progresses, inelastic fibrous cells replace elastic functional tissues. If there is any process that can be called simple aging as distinguished from disease, that is perhaps this process of loss of elasticity of tissues.

The other general effect of aging is to reduce the respiratory efficiency. The capacity of lungs to hold air and expel the spent up air is reduced. Although poor oxygenation is known to be a factor in biological aging, its immense impact does not seem to have been fully recognized by modern geriatrics and medical profession.

In a number of studies conducted on the relation between physiological variables and aging, it was found that rate of blood pressure, hearing loss, and reaction time are positively correlated with aging (they move in the same direction) while lung capacity, forced expiratory volume, visual acuity, and grip strength are negatively correlated with aging (they move in opposite directions)[15]. Of all these variables, lung capacity has the highest negative correlation (-0.77) indicating that preserving and increasing lung capacity is of vital importance in retarding aging.

What drives Aging?

Why do all these undesirable things happen? Molecular biologists were never sure of why these events happen. They used to propose a number of theories from time to time. It may be due to a number of reasons like fall in protein turnover rate, free-radical damage, weakening of immune function, or inability of system to repair the DNA damage or a combination of some or all of them. However, the question arose why should there be more free-radical damage and why should the immune system weaken, as the time of an individual passes? Why should the system gradually fail to repair the DNA? While it may be true that all these may be the proximate causes and play their part in the expression of aging there should be something else that forces them to do well when one is young and yield gradually as one ages. Who drives these vehicles?

Germ cells do not age while somatic cells do age. Germ cells do not age because they are needed for perpetuation of life and Nature has a stake in them. Somatic cells have the important function of supporting the germ cells, yet they age and die! According to official theory of evolution, Nature does not care about you after you reproduce. Your genetic system supports you until you have reproduced, and have assured your offspring their own chance of reproduction. After that, Nature turns against you. Evolution has a stake in giving you a fair chance of passing on your germ cells. After that is done, you are of little use from the evolutionary point of view. So, Nature abandons you. It does not just leave you alone. It turns against you, orchestrates, and enforces your aging and death.

One may wonder: if we are so sure of our subjugation to evolutionary force that orchestrates death, why should we worry about the inevitable end of the physical body? Is it not better to give up our resistance and make the best use of the time available to us, living in harmony with nature? Molecular biologists were not satisfied with the blanket explanation of evolutionists that the amorphous entity called 'Evolution' wills it that way. So their quest continues.

They say that instead of adopting the pessimistic view that some imperious outside agency governs the aging processes, why not we adopt the optimistic attitude that the mechanisms that control these processes are within our own bodies. Those mechanisms that drive you into old age are your gene expressions that drive aging. None of them is however, the root cause.

Something else triggers these gene expressions. If we can locate that cause, we can hope to arrest aging.

Is Aging a Disease?

DIFFERENCE BETWEEN DISEASES AND AGING

Disease means disruption in ease of living. People age without much discomfort and meet their end in many cases. If that is the case, aging is not disease because aging implies passage of time with or without discomfort. Studies have revealed that even if all killer diseases were eliminated, life expectation cannot be raised by more than fifteen years. Seeking a cure for aging therefore means seeking cure for development of an adult.

The differences in life spans and aging of different species give us some clues. Aging in multi-cellular organisms is universal while there is variability in simpler organisms. Bacteria and viruses do not age though other unicellular organisms age.

How is it done? The onset of old age is not sudden. You do not get up on one morning and suddenly find yourself with gray beard, shriveled skin, stooping shoulders and failing vision. You cannot find the difference between yesterday's body and today's body. Old age sneaks in gradually and imperceptibly as years are counted. You can only find the difference in your bodily states between now and years ago. Therefore, we presume that there is a timing mechanism in our bodies, which runs parallel with the clocks. For a long time biologists suspected that this might be in the pineal gland in our body but no progress could be made in establishing that idea. Recent researchers have however gone in a more promising direction and made the startling discovery. The clock is genetic.

Every aspect of human life undergoes some change as time marches on; the structure and functions of the body gradually get impaired; motivation, perceptions and emotional stability change. About a quarter of our lives is spent in growing up and the remainder is spent in getting old.

But, why do different species have different maximum life spans and why do even members a species die at different ages? There appears to be some timing mechanism in every organism that tips the balance in favor of entropy in the fight between entropy and homeostasis. As the final time approaches you are abandoned and entropy takes you over and destroys.

Why, why does it happen so? We do not know. It is steeped in mystery. Science has not yet succeeded in unraveling this mystery of life. The failure of science to solve this mystery of life can perhaps be traced back to its persistence in sticking to the mechanistic model of the universe.

Physicalist vs. Vitalist Controversy

In nineteenth century, thinkers who were wrestling with the ideas of matter and life were divided into two streams. Those who started a new way of thinking which sought to explain away everything in the world by mechanical principles were known as "Physicalists." This omnibus extension of mechanical explanation even to the living organisms was not acceptable to many naturalists. Those who protested against the wholesale mechanization of the world picture and contended that some vital principle must be introduced to explain the working of living things, were known as "Vitalists." Physicalist movement had the merit of reducing the influence of the then prevailing magical thinking but it suffered from the demerit of over extension of its sway into matters involving life. They were never able to account for the phenomenon of life with their physical principles. The few explanations they offered were vacuous and often bordered on the ridiculous.

In spite of their inability to explain the phenomenon of life, Physicalists were able to suppress the Vitalist objections, riding on the prestige of the discoveries made by Galileo, Kepler and Newton. Curiously, Newton himself never repudiated the vital principle. He respected 'the hand that hurled the celestial spheres into their orbits.' The success of Darwin's theory in penetrating the domain of science in spite of many gaping holes in it, has strengthened the Physicalist movement. Darwinism was accepted in the halls of science, in spite of its fragmentary nature and its failure to account for the phenomenon of life. It is curious that Darwin's book titled "*Origin of Species*" goes on explaining the mechanics of evolution without justifying the title of the book. It raised a storm of protests but its acceptance was facilitated by Darwin's friend T. H. Huxley's defense of it. Huxley, who called himself the "bulldog of Darwin," silenced his contemporary critics with his dialectical skills, without plugging the holes in the theory and without accounting for the central problem of the controversy, namely, the origin of life. Huxley conceded that some sort of evolution might have occurred but rejected the mechanical theory of natural selection. His position on the force that drove evolution was 'agnostic,' a term which he himself coined.

Another idea worth mentioning here deals with the physicalist idea of proving their theory of non-difference between inanimate matter and living things by showing that inorganic compounds can be converted to organic compounds. In 1828, a young chemist named Frederick Wohler discovered that he could convert ammonium-cyanate (inorganic substance) to urea (organic substance). The Physicalists paraded the single experiment in which one inanimate substance was converted to another inanimate substance as a proof of their theory. However, during the one and three quarters of centuries since Wohler's experiment nobody has ever succeeded either in putting life into an artificially created organic molecule or in making a cadaver stand erect though it contains the same chemicals as a living body.

Coming back to the topic of life extension, the present cleavage looks like a hangover of the earlier controversy. The unlimited-life-theorists, who reject life principle that is beyond our reach, appear to be taking the side of the Physicalists who claimed that they could create life. Limited-life-theorists appear to take the side of Vitalists, having realized the futility of decades of research in coming up with any tangible results that will enable us to keep our bodies alive forever.

Regardless of the historic controversy, we should consider the question of how we can deny the idea of life as the vital principle in academic discourses and then hanker after it when it concerns our own existence? Vitalism cannot be dead as it is linked with our very existence in this phenomenal world. By dismissing ideas simply because they smell metaphysical, as if all metaphysics is trash we will only succeed in retarding our journey towards our cherished goal of finding the mystery of life.

If the followers of Huxley tradition remember that Huxley himself said that science and religion are twin sisters, they would perhaps be receptive to other ideas. Though Huxley conceded that evolution might have occurred, he did not support the theory of 'natural selection,' and declared himself an agnostic, which meant that he did not know the mystery of the spiritual. Scientists who are wary of using the 'G' word and readers who are interested in getting a glimpse of that missing link that baffles biologists should not stop at elder Huxley's writings. They should proceed to study and discover that link in the writings of his grandsons, Aldous Huxley and Julian Huxley.

Reductionism

Physicalists view a system as fully explained if it is reduced to its minutest parts and these parts are fully described and explained. They are not concerned with their integration into the whole.

Even this physicalist-reductionism has been undergoing change. What physical and chemical laws does the embryo follow in its flawless effort to develop from the stage of a zygote to a fully developed baby? What physical and chemical laws can describe the wonderful coordination displayed by mother's body to deliver the baby? Coordination and self-propulsion are the characteristics of life that cannot be accounted for by purely mechanistic explanation. One may explain the motion of molecules in lifeless bodies by Brownian motion but living systems are not piles of molecules to be explained away by Brownian motion. Movement of molecules in living systems is characterized by **direction**. Who provides that direction? That is a question, which cannot be answered by purely mechanical explanation. In living systems there is integration of parts at every stage from the level of cells to the level of whole organisms. A living system displays this integration and coordination in its biochemical, developmental and behavioral aspects. There is something, call it what you wish, that coordinates and manages the various parts of a living system into an organized whole and gives it the characteristics of life as we know it.

The cleavage between science and philosophy has arisen mostly because some scientists and philosophers overreach their respective domains. In the middle ages, the Christian church did irreparable harm to the cause of philosophy by exerting itself in an area in which it was not competent to pass judgments. Some modern scientists tend to make the same mistake of applying science to ideas in which scientific method is irrelevant or inadequate.

Even the Physicalists, if they claim to be the followers of Cartesian model, must admit the Cartesian postulate '*Cogito ergo sum*', as all knowledge starts with self-awareness or consciousness.

Living things have consciousness and the ability to interact with their environment. This self-awareness or consciousness is the seat of self-knowledge. A scientist, who uses consciousness as a device for extracting information from telescopes, microscopes, and spectroscopes, cannot refute the claim of those, who are in search of truth beyond the domain of science, that consciousness is the vital principle on which all activity depends.

After all, machines are not self-reproducing-automata. No machine has ever built itself, replicated itself, programmed itself or procured its own energy without human intervention. How then could it be said that machines are living things?

Of late, some scientists have been recognizing the deficiencies of mechanical reductionism. New discoveries in biology, especially genetics have enabled the surviving Vitalists to face the challenges thrown earlier by the Physicalist movement.

The eminent mathematical biologist J.B.S.Haldane questioned the validity of purely reductionist explanations of life. Short of taking a metaphysical position, he began using new words like 'holism' and 'integration' to explain his position. At least this change enabled scientists not to look upon the physicalist reductionism as sacrosanct and look for alternate explanations. As a result, a new paradigm known as "Organicism" has been slowly emerging. It may not be the same as earlier Vitalism but it is definitely an ally. Organicism is characterized by the dual belief in the importance of considering the organism as a whole and the conviction that this wholeness is open to analysis.

Every system loses some of its characteristics when dissembled just as a car ripped apart. Many of the interactions of components of a living organism do not occur at the physical or chemical level but at higher level of integration. What integrates the activities and development at each higher level of integration is the genetic program of the individual, containing historically acquired information. How this information was originally obtained cannot be answered without postulating an independent entity called soul or consciousness. The eminent mathematical physicist Von Neumann showed how quantum mechanics implied that consciousness of observer is distinct from all aspects of observer's body and brain.

A number of properties distinguish the living systems from the inanimate. Ability of living systems to reproduce themselves and capacity for metabolism are well known properties found in basic textbooks. Some more important properties that are gaining acceptance are:

- Living organisms are subject to evolution.
- They are self-regulating systems maintaining steady state (feedback mechanism via homeostasis).
- Historically inherited genetic programs characterize them.
- They exhibit a dualism in their functioning. We say it is dualism of body and Soul. Those who feel shy of using the word Soul may substitute a physico-chemical nomenclature and say that it is a dualism of phenotypes and genotypes; but then they have to shoulder the responsibility of tracing each genetic program, unique to each individual, to its origin.

These characteristics distinguish the living creatures categorically from the inanimate matter.

The present confusion in aging research seems to have arisen more or less as a legacy of Darwinism, which looks upon man only as an animal interested in sex and reproduction. Even as it dismisses teleology in evolution it attributes a purpose, the all-important purpose of regulating the duration of man's life span, to the amorphous entity called "natural selection." Darwinians say that evolution is interested (notice the word 'interested'- it has purposes too) in protecting the individual only up to the time of reproduction. This conclusion of Darwinians is particularly intriguing as most of the people live longer after the reproductive stage than before it. Darwinian scientists have no explanation for this phenomenon.

Evolution, in its full sense, envisages mental as well as moral development. Rationalism is only a stage in human evolution. Evolving into supra-rational human beings is the next stage in the journey. The time in excess of that required for reproduction and rearing children is meant for making efforts to evolve into better human beings, for spiritual evolution. The evolutionary link is established by the transmigration of the soul that carries with it the fund of its old experiences.

The Western thinkers of the nineteenth century developed a very rigid frame in which the concept of reality applied only to things or events that could be perceived by our senses either directly or through the medium of tools that we have devised. Developed in this framework, the physical comforts produced by nineteenth century physics and chemistry appealed to the great masses of general public who were already under the spell of utilitarian philosophy. No wonder then, matter became their primary reality and scientists began adoring the objective world so much that they forgot themselves until they were jolted by the problems of quantum mechanics. The contradictions of the quantum theory forced them to recognize that they cannot be reconciled unless they recognized the need to incorporate their own consciousness as a part of their model.

The question of consciousness has arisen in modern physics in the context of study of sub atomic phenomena. Quantum mechanics has made it clear that these phenomena can be understood only as links in the chain of processes that terminate in human consciousness. Scientist should no longer think that he is a merely spectator. He should regard himself both a spectator and an actor and include consciousness in configuring his models whenever they are deemed necessary.

The phenomenon of life is not inherent to matter. Matter is animated by an entity called "*prāṇa*" or "soul" as it is known in sentient beings. The working of the brain, the seat of mind, cannot be satisfactorily explained without reference to the soul concept.

As long as we stick to the position that our corporeal existence is autonomous, that evolutionary changes are limited only to mindless body, there is little hope of understanding the mystery of life. Understanding this mystery is at the root of the problems of old age and death. Our success or failure in solving these problems depends on this knowledge.

Should we therefore resign to the inexorable old age and

death? Not necessarily. We have to compromise to the extent that we cannot maintain the same physical identity forever but we should be able to extend the present life expectation to the full length of the known maximum life span of humans and live it vigorously, and perhaps may be able to extend it within reasonable limit. If you are bewildered that your youth melts away before your eyes and senescence overwhelms you before you are contented, yoù are not alone. In the inimitable words of Oscar Wilde:

> There is such little time that your youth will last - such a little time. The common hill-flowers wither, but they blossom again. The laburnum will be as yellow next June as it is now . . . but we will never get back our youth. The pulse of joy that beats in us at twenty becomes sluggish.

Therefore, a more rational attitude would be aiming at post-poning death to a reasonably distant time and at living a healthy and vigorous life until the end. Our motto should therefore be: '*Stay youthful until the end.*'

That is why we find in the Vedas frequent prayers to grant us not only the full, allotted span of life but also firm limbs and acute senses until the end.

> Om, O worshipful Ones, may our ears hear what is auspicious. May we, efficient to worship, see with our eyes what is auspicious! May we, who sing your praise, live our allotted span of life in perfect health and strength![16]

Chapter Five

An Interlude

Background Theories to Help Lay Readers

> ... nature of the Universe delights above all in changing the things that exist and making new ones of the same pattern. For every thing that exists is the seed of that which will come out of it.

---Marcus Aurelius

This book is meant for everyone- general public who have a stake in longevity research, sincere seekers of longevity, health buffs, students of biology, and anyone who wants to unravel the mysteries of life and death. Even professionals may find it useful to be introduced to some refreshing ideas.

As the topic of the book is scientific in nature, understanding the background on which the themes are developed gives a better understanding. As all readers may not be familiar with such theories, this chapter is designed to help lay readers by explaining a few background theories briefly, and to save them the trouble of referring elsewhere. Even readers who may be familiar with the theory of evolution, but are not quite happy with the official version may find some new ideas in the presentation of the "Theory of evolution." Others may skip this chapter and proceed to the next without loss of continuity.

Even if one tries to render explanations in non-technical language, use of some jargon is unavoidable if one has to be faithful to the central ideas of the subject. Glossary is provided at the end of the book, which explains most of the jargon.

1.Theories of Evolution

When you ask a man in the street, what evolution is, you need not wonder if he replies; 'Isn't it the theory that man has descended from monkey?' Many people might not have even heard of the theory of evolution but whoever has not heard, about the famous (or infamous?) monkey trials of nineteenth century? Monkey trials might have become popular because of newspapermen but that does not mean all the newsreaders understand the theory of evolution and the many variations of it. Most of the readers might have heard about the Theory of Evolution but not all of them are familiar with the controversies surrounding it and the subtle differences among different versions. Essential features of these theories are presented here to help such readers.

The idea of evolution has been around us since ancient times but it has acquired prominence after the publication of Charles Darwin's "Origin of Species".

The classical writings of the Greek philosophers Thales (624-548 B.C.) Anaximander (611-547 **B.C.**) and Empedocles (495-435) contained the idea of evolution. Centuries before that, Indian philosophers elaborated their theory of cycles of evolution and involution in Vedic literature.

With the Renaissance in Europe, interest in natural sciences picked up speed and many possible mechanisms were proposed for the phenomenon of evolution. First, it was thought that species are immutable. Most of these fall into two categories, Creation with intelligent design and programmed evolution.

Creation and Intelligent Design

Intelligent design argument encountered difficulty when some fossils of animals, which were no longer existing, were found, because how could an intelligent designer create species that would become extinct. Georges Cuvier, a French biologist and paleontologist offered catastrophic hypothesis as a way out. A series of natural catastrophes removed a number of species originally created. Buffoon (1707-1778) suggested that the Supreme Being had created perfectly designed types embodied in the original species, but new species arose from them by a process of hybridization and degeneration. Species were no longer considered immutable.

Programmed Evolution

Some medieval scholars considered the story of creation in *Genesis* to be a myth. Augustine likened the work of the Creator to the progressive growth of a tree. The explanation given by St. Thomas Aquinas (1225-1274), is perhaps more plausible. He said creation unfolded by the powers given by God to the matter.

Less Rigid Program:

Erasmus Darwin (1731-1807), Charles Darwin's grandfather proposed that all life arose from primordial protoplasm. In his system, evolution was determined based on the acquisition of new characteristics by the animals themselves.

Jean Baptiste De Lamarck (1744-80), one of the best-known evolutionists, formulated a strikingly similar theory. Like Erasmus, Lamarck believed that some innate power propels an organism to perform actions sufficient to meet the needs of changing environment, but environment does not cause the change. Lamarck thought that acquired characteristics are inherited but his theory was dismissed rather hastily based on some weird experiments.

Darwin-Wallace Theory.

Darwin used his extensive observations of flora and fauna to conclude that evolutionary changes did occur. To answer the question how the changes took place, he borrowed the principle of struggle for existence and survival of the fittest, from the economist T.R. Malthus and arrived at the conclusion that organisms with unprofitable characteristics would perish and those with profitable characteristics would survive and spread their characteristics. As the instinct for survival and the involved struggle for existence are natural for all living organisms, this process is mechanical and automatic.

He took for granted the existence of abundance of flora and fauna in nature and the enormous variations within species. Then he assembled a theory of how one species gradually evolved into another species over millions of years by slow and imperceptible degrees, as set out in the following mechanism.

- Individuals within a species show considerable variation in form and physiology.
- This variation arises in a random fashion and is heritable.
- Populations of animals and plants have the potential to multiply themselves rapidly.
- Since resources are limited, individuals in a population struggle for their own existence and that of their offspring.
- In that struggle for existence only the 'fittest' (those who have more favorable characteristics to adapt to the changing needs) survive and leave offspring with their traits.

Darwin called this tendency of choosing individuals with favorable characteristics for survival, "selection." Since the choice was not done by any human agency as in artificial breeding, but by the natural instinct of the species to struggle for survival, he called it 'Natural Selection.'

Nature automatically selects the profitable characteristics (profitable in the sense of reproduction) and eliminates the unprofitable. The weaker in the struggle are eliminated by the stronger and the profitable characteristics of the survivors are preserved and passed on to the next generation. Changes in the characteristics occur by accumulation of innumerable slight variations over millions of years, leading to the emergence of new species. He contended that evolution of new species occurred automatically without the conscious action of any outside agency, or spirit.

One important implication of this theory is that Nature is interested in perpetuating a species but not the life of any individual member of it. Individuals have to die in order that new species may evolve. However, Nature protects the individuals until their reproductive period ends as it helps to continue the lineage of the species. After that, the individuals are abandoned.

The tumultuous story of post-Darwinian evolutionary theories can be roughly demarcated into three periods.

The Romantic period (1860-1903):

The romantic period was characterized by extreme enthusiasm for Darwinism together with an uncritical acceptance of whatever data were presented to support Darwinism. Contrary evidence was given no or little weight while illogical extremes of interpretation in order to make observed facts fit Darwinian theory, were quite common. Leaders of this group included Thomas Huxley, Herbert Spencer and George Romanes in England and David Jordan and A. S. A. Gray in America. As a group, they went to interpretative extremes, reading adaptive significance into every organic structure, even on the most imaginary evidence. They cited good anatomical and taxonomic evidence but the experiments to test the adaptive values were either unusual or flimsy. No doubt, they were all scholars in their fields but the atmosphere in which Darwinism got acceptance was extraordinary.

The theory of evolution arrived at a time when the climate was dominated by the materialistic determinism of physics. It was a time when mechanistic worldview was spreading as a revolt against the bigotry of medieval Church. Riding on the success of Galileo, Kepler and Newton in formulating certain astronomical and physical laws and the emergence of industrial revolution, utilitarian philosophy dominated the intellectual climate of England. Truth was measured in terms of utility. In such an atmosphere in which man was arrogating to himself sovereignty, explanations of phenomena without the role of spirit were readily welcomed. It is no wonder then that Darwin's 'Natural Selection' came as a handy answer to the materialists who were looking for arguments to confront Paley's intelligent design argument.

Agnostic Period (1903-1935)

This period witnessed a lot of skepticism and disillusionment. Many factors converged to cause this change, one of which was Gregor Mendel's enunciation of the laws of heredity, because they seemed to raise fearful obstacles to the origin of new species. Consequently, the science of genetics was regarded as a blind alley at the end of which stood the sign post 'Gene, the dead end'.

Another factor was the conclusion drawn by another scientist Johannsen from his experiments on the size of beans that selection could be effective in an organism with hereditary variability without being influenced by environment.

Finally, Many additional mechanisms were advanced besides natural selection. Moritz Wagner (1868) emphasized the effects of different environments on isolated forms. Gulik pointed out the unavoidability of divergence even without environment.

Karl Nagel suggested an inner directive guides the course of evolution mainly, independent of environment.

Followers of Mendel like DeVries played down the importance of natural selection, emphasized the importance of Mendelism, and set out to reconcile it with the theory of evolution.

Because the results of their efforts went beyond the mechanism that Darwin proposed, the synthesis that emerged is known as Neo-Darwinism.

Further the interest of some neo-Darwinians shifted from considering consequences of breeding from pairs of individuals to the consequences of breeding among individuals in populations. The Mendelian ratios then became frequencies of occurrence of genotypes in population. Genetic and evolutionary changes were defined in terms of changes in relative frequencies. In 1908, the British mathematician Hardy and the German physician Weinberg concluded that Mendelian mechanisms do not themselves cause changes in gene frequencies and that they cannot be considered to be an important element of evolutionary change. Subsequently, statistician R. A. Fisher and the mathematical biologist Haldane expressed the intensity of selective force in terms of changes in relative frequencies of genes. Those alleles that increase in frequency relative to others are said to be fitter. Therefore, the changes in these relative frequencies measure the neo-Darwinian fitness. Fisher postulated this idea in his fundamental theorem of natural selection, i.e. the rate of increase of fitness of any organism at any time is equal to its genetic variance in fitness at that time.

Modern Synthesis period

The synthesis of a coherent theory of evolution which takes into account all the pertinent facts of modern biology has been the work of a number of biologists during the past fifty or sixty years.

Julian Huxley first used the term modern synthesis in 1942. This synthesis was an amalgamation of the data and concepts of genetics, taxonomy, embryology, biogeography and other disciplines. In 1942, Ernst Mayr argued that amalgamation of changes induced by environment, mutational limitations, epigenetic limitations, random events, and natural selection have all played their roles in evolution.

Present State

At present five basic types of processes are recognized in the evolutionary change. Gene mutation, changes in chromosome structure and their number and genetic recombination provide genetic variability while Natural selection and reproductive isolation work to bring about the change.

Punctuated Equilibrium Theory

This theory proposed by paleontologists Stephen J. Gould and Eldridge is the modern variant of Darwinian type of evolution with the caveat that selective change occurs in bursts, rather quickly, not in millions of years as conceived by Darwin. The change having occurred in bursts, the new state known as stasis persists for long periods. It changes, stops, and changes again quickly instead of changing slowly and imperceptibly. They called this process, 'Punctuated Equilibrium'. Many biologists consider this a plausible explanation as it explains the missing links in the fossil records.

Epigenesis:

We cannot give the exact definition of this word, as it comes from an area yet hazy. Chambers dictionary gives the meaning of the word as 'development of the organism by growth and differentiation of a single germ - i.e. by division or segmentation of a fertilized egg'. Webster's new world dictionary of American language says it is 'the theory that germ cell is structure less and that embryo develops as new creation through the action of environment on the protoplasm'. But what biologists consider now as 'epigenesis' is the phenomenon of changes brought about in the development of an organism, by influence of environment, without changing gene sequence.

Genes have generally been regarded as instruction booklet for the body, but it is now realized that genes themselves need instructions to do their work – what, where and when? These instructions are not found in the letter sequence of the genes. Instead, they reside as chemical markers on the double helix.

The collection of all chemical markers lying along the length of the double helix is known as epigenome. These perform the function of switching on and off the expression of various genes. It has recently been discovered that the epigenome is sensitive to signals from the environment. If the signals commanding the gene action could be altered by environment, it amounts to saying that environment can influence action of genes without altering their code and produce changes in the phenotype. A more surprising discovery is that the altered characteristics so produced can be inherited. This is a far-reaching conclusion that may upset the apple cart of hard core Darwinism.

2. The Story of the incredible Cell

Cell-The Incredible Storehouse

A cell was originally defined as the smallest unit of an organism that is alive. For today's biologists, a cell is nothing more and nothing less than a special collection of complex molecules, enclosed by a membrane and having the very special ability to reproduce itself from the much simpler molecules available in it. A chemist defines it as a self-replicating collection of catalysts - most of which are proteins. A living cell is composed of molecules, many of which are constantly converted into other molecules through a series of steps called metabolism. The structure of the cell is very complex which is a miracle of design.

Under the outer cover, there are sheets of membranes called endoplasmic reticulum that serve to transport proteins manufactured inside the cell. Between these sheets and the core structure called nucleus, there are structures called ribosome, mitochondria, and centriole. Ribosomes are structures on which proteins are assembled using amino acids. Mitochondria are production centers for ATP, the molecules that supply energy to the cell. Centrioles that are close to the nucleus play an important part in cell reproduction. The organization inside the cell is like that of a small city-state wherein factories produce proteins, powerhouses generate energy and complex transportation systems guide specific chemicals from place to place within the cell and beyond. Sentries at the barricades control the exports and imports and guard the cell from outside danger. Disciplined biological armies stand ready to repel invaders while a centralized genetic government is in charge of general administration. The structure of a cell is indeed a miracle.

Cells - Varieties of

Cells are classified by the function they perform. There are varieties of cells known as germ cells, somatic cells, stem cells etc.

PROKARYOTIC CELLS

Prokaryotic cells are believed to be the progenitors of eukaryotic cells. They are seen in primitive organisms like bacteria and blue-green algae. A distinct nucleus with nuclear membrane is absent. Instead, all the genetic material is concentrated at a single place in the cell called the Nucleoid or Bacterial Chromosome. It is devoid of any membrane surrounding it. The DNA present in the nucleoid is naked. Reproduction takes place by amitosis and fission.

EUKARYOTIC CELLS

Eukaryotic cells are those enclosed in a nuclear envelope and having one or more membranes bounded by organelles in their cells. These cells make up all multi-cellular organisms like algae, fungi, plants, animals and some unicellular organisms like protozoan. They have a distinct nucleus with a nuclear membrane. Reproduction takes place by mitosis in somatic cells and by meiosis in germinal cells.

SOMATIC CELLS

The word somatic refers to body. Somatic cell means a cell pertaining to body but not a germ cell.

STEM CELLS

The origin of stem cells is the fertilized egg known as zygote. We start our lives as zygotes or single fertilized eggs. In time, this divides into two cells, two into four, and so on in geometric progression to become the embryo. At this early stage of development, the embryo has no resemblance at all to what we are, but the DNA in the nuclei of these cells contains the programmed information for developing gradually into coordinated limbs and organs of the body in course of time. These cells are called stem cells because they are like plant stems that spread out into various branches. Out of these, emerge other cells that go to make other organs.

GERM CELLS:

Germ cells are a particular type of cells, which are responsible for creating the next generation and are known as sperm cells and egg cells. These are also called gametes.

Cell Division

MITOTIC DIVISION

All living matter consists of cells. Growth of living matter takes place by cell division. The first cell divides into two, these two divide into four, each of them divides into two again and this process continues for some time. This process is known as cell division or proliferation. Although organisms exhibit wide variety of reproductive patterns, they can be divided into two basic categories, sexual and asexual. In asexual reproduction, a single organism gives rise to offspring by division, with characteristics essentially identical to those of parent. This is known as mitotic cell division.

MEIOTIC DIVISION

For sexual reproduction, another kind of division is necessary. This involves production of specialized cells known as gametes. In animals, female gametes are called ova or egg cells. Male gametes are called sperm cells. A new individual of the species is conceived when a male gamete fuses with the female gamete resulting in a single cell known as zygote or fertilized egg. From this stage, mitotic process continues again.

Meiosis is a process in which two chromosomes derived from a father and mother are paired, genetically recombined and segregated to form four gametes by two successive cell divisions.

A human adult consists of some 100 trillion cells. Consider how you, a complex of 100 trillion cells came about. You began as a single cell when the sperm of your father united with an egg cell of your mother. At that time of union, plans were drawn up within the DNA of the newly formed cell to produce what eventually became you. One cell became two, two became four, they became eight and so on. This process continued for about 9 months during which time thousands of millions of cells of different kinds developed in the body of your mother and eventually you were born as a beautiful human baby. It happens as though the first cell had a shelf full of blueprints containing instructions on how to make you as a unique individual with specific physical and mental features.

3.The Story of DNA:

DNA is an acronym for Deoxyribo Nucleic Acid, a long and thin fiber-like chemical substance. This stores all the information necessary to control the chemistry of the cell. DNA fibers are found in all cells except the red blood cells. These fibers dictate the behavior of the cells. Thus, DNA controls our body by controlling the chemistry of our cells.

DNA and its connection to Inheritance

The early investigators of inheritance thought about it, naturally, as a mixture of characteristics drawn from the mother and father; but actual experiments proved they were wrong. When two different dogs were crossbred, puppies were produced which resembled either this or the other, but not both. Similarly, plant breeders realized that when they cross breed two different varieties of peas the outcome may not necessarily a mixture of both; it may be a plant resembling completely the one or the other. This puzzle was not resolved until a Moravian monk Gregor Johann Mendel drew his conclusions from his systematic experiments. He explained that inheritance occurs by drawing together of what he called 'primal characters' from each parent but sometimes one of them is obscured because of the dominance of the other. This means, for example, the child of parents one of who has a squint eye may have normal eyes, because this squint factor is obscured by the dominant factor. The squint factor may however acquire dominance in subsequent generations and squint-eyed children may appear again. Thus, Mendel concluded that living things are a mosaic of independent factors.

Subsequently these factors have come to be known as genes after the structure of cells was deciphered. While Mendel was working in obscurity, biologists were working in their labs on cell structure. With the aid of powerful microscopes they discovered that the cells contain within their nuclei, a dark, long rod-like structure. When a somatic cell divided, these rods split longitudinally bestowing on the daughter cells similar assortment of rods. But when a germ cell (sperm cell or egg cell) divided, this mechanism differed; only half the number of rods was passed on to the daughter cells. When the sperm cell and the egg cell fuse together, the total number of rods found in ordinary cells is restored. Thus, the two halves of rods in both varieties of cells are perfect complements to each other in producing life. The dark black rods in the nucleus of the cells are called chromosomes. So, these chromosomes are the material that carries heredity.

The next stage in the story of DNA consists of one the most seminal experiments ever conducted in the field of biology. Oswald Avery, an American chemist, and others conducted experiments on two strains of pneumococci, one with a smooth coat and the other rough coated. He extracted some substance from the smooth coated, placed the rough coated in that substance and allowed them to divide. He observed that some of them acquired smooth coats and then when they divided, in turn transmitted that characteristic to their offspring. He recognized that this substance caused the transmission of characteristic from one generation to the other and purified it. On purification, he found that it was an obscure acid called Deoxyribo-Nucleic Acid, which is now known by its acronym DNA. A decade later, it was shown that this DNA was a complex molecule with two strands twisting together like a helix.

How is Information stored in DNA?

Different regions of DNA contain different bits of information. Specific regions of information are called Genes. DNA can be thought of as a motion picture film. Like the film, DNA is subdivided into frames that make sense only when viewed in proper order. Frames in the motion picture film correspond to genetic letters in the genetic code. In cinemas, film editors create scenes by organizing a specific combination of a number of frames. Similarly, Nature organizes combinations of genetic letters into distinct units of what are called genes. Specific combination of a number of letters yields specific genes.

Information is used in the manufacture of proteins. Proteins are chain like molecules that fold in a particular way to form specific structures. Some proteins contribute to the architecture of the cell while some control the cell chemistry. It is the specific information in the genes that makes a human being different from a bird or a monkey. Information in your DNA makes you different from anyone else on the earth, unless you have an identical twin. It is well known to the readers that DNA has come to be used as clinching evidence in criminal trials. It is this uniqueness property of DNA that has made it a criminal investigation tool.

The Structure Of DNA

In 1953, James Watson and Francis Crick proposed a structure for DNA, which is now regarded as one of the most important scientific findings of modern biology. DNA is a long thread-like molecule composed of sub units called nucleotides. It resembles a two-stranded beaded necklace. They called this structure as a double helical structure. This structure has guided the biologists to such an extent that the chemistry of heredity can now be clearly explained and manipulated. DNA molecules can be very long, sometimes containing more than a 100 million nucleotides. Four rules govern the DNA structure.

- DNA strands do not have branches. This makes it possible to have a helical structure.
- The ends of DNA strands are chemically different.
- DNA strands have directionality.
- The two strands of DNA are complementary.

There are four different nucleotides called Adenine (A), Guanine (G), Cytosine(C) and Thymine (T), abbreviated by their first letters. Genetic information is stored in DNA through the specific order of their nucleotides. Nucleotides in the two strands pair with each other so that an A is always opposite to a T and a G is always opposite to a C. The two strands are strictly complementary. One can imagine that they fit like electrical plugs and sockets. The two strands of DNA tend to stick together tightly because of millions of tiny bases fitting together and stacking on top of each other.

Organization of Information in DNA

The sub units or nucleotides of DNA are the chemical basis

for storage of information in DNA. The genetic letters mentioned earlier A, G, C and T represent the abbreviations of chemical base names. Thus, information is stored in DNA by the specific sequence of a 4-letter alphabet. A *gene* corresponds to a scene in a motion picture film. The information in genes is arranged in words. Genetic code is read in triplets of nucleotides. A group of Nucleotide-Letters is known metaphorically as a Gene-Sentence. A group of Gene-Sentences makes up a Chromosome-Book. A group of Chromosome-Books makes up what is called the Genome-Library.

It should be understood that the words 'letter,' 'sentence,' 'book ' and 'library' are metaphors for nucleotide, gene, chromosome and Genome respectively. Every cell in your body has a reference library from the past. You can think that half of the books in this library were donated by your father and that the other half by your mother as you inherit one set of chromosomes from each of your parents. All the books were shelved together in a common library called human genome.

THE GENOME-LIBRARY:

1.DNA Base - - - --Letter

2. Gene - - - -- - --Sentence

3.Chromosome- - -Book

4.Genome - - -- - -Library

DNA library is divided into 46 chromosome-books or 23 paired volumes. The library has two main functions - development and maintenance. The development aspect tells the cells what to become as the cells divide- whether to become a blood cell or a nerve cell or something else. The library also directs the housekeeping chores of the cell, viz. repairs and maintenance continuously throughout life.

DNA Replication:

Replication in general means repetition. In the context of biology, it is the technical term for reproduction. Reproduction is defined as the duplication of information content of the cell, followed by segregation of information into two newly formed daughter cells. Since information is stored in DNA, knowledge of how DNA duplicates is essential for understanding reproduction. The purpose of DNA replication is genetic continuity. To ensure this continuity, DNA must not only be stable but it must also be accurately copied. If not, characteristics of successive generations will not be the same. Copying is accomplished with amazing fidelity.

As the DNA strands begin unwinding, they form an inverted V-like fork. The inner walls of the V-fork are now used as templates for forming new strands. As the fork is separating, an enzyme called DNA polymerase mediates, linking the free nucleotides to the growing new strand.

The template walls are now filled with the nucleotides using the complimentarity rule. Each daughter cell then consists of one old strand with the same sequence as that of the mother and the other with the complementary sequence. The replication fork moves through the double stranded molecule like an unzipping zipper. Thus, each new strand becomes a copy of the mother strand.

CHAPTER SIX

Battles to Beat Aging

The State of Aging Theory

> Death, be not proud,
> Though some have called thee
> Mighty and dreadful, for thou art not so;
> For those whom thou think'st thou dost overthrow
> Die not, poor Death; nor yet canst thou kill me...
> One short sleep past, we wake eternally,
> And Death shall be no more; Death, thou shalt die!

------John Donne (Sonnet on Death)

Competing Theories

You heard the joke that only two things are certain in life and they are taxes and death, but now biologists are saying that distinction should belong only to taxes. Like John Donne, they throw a gauntlet at death though Donne's challenge is figurative.

The ageless quest for postponing aging and to stay alive long and healthy appears to be approaching a watershed. However, gerontology, the science of aging is now in a state of ferment like any other science undergoing revolution. Many sound theories are competing with their interpretations of aging phenomena and observed experimental facts.

One theory says that aging and death are pre-programmed by Nature and they unfold, in time, as they are programmed. Another theory argues that aging is the consequence of the deterioration of the immune system. Yet, another theory holds the imperfections of the DNA repair mechanism responsible for aging. Some other minor theories have different postulates, but no theory seems to be in a position to assert its supremacy, at least for now. That is how scientific revolutions occur.

In the ferment that precedes scientific revolutions most of the theories appear to be well reasoned, well documented, self contained and all appear to be reasonably likely. All of them seem to explain the same set of facts more or less satisfactorily. All work within the existing paradigm. After a period of intense competition in which there is no decisive winner, suddenly one of the competitors stumbles on some experimental result or achieves some special illumination that clinches the issues; or one or more of the theories are synthesized to produce the sought after result.

Until recently our knowledge of aging has been like a puzzle spread out before us with many missing pieces. The individual pieces are crisp, well-defined, patterns but they do not make up a cohesive picture. A few tried to find the missing blocks while some tried to assemble the puzzle but there were no clues as to how to fit together all the pieces to form a picture. Nobody knew how to complete the puzzle. Things started to change recently as scientists labored incessantly to supply the missing piece. As the new pieces are fit one by one into the gaps of the puzzle, hints of the shape and nature of the whole picture have begun to appear. The theme of the picture appears to be cellular.

As there are many theories of life, aging and death making rounds, we may classify them first into four broad categories:

(A) Theories based on the idea of Evolution,

(B) Those based on Molecular Biology,

(C) Replacement Theory,

(D) Esoteric phenomena.

These can be further subdivided depending on the nature of the problems and the branches of knowledge that try to tackle them. As there are various overlapping theories with numerous branches, the classification should be viewed only as device to organize and understand and not as watertight compartments.

Among the theories, explaining the phenomenon of aging, the theory of evolution is the oldest and has been firmly established on logical grounds, but it doesn't offer any solace to the gerontologists as it considers death is immutable. External program theory and the entropy theory come under this category. Other theories that have been in quest of immortality are in a state of disarray.

(A) State of the Theories based on Evolution

The theories of life and death based on the theory of evolution have not budged a bit from their original position as they are based on logical grounds and mathematical demonstrations.

(A-1). External Program

As you may know, the theory of evolution is based on the idea of natural selection. Natural selection is the name given to a force that operates in nature to eliminate the weak and to promote the survival of the fittest. Evolution is said to be interested in perpetuating species. To promote this objective, it selects and protects characteristics useful for reproduction. Therefore, heritable traits prevail in a population while the opposite traits are gradually eliminated before the reproductive age. Thus most useful traits result in the most offspring and hence the greatest perpetuation of the genes controlling these characteristics. Similarly, the traits that are injurious to survival in youth become less common because the carriers of those traits do not live long enough to reproduce.

In contrast, those genes that are antagonistic to later life accumulate as their carriers pass them on to the next generation before they interfere with reproduction. However, the later these genes lead to disability in life, the more widespread their prevalence becomes because their carriers reproduce longer. This bias of Nature towards reproduction fades out after the reproductive period is over and it acts against the animal with the same fervor, as it showed against the non-heritable traits in earlier years. This results in accumulation of genes in a population that act against the individual in later years.

Aging thus creeps into population because the process of natural selection, which was favorable to them in young age, has built-in bias against later years. Thus, the guardian angel that protected you before you reproduced becomes the monster afterwards and robs you of your vitality. This idea is illustrated in Fig. (6-1).

Evolutionary biologists say that neither there is a single reason that is responsible for aging nor is it derived from any genetic program. It does not occur because of a single universal defect in all cell types.

Fig.6-1: Adaptability to Natural Selection with advancing age

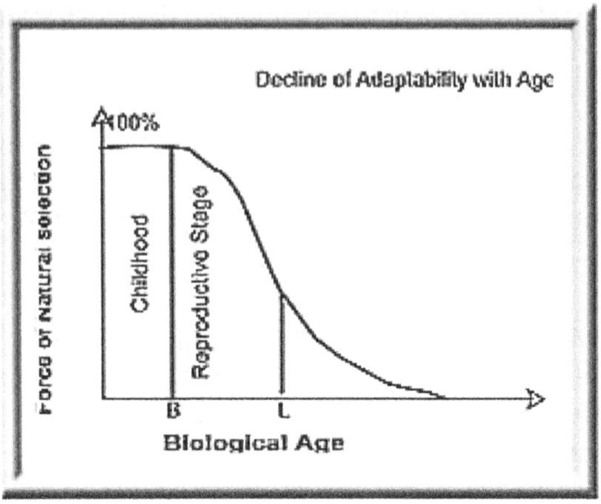

Adaptability to Force of Natural selection begins to drop after the earliest age of reproduction, say position B in the diagram is reached. It continues to drop almost vanishing after age L.

If some single flaw causes all systems to fail, no animal can escape aging, but some animals do escape. Some primitive creatures are virtually immortal. Sea anemones kept in aquariums live on for decades looking hale and healthy. Among the more complex organisms are the jellyfish produced from a stub known as schyphistoma larva. Though each animal produced from the stub has a limited lifetime, the stub appears to carry on indefinitely. Therefore, they are regarded as immortal.

The argument of the evolutionists is that senescence is the byproduct of natural selection, which afflicts human beings and other vertebrates but not asexual creatures like anemones. They say that death has become part of Nature's plan of evolution that favors creatures that die after reproduction in order to enable it to modify the species by natural selection to meet changing conditions. Species of creatures, which live forever without reproduction, become extinct. A new species, for better or worse, can start only with new life. Therefore, they concluded that death is an evolutionary necessity. That, however, does not mean that molecular scientists gave up.

That explanation, naturally, raised the question: if the ability of the body to adapt to the force of natural selection (force of natural selection really does not decline as some people erroneously express), why not we delay the beginning of reproduction which expands the life span in its youthful stage. If the active phase of natural selection could be progressively increased over a number of generations by postponing reproduction, it is predicted that aging can be postponed. This prediction was tested on short living organisms like fruit flies (Drosophilae) and found to be true. By manipulating to delay reproduction in these flies for over 10 or more generations scientists have now gathered stocks of flies that would live healthily three to four times the normal span of their life. The flies that displayed postponed aging were said to be 'surprisingly perky' and sustained normal biological functions for longer periods and showed superior capabilities at all adult stages

Can we solve the problem of aging in human beings in the same way? It depends on the willingness of people in several generations to postpone child bearing. It is not known clearly whether it means sexual abstinence. If it means that, the chances of enforcing this method in human beings are very low it may remain a utopia.

(A-2): Disposable Soma theory:

The name of this theory is derived by the analogy with the disposability of short-term durable goods. Soma is the generic name used to distinguish body parts other than those that are involved in germ line — the parts engaged in production of sperm and ova.

The theory is based on the economic idea of optimal allocation of scarce input resources in their alternate uses. The fixed inputs in this case are some indefinable life maintenance forces. Alternate uses are reproduction and continuation of species on the one side and body maintenance and longevity on the other side. The theory says that because Nature is more interested in perpetuating species than in body maintenance, it allocates more resources to the former to the detriment of the latter. Thus, Nature's bias disposes off the soma as it considers them as disposable. This apathy of Nature causes aging and death. As the inputs here are not quantifiable, the theory is not testable.

(B). State of Theories based on Molecular Biology

In contrast to the macro-view of evolutionists, molecular biologists take a micro-view and try to look for the aging mechanism inside the body itself. They argue that if there are certain causes of aging, removal of those causes should stop the aging process or at least decelerate that process. That hope has awakened the slumbering field of gerontology. Scientists in life sciences community are now agog with anticipation of some important breakthrough that would lead them to the elixir of life and they have brought the field of gerontology to the forefront from the backwaters where it lay for a long time. They seek to tackle the problem by one or more of the several methods given below.

- Discovering the identity of clock-metaphor and trying to stop it or reset it.
- Reprogramming the genetic code.
- Damage control by developing physiological methods for limiting damage occurring to DNA and / or increasing the rate of repair to DNA.
- Protecting the autoimmune system of the body.
- Achieving control over the metabolic rate.

This area of investigation has been a jigsaw puzzle, in which some blocks were not available or it is not known how the available blocks fit in to complete a cohesive picture. However in recent times some missing blocks have been coming up, and insights are developing as to where these blocks fit in to form a picture, as many scientific disciplines like biochemistry, cell biology and gerontology have been coalescing. We will now review the several methods under consideration for solving the mystery of old age.

(B-1) Internal Program

Program theories envisage that the march of events like childhood, adolescence, youth, old age and death follow a predetermined course and unfold gradually according to the instructions contained in the program. Molecular biologists hold that this program is internal to the body and vow to unravel it.

Program theories are metaphorical, the underlying presumptions being that our bodies are organized like computers with stored programs, that the code is written in genetic language and that survival, aging and death take place as per the commands in the genetic program. It is the software that is important here, not the hardware. According to program theory, aging is intrinsic. Childhood, adolescence and adulthood are all preprogrammed for the body-computer and controlled by the genetic code in the DNA tape that influences the metabolic rate or the fundamental cellular processes or both. Who wrote the code? Materialists avoid this question, as they cannot answer it. You may say 'God' if you wish!

(B-2) Genetic Manipulations

Genetic Manipulations envisages deliberately inserting specific desired factors (genes) into the genetic material or removing undesirable factors or a combination of both.

Actual genetic manipulations are complex laboratory techniques involving isolating certain genes, cutting DNA and inserting them again into the genetic material with viral carriers.

Some scientists have been skeptical of such possibilities. On the other hand, there are optimistic scientists who foresee great strides in genetic engineering. Letting alone the theoretical objections voiced above, the major problem has been to send the genes into the human body and make them behave, as we want while avoiding dangerous side effects.

Gene cloning

Classic gene therapy is based on a simple idea. Since genes direct the assembly of every cell in body, it should be possible to treat chronic diseases by inserting corrective genes into patients. It may be interesting to the reader to understand a little about the method of gene therapy. The relevant question here is even if we identify the particular gene needed to treat a given disease, how can we send that gene into millions of cells of the body to do our job? The answer comes from a simple extension of our common experience. When some disease-causing virus attacks us, the virus spreads rapidly and cripples the whole body. Taking cue from this idea biologists have hit upon the idea of using the viruses themselves to do the gene insertion job. Broad details are like this: A virus which is known to be relatively harmless is genetically altered to render it as safe as possible. Then its DNA is extracted. Into this DNA, the therapeutic human gene is inserted and this compound DNA is inserted back into the virus. The virus is then injected into human body. As the virus has the quality of rapidly multiplying itself, it is expected that the virus carry the modified genes into various cells of the body. With everything working in our favor including some luck, the modified gene is supposed to spread all over the body and do its job of curing the disease.

Scientists have developed techniques of isolating required strands of DNA and splicing them into vehicles (or 'vectors' as they are technically called), which can penetrate cells. But getting the body to accept these vehicles and give expression to the therapeutic genes has been a tremendously difficult job. When the vector is introduced into the body, it sets off an immune reaction in the body that destroys the needed gene.

Gene Therapy

There is some confusion between the phrases "Genetic engineering" and "Gene therapy" due to indiscriminate use of words. To avoid such confusion, we use the phrase "Genetic therapy" to refer to treatment of specific diseases by genetic manipulation; while we reserve the phrase "Genetic engineering" to refer to producing large scale changes affecting species characteristics. Genetic therapy may be viewed as a piece-meal operation aimed at curing specific ailments by trial and error process. This involves identification of gene-characteristic relationships and involves either introducing appropriate genes into human cells or knocking out bad genes. The conceptual basis of gene therapy is attractive. Instead of attempting to treat diseases with small molecules (drugs) that temporarily ameliorate metabolic disorders resulting from genetic defects, gene therapy proposes to correct the genetic defects themselves. Despite the promise of this concept, this field has utterly disappointed the scientific community for the past two decades in spite of occasional hints of success.

The foundations of gene therapy were rooted in the advances made by molecular genetics, cell biology and virology during 1950's and 60's. A few scientists suggested the possibility of obtaining therapeutic effects by introduction of genes into mammalian cells.

The aim of genetic engineering is to produce a desired form of creatures by tinkering with the genetic material. This may involve deletion of some undesirable gene(s), introduction of a desirable gene(s) or modification of some existing genes to yield desired result. Some gerontologists are very enthusiastic about the research that is going on in engineering genetic changes.

Curing Specific Diseases with Gene Therapy

Gene therapy was already used with success to cure a condition called SCID (Severe Combined Immune Deficiency). A four-year young girl was afflicted with immune deficiency and was exposed to all sorts of infections. As she inherited a defective gene from her parents, her immune system was shut off as the defective gene was unable to produce the enzyme called adenosine deaminase, required for proper functioning of immune system. A team of specialists at The National Institute of Health used genetic therapy in 1990 to treat the girl and restored her to normal health[17]. They removed the white blood cells of the immune system from her body, inserted normal and healthy copies of the defective gene into the cells and returned the modified cells into her blood stream. After four months of treatment, the girl is reported to have returned to normal health and started enjoying life in full bloom.

If proper corrective genes can be identified and safe methods of delivery to the targets are perfected, patients suffering with other diseases like cancer, cystic fibrosis, hemophilia, Hunter's syndrome, Rheumatoid arthritis and AIDS will soon become candidates for this therapy. Medical doctors will be able to inject the needed genes directly into blood stream. The genes contained in viral carriers will seek out targeted cells and unload the genetic material that will then produce necessary proteins to help fight diseases.

At present understanding of the precise genetic basis of many diseases is sketchy. After the recent unveiling of the Genome, these prospects have increased, as scientists now know the DNA code of more than 99% of active human genes. As research on uncovering the function of each gene progresses, rapid progress in gene therapy can be expected if only safe delivery methods are also perfected simultaneously.

Identification of Specific Genes

The prerequisite for obtaining the desired effect is to know what gene causes what effect. As already mentioned a single gene may not cause a single effect. Unfortunately, there is no such one to one relationship. There may be one to many relationship or many to one. That complicates the matter and makes it hard to identify the relationship between genes and aging process. Then, hosts of practical difficulties have often plagued the scientists.

A Theory of Yeast Genetics

Even as the evolutionary biologists have been cautioning that any prospects for genetic intervention should be explored with circumspection, molecular biologists have been expressing hope that they will be able to identify the mechanisms underlying the aging process with their empirical work. They claim that the first validation of the role of genes in life span extension came from the extended longevity of lower organisms like fruit flies but in this case, the role of genes was conjectural as no specific gene or genes were identified. The second and more important validation came from extension of the life span of the round worm as a specific gene was identified as responsible for modulating the life span. This is the first identification of a longevity gene.[18, 19]

Recent reports claim to have discovered a few such genes in nematode roundworm (*Caenorhabditis elegans*) and the fruit fly (*Drosophila melanogaster*). Some genes capable of lengthening life by about three to four times were reported to have been identified in *C-elegans*. If that could be done to humans, they say it would mean, adding 200 more years.

The other report is from MIT[20] where experiments conducted by Dr. Lenny Guarente and his staff, on yeast is reported to have yielded similar results. The report claims to have found a fundamental defect in the nucleus of yeast and they believe it is the cause of aging in yeast cells. According to their work, the genetic material called ribosomal DNA seems to hold the key to aging in yeast. Yeast has a string of vital DNA called ribosomal DNA (r-DNA for short). Every now and then, the report says, the r-DNA gene pinches off a small circle from the chromosomes (we don't know why) and the circle replicates several copies of itself eventually overcrowding the small area called nucleolus disabling it in its function of protein manufacture. That is the defect, which causes aging in yeast. They say the culprit is the ribosomal gene.

This argument cannot be applied to humans, argues Dr. George M.Martin of the University of Washington at Seattle, because unlike yeast, humans have two kinds of cells-those that divide often during lifetime, like skin and intestinal cells and those that never divide after maturity, like brain cells and muscle cells. Even if the yeast mechanism is found to occur in the dividing type of cells in humans, it cannot tell us anything about aging in muscles and neurons. This argument brings us back to the round worm whose cells never divide again after maturity.

Alternative Theory of Yeast Genetics

The most recent theory coming from the MIT scientists claims to have found a gene that governs the rate of aging in yeast cells. This gene's role is to silence the DNA in yeast cells. Silencing means, this gene switches off some groups of genes in the chromosomal region and prevents the cell machinery from gaining access to gene transcription machinery. This DNA suppression in turn leads to suppression of production of waste genetic material, which clogs up the cells. According to Dr. Guarente and his colleagues[21] at MIT, this silencing of DNA is said to have been accomplished by shaving off a chemical group called acetyl group, from the sites around chromatin, the dressing material of DNA. Loss of this chemical group brings the chromatin closer to DNA blocking access to the cell's gene transcription machinery. It has been known for some time that this chemical group acts like a gatekeeper opening and closing the chromatin. Adding the chemical opens the gate and withdrawing it results in closing it. It was not exactly known what the protein is that accomplishes this task, though protein called SIR-2 (Silent Information Regulator No.2) was a leading candidate. Dr. Guarente and his colleagues have now confirmed that SIR-2 is indeed the silencer and its silencing activity results in extension of life span of yeast. They knew this since 1997, but the present discovery is the mechanism through which this life span extension occurs. That is, this SIR-2 protein cannot accomplish the task of silencing alone; it can do it only in the presence of a complementary chemical called NAD (Nicotinamide Adenine Dinucleotide).

This discovery provides an important link to the association between caloric restriction and longevity. Many empirical studies have concluded that caloric restriction prolongs longevity but the biochemical pathways through which it is done are not exactly known. Perhaps it is done through the energy metabolism path of the cells' NAD since NAD is an energy-carrying chemical. Perhaps it's concentration increases when the cells are in a state of low energy production mode. These MIT scientists suggest that increase in NAD spurs the activity of SIR-2. That link completes the chain connecting caloric restriction to longevity, as the increased activity of SIR-2 is known to prolong longevity.

Shortcomings of the Genetic Manipulation Method

Notwithstanding the enthusiasm of a few biologists, there is a serious problem in genetic manipulation efforts. The relationship between genes and characteristics of living beings is not so simple as to have a one to one correspondence. A particular gene may produce a number of characteristics like color of skin; texture of hair and height of body while a group of genes may be responsible for a single characteristic like intelligence. For example, if you manipulate a gene with the intention of changing the color of the eyes, it may not only change the color of the eyes but also may damage vision. If you remove or repress a group of genes, with the hope of eliminating a disease you may succeed in that aim but you may also end up producing a stupid. It is important to note that we cannot eliminate an undesirable characteristic by eliminating one or two genes.

The characteristics and their causative factors are so overlapping like the warp and woof of a fine colorful fabric that you cannot manipulate the genetic material at will without causing damage to the very fabric of life.

The Safety Controversy

Safety controversy revolves around two aspects. The therapy, instead of being useful to the patient, may endanger the patient. The bacteria modified in the process may become pathogenic and accidentally escape into the environment. In such an event, lives of others who are in no way concerned with that process may be endangered.

If the cloning vehicles are not right, they may create side effects that can prove to be fatal to the patient.

This danger is highlighted by the death of a volunteer due to adenovirus side effects in the gene therapy experiment of the University of Pennsylvania. This has led some experts to stop such dangerous experiments. Yet, in another blow to the gene therapy enthusiasts, another death involving gene therapy has come to light very recently. The Food and Drug Administration has accused the researchers at St. Elizabeth's medical center in Boston, of having failed to report the death of a patient under their care for gene therapy.[22]

After success with their gene cloning experiments in 1970's, scientists realized that this type of genetic manipulation might result in health hazards. Such hazard was not observed in laboratories but several scenarios in which the frightening situation can occur could be imagined. For example, take the case of botulism. Botulism is a food poisoning condition caused by toxins produced by a kind of bacillus. Digestive tract is not the normal habitat for these organisms but once they go into the body, they kill their victims fast. Digestive tract is the normal habitat for the bacteria used in many gene-cloning experiments. Suppose now that the gene containing information for botulism toxin is inserted into some harmless bacteria and they escape accidentally into the environment. Disaster strikes. Anyone infected with these toxin-producing bacteria soon succumbs.

This is not to say that the scientists who developed the gene cloning were not aware of this danger and did not recognize the need for strictly controlling the laboratory operations to prevent such accidents. They do understand the danger of such accidents and observe all precautions to prevent the escape of bacteria from their jars and test tubes, but as the gene cloning process proliferates and hundreds of scientists all over the world perform the same kind of experiments the probability of such danger increases. A few scientists from any particular laboratory will not be able either to prevent such an accident or to arrest the spread of the effects of such an accident.

However, most scientists do not view the situation with alarm for various reasons:

- First, recombination of DNA occurs in nature and therefore the related technology cannot be viewed as hazardous.
- Second, recombined DNA, which sits in isolated test tubes, is transferred to living cells only under strictly controlled conditions. Hence, it has no scope to escape from its premises. Even highly developed strains of pathogens have been successfully contained in the laboratory.
- Third, the strains of bacteria used for cloning are generally weak and require special nutrients for their growth. So, they cannot survive outside the laboratory.
- Fourth, the strains of bacteria used for cloning do not automatically convert themselves into pathogens, as pathogenic organisms have sophisticated mechanisms to infect their hosts. For these reasons, the protagonists argue, the procedures are not dangerous if prudent care is taken. After the University of Pennsylvania incident, biologists have tended to be cautious though they are proposing safer vectors.

That is all fine as far as we are concerned with dangers through accidental release of the pathogens. There is different scenario that causes concern. What would happen if an aggrieved scientist or a weird lab-worker releases them intentionally? In addition, the problem of terrorists taking control of these pathogens and terrorizing the world can never be ruled out. There is no defense against such an event. The only protection the public has against such a contingency is the honesty and diligence of the scientists who agreed in the Asilomar conference of 1974, to regulate themselves in safe pursuit of this technology. We do not know how effectively they can regulate themselves.

Of all the problems created by scientific research, the imminent power human beings to alter the genetic structure of their own species is the most serious. Attempting to crack the human genetic code and tinkering with it is a challenge to creation itself and may end up as a self-destructive process for the humanity.

Reprogramming the Genetic Code

This theory postulates that the program of unfolding of various bodily stages does not lie somewhere outside the body but it is internal to the body. That program is written in the special programming language of the body but the dialect of that programming language is not English. It is a chemical language. If it were so, the biologists hope to reprogram it to serve their will. Genetic programming should imply an integral plan of all the steps and their proper sequence, which when let go will unfold in time without further intervention and leads to the desired result. While we do not go into the details of how the desired changes are engineered, we will consider various possibilities and the consequences of genetic manipulation. But what is the genetic code anyway?

Computers do not understand the English language as we use it for speaking and writing or do they. They work with English but its syntax and grammar are entirely different. The language with the special syntax and grammar they can understand is called programming language. Programs are written in that language to achieve a specified objective. The sequence of statements written in that program is called the source code of the program. If somebody wants to modify the objectives of the program, they have to change the source code. Similarly, if somebody wants to change the functioning of the body, they have to change the code of the programming language of the body. The dialect of that language is chemical and the sequence of letters written in that chemical dialect is called genetic code. It is called genetic code because the material with which the letters are formed is neither chalk nor ink but genetic material hidden in the cells of the body. Therefore, genetic code is the metaphor for the sequence of the names of chemicals hidden in the DNA tape.

If we want to change the objectives of a computer program, first we should know the source code of the program. Similarly, we should know the code of the original program of the body to enable us to make any changes. The code of that original program of the body is genetic program.

The length of the program is enormous surpassing any known computer program as it runs into billions of characters. The totality of all these letters is called Genome. Using a string of metaphors each chemical base of DNA is called a letter, gene a sentence, chromosome a book and the genome a library. This genome has to be known first to think of any corrections to the great program.

The Genome

Deciphering the code of the great books of life in the genomic library gives us a lot more information than piece-meal identification of a few genes here and there. It is expected that knowing how the genome works will place us in a better position to understand how the great edifice of life is built, how the various structural elements complement and cooperate with each other at the commands of a master architect, how the structure endures the ravages of time, and how it eventually crumbles. It is like having the blue print of a big ornate palace in your hands. It is expected to enable us to prevent and cure many diseases. Best of all, it may provide a clue for finding the fountain of youth and immortality.

With such fond hopes and expectations the most ambitious project in biology, the project to decipher the genome started 13 years ago. The goal of the project has been to determine the sequence of every letter that constitutes the DNA in every cell of the human body. With the work of more than a thousand biologists, computer scientists, and technicians spread over 16 laboratories in six countries and with a huge expenditure of $250 million; the human genome project has recently been completed. An estimated 3.2 billion characters have been sequenced. With this blueprint of human genome, biologists hope to change this great program to suit individual needs.

Dissents:

Junk DNA: - One criticism based on pragmatic grounds is that about 97% of the human genome-about 3.1 billion of the total 3.2 billions chemical letters-does not spell out a gene. Why then, should we bother about sequencing the junk DNA? (Being defective, these DNA fragments are not of much use to the host cell in which they reside. Critics question the usefulness of junk DNA)

DNA is not your destiny: The other is about the rigid determinism implied by the genome model. Your genes determine all your health, your behavior and your future. In short, your genes preordain your life at birth. We cannot accept the implication that your genes determine everything in your life. This kind of philosophy robs the individual of free will and leads to fatalism. This attitude may deprive the individual of initiatives in life. Yet, this is not the way genes work. For example, if you have a gene for say, schizophrenia it means that you have the risk of becoming schizophrenic if other conditions also encourage it. A particular gene can produce a different effect depending on environment. The presence of a particular gene associated with a particular trait makes you vulnerable to contracting that quality; it does not indicate inevitability. It is the notion of the primacy of the genes that frightens us.

Such a deterministic argument, which leads to fatalism, is really out of place. Genes do not work in isolation. Just like hormones, they are regulated by chemical messages from other parts of the cell or from other cells of the body. External environment plays a critical part in switching the genes on and off. For example carcinogens work by getting the cells, binding to the DNA and turning on the genes that cause uncontrolled growth of the cells known as cancer. It is experimentally verified that intense stress causes suppression of immune function of the body and leads to bad colds and related ailments. In primate, the mere smell of a female in 'heat' activates the reproductive genes of a male. Why go that far, don't we know that our sex genes are activated when we see a beautiful person of the opposite sex! Isn't it true that this activation becomes faster with tactile sensation? Therefore, we cannot dissociate genes from environment. Genes are not every thing. They work in conjunction with environmental factors. DNA is not your destiny.

This viewpoint has been upheld by recent researches in the new branch of biology called epigenetics.

(B-3):Cellular Process Theory

DISCOVERY OF THE CLOCK-METAPHOR

If life is a temporal program taking its commands from a preprogrammed code, it should have a clock that controls the timing of various temporal events like development of various organs, growth from childhood to adolescence, to puberty, to adulthood, and decline into old age eventually reaching the end of the program with death.

Aging therefore is a pre-programmed process in which a genetic clock, ticking away in the body, determines when the old age sets in. The time it keeps is a physiologic time, not Greenwich time. This clock controls metabolism, biochemical changes and cellular processes in the body. Aging proceeds from the cellular level to the grosser level.

HAYFLICK PARADIGM:

In 1912 a biologist named Alexis Carrel claimed to have grown cells of chicken heart continuously by feeding the culture with necessary nutrients for an indefinite period in his laboratory and that this continuous growth of normal cells indicated that they are immortal. Though some of the contemporary scientists were skeptical of the claim that the culture was doubling without limit they remained quiet because of the authoritative position Carrel was occupying at that time. However fortunately for the scientific world, a chance revelation by one of the staff of Carrel led to the discovery that the nutrient Carrel was feeding was not clean of the subject cell and at once the scientific world dumped the immortal cells theory. This misadventure can be said to have delayed the discovery of the truc nature of the cells and the progress of the related sciences by about three decades.

Dr. Leonard Hayflick conducted some experiments in 1960's to study the behavior of human cells and found that human fibroblasts in tissue culture would not divide more than 50 times. He put human tissue cells, after washing them free of enzymes, into bottles containing nutrient solution and kept them at body temperature. The cells went on dividing and filled the bottle. When the bottle was filled with cells, half of them were discarded and the remaining half was maintained in the nutrient solution at body temperature. Again, the cells divided and filled the bottle. He called each refill "Doubling". As the 50th doubling approached the rate of cell division decreased and microscopically visible, senescent changes were observed within the nuclei and the fluid space. Finally, the cells stopped doubling after the 50th count. Hayflick also found that, if the cells were frozen, say after 30 doublings, they remembered that they had 20 more to go when they were thawed and re-fed. Thus, he concluded that aging is fundamentally a cellular process.[23, 24]

A number of cell cultures grown in the laboratory have a finite ability to proliferate. After repeating a number of population doublings, they enter the state when they no longer divide. This state is known as replicative senescence. The degree of senescence is determined by the number of times the cells divide, not by the calendar time. Senescent cells accumulate in the culture. Gradually, the percentage of senescent cells increases until all the cells in the culture become senescent. The inability of cells to continue dividing is known as cellular senescence. A model in which the aging of cells in the laboratory is described is called 'in vitro' aging model.

The growth capability of fibroblasts (cells found in connective tissue), vascular endothelial cells, T-lymphocytes and a variety of epithelial cells was shown to decrease with increasing donor age[25, 26].

Cells from patients of Werner's syndrome, a premature aging disorder achieved fewer divisions than normal cells[27, 28, 29].

Proliferative capacity of fibroblast cells is also directly related to known Maximum Life Spans of the species, from which they are derived[30, 31,] For example cells from Galapagos tortoise achieved up to 130 doublings whereas those from the mouse with a maximum life span of just three years could undergo only 10 or fewer doublings before becoming senescent.

From these experiments, it was surmised that aging is a preprogrammed process in which a genetic clock, ticking away in the body, determines when the old age sets in. The time it keeps is a physiologic time, not Greenwich time. This clock controls metabolic, biochemical changes and cellular processes in the body. Aging proceeds from the cellular level to the grosser level.

The important question, then, is whether this 'in vitro' (in lab) model can explain the process of human aging in vivo (in real life). The answer is yes. Many investigators have established a link between in vivo aging and proliferative potential of cells in culture.

CELLULAR AGING VS. BODILY AGING

Aging starts at the minutest level and proceeds to grosser levels in haphazard abandon. Cell division decelerates with Aging and eventually the cell dies. Even if cell death cannot be prevented, years can be added to life by replacing body parts just as we extend the life of a car by replacing it's worn out parts. Implantation techniques provide some relief for a limited time but the real problem is prevention of senility. The current state of research is to go to the root cause and attempt to remove it.

IDENTIFYING THE GENETIC CLOCK

After a long struggle biologists have veered round the view that, more than anything else, Aging is controlled by some sort of clock hidden somewhere in the body. But, how do we find this clock? How does it look like? Is it silent or can we hear the sound of its ticks? Does it have a dancing pendulum? Does it have a dial to watch the time? If not, how is time reckoned?

For a long time, cells were suspected to contain the clock that controls Aging but no definitive results came forth. Recent researchers in molecular biology have settled down on the cells of the body as possible locations of the Aging clock. They have now come up with the proposition that a part of chromosomes themselves act as biological clocks. The end segments of chromosomes are identified as these age-clocks. What is a biological clock anyway? It is a biological analogue to a time machine for measuring Aging process, just like your grandfather clock, which is a mechanical analogue for measuring time. In a grandfather clock the circumference of the dial is taken as a day or night and the graduations on the dial are used to represent hours and minutes as the short and long hands rotate around. Since the movement of hands is taken as an analogy for the passage of time, this mechanism is called an analogue clock. Biologists were in search of such an analogue mechanism in the human body with the hope of manipulating that mechanism in order to slow down or stop aging. What should be the characteristics and qualifications of the candidate for this bio-clock?

It is agreed that in order to be recognized as the clock that controls aging, it should have certain characteristics and qualifications.

- It should be part of your genetic library
- The clock should be active and not passive, i.e., it should be playing an active role in the life process.
- It should be directly involved in stopping the replication of cells after a certain number of cycles.
- It should be related to the process of causing senescent expression in the genes responsible for dealing with the drivers of aging, viz., free radicals, protein-turn-over rates and the enzymes that repair DNA.

The candidate process should be unidirectional like the arrow of time, be capable of running down, and come to a stop after some time.

Let us see what part in our body satisfies the conditions set forth above.

The Telomere Model

Over the years, since Alex Comfort predicted that such a clock would emerge, a number of suggestions were made to be that candidate but none could satisfy all the criteria set forth. At last, the prediction came true after two scientists Hayflick and Moorhead demolished the Carrel 's theory of cellular immortality and firmly established that normal human fibroblasts have finite life span in labs and a Russian scientist named Olovnikov, proposed a mitotic clock in the chromosomes. He suggested that the cause of cellular senescence is the gradual truncation of chromosomal ends called telomeres[32]

Recent interest in telomeres has its roots in the independent experiments conducted in 1930's, by two genetic scientists Barbara McClintock and Herman Muller (then in the Universities of Missouri and Edinburgh respectively) who discovered that chromosomes bear a special component at their ends which prevented the ends from splicing and sticking together. Muller coined the word "telomere" from two Greek words *Telos* meaning "end" and *Meres* meaning "part." These end parts are functionally like fountain pen caps or like the metal pieces on the ends of shoelaces, which prevent the fraying of threads.

Telomeres, simply stated, are the end segments of DNA on chromosomes that act like caps. These molecular caps consist of nucleotide 6-unit phrases 'TTAGGG'. These telomeres serve the purpose of preventing errant recombination of the frayed and exposed ends of chromosomes.

How Do the Telomeres Shorten and Why?

The Non-Replication Problem

Before a cell can divide, it should first replicate the double-stranded DNA in its chromosomes, but the cell has a problem in doing so at the ends of chromosomes. In a normal somatic cell, the internal machinery copies the bases, but then it cannot copy the last few bases of the telomere region of one of the strands of the DNA. As a result, these telomeres shorten with each round of DNA replication. What is the replication machinery that does the job of copying the nucleotide bases all the way from the beginning but fail to do so at the end? It is an enzyme called DNA polymerase. This inability of the conventional DNA polymerase to copy the bases until the end of the chromosomes was discovered by James D. Watson in 1972.

As pointed out by him, every time a normal linear chromosome duplicates, it shortens. He called this 'end replication problem'. This problem is really end non- replication. To understand why it can't be replicated we have to understand the process of replication. The key point here is that DNA strands can be copied only in one direction. The process of copying is started by a 'primer', which latches on to the end of a chromosome. The primer moves in one direction priming the enzymes that do the copying. The primer separates as its job is finished but never turns back. Therefore, the end portion of the chromosome where the primer sits initially is never copied. At each replication, the chromosome is truncated at the end and thus the chromosome shortens as replication is repeated with cell division.

All living creatures face the same problem of maintaining their telomere without shortening. With progressive shortening, the chromosomes tend to disappear altogether. How then life has been continuing for millions of years. How could we have inherited the genes from our forefathers?

Nature solves its problems with its own expert solutions to maintain its realm. If chromosome ends are creating problems in maintaining the natural order, why do they have the ends at all? Linear forms have ends. So, avoid linear forms and replace them with forms without ends! One of such forms is circular (not necessarily accurate circular shapes but closed loops). That is what bacteria and certain viruses do. A second method used by Nature to attack the end-non-replication problem is to have protein locks at ends of chromosomes. When there is a protein lock at the end of chromosome, copying is done faithfully from the lock without loss of end segments. Some bacteria remain immortal that way. Yet there is another solution which is used by almost all the organisms with nuclei and that is to allow the shortening and replenish the lost pieces. That is exactly how the germ cells remain immortal maintaining their chromosomes without ever shortening. It astonishes that yeast cells which have survived billions of years since the beginning of creation, they too succumb to Aging and death when conditions are not favorable. The ageless and immortal yeast too dies when their telomeres shorten accidentally. From this it is clear that shortening of telomere triggers aging and death of cells.

Why does Nature work to keep the telomeres short? The reasons implied in shortening of the telomere are one more example of the wondrous and inimitable designs of Nature. What interest has Nature in keeping you alive for a long time? Its only interest is in perpetuation of the species and not in keeping you alive to pamper your endless desires.

In this process, an enzyme called telomerase rebuilds the lost base pairs. Telomerase acts like a rubber stamp. It stamps the lost TTAGGG sequences restoring the original telomere length. Germ cells and many unicellular organisms remain immortal by faithful operation of this mechanism. Yeast has remained immortal by the successful operation of this mechanism. Even among the yeast, there is a mutated variety in which telomere shortening leads to aging and death. It is astonishing that even yeast cells that have survived billions of years since the beginning of creation, they too succumb to aging and death when conditions are not favorable. The ageless and immortal yeast too dies when their telomeres shorten accidentally. From this it is clear that shortening of telomere triggers aging and death of cells.

Anyway, how short is short and how long is long for a telomere? Nobody knows. If a telomere is very long, it wastes cellular resources. A cell may become immortal by spending a lot of metabolic energy to build a long telomere and yet may succumb to the very process because long telomeres may result in haphazard and dangerous growth. On the other hand, a cell that builds its telomere sparingly may risk premature aging and death. It is like a profligate man consuming himself due to overindulgence or a miser dying in the midst of plenty due to his niggardliness.

Cells have to evaluate the signals received by them carefully and continuously and make decisions regarding balancing the length of the telomere. In unicellular organisms, this balance may not be very important but in multi-cellular organisms, maintaining this balance carefully is crucial to their survival. If somatic cells recover the lost telomeres, they grow unchecked and turn cancerous. In fact, mere cell growth is not as destructive as the aberrant behavior of the growing cells that cannot cooperate with the rest of the cells in the body. Cells that do not respond to the organizational cues become dangerous by threatening the neighboring cells. These anti social cells multiply rapidly and invade the organs, take away their resources and prevent their function. That is cancer.

The body tries to prevent cancer. How does it do? Normally, in a healthy body, needed cooperative responses are produced in cells in response to messages received from other cells. If responses detected are not favorable, DNA damage is suspected and brake is applied preventing cell division. If the brake fails due to any reason, then the second line of defense that is the telomeres come into play and stop the cell division. The importance of this mechanism to the organism is obvious, as it requires enough healthy time to reproduce and raise its progeny. Therefore, the body is designed to do whatever is necessary to prevent unnecessary cell division and outbreak of cancer in the young. If longer telomeres obstruct this process, they have to be kept short. Beyond the reproductive age, nature is no longer interested in you. Evolution is concerned about maintaining the line of your species and not you as an individual. It does whatever is needed for maintaining your line and then abandons you.

The suggestion of Soviet biologist Olovnikov in 1970's that the length of these telomeres is associated with the aging of cells paved the way for the concept of mitotic clock. He linked the end-non-replication problem of the chromosomes to the cessation of cell division. He suggested that the shortening of the chromosomes that occurs in human somatic cells might not be corrected. That is the idea behind the concept of mitotic clock.

Resetting the Clock

Therefore resetting this genetic clock implies re-lengthening the shortened telomeres. How can it be done? Single cell species and the germ cells and early embryonic cells in multi-cellular organisms overcome the problem of shortening of telomeres.

Investigators found large variations in the lengths of telomeres in different animals, humans, and even in different cells in the same animal or human. This observed heterogeneity of telomere lengths led to the surmise that some of the cells might somehow be re-lengthening their shortened telomeres. This surmise, in turn led to the suspicion that it might be some enzyme doing the trick.

In 1984, Carol W. Greider and Elizabeth H. Blackburn, then in the University of California at Berkeley, discovered that when synthetic telomeres were mixed with *Tetrahymena cells,* the telomeres gained additional sub units. That confirmed the suspicions that a re-lengthening enzyme existed. It was identified as a complex of proteins and RNA. The RNA component of this complex contained a template on which TTAGGG repeats at the ends of DNA are synthesized. As this enzyme accomplished the task of adding sub-units to the telomeres, (which could not be done by normal DNA polymerase) it was named telomerase. In 1989, Gregg B. Morin of Yale University discovered it in the immortal cancer cells. These two pieces of work, established the position of telomerase as the remedy for repairing the telomere attrition.

What, then, is the meaning of the metaphor 'Resetting the genetic clock'? It means re-lengthening the shortened telomeres by application of telomerase.

Now it is well established that all organisms with nucleated cells synthesize telomerase. Surprisingly, human somatic cells lack that ability. Somatic cells were found to have shorter telomeres than the germ cells. Germ cells maintain the long lengths of their telomeres by the activity of telomerase whereas this enzyme is repressed in somatic cells resulting in telomere loss.

Telomerase is active in the germ line, but is repressed in most normal somatic cells resulting in telomere loss in dividing cells. The gradual shortening of telomeres in human somatic cells that lack telomerase provides a good marker of replicative capacity. Available evidence suggests that telomere loss initiates the retardation of replicative process and eventually brings it to a halt at a point known as Hayflick limit. This process can be characterized as shown in Fig. (6-2).

Fig.6-2: Schematic View of Telomere Hypothesis

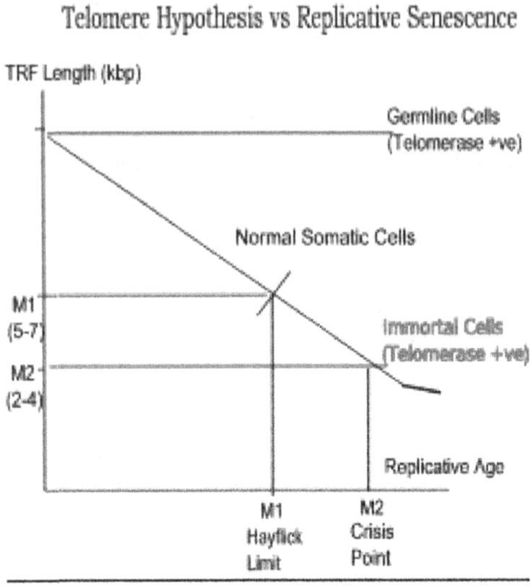

At M1 (the Hayflick limit), there is critical telomere loss in a few chromosomes signaling cell cycle arrest. When telomeres become critically short on a large number of chromosomes, cells enter crisis stage (M2)

Source: Richard C. Allsopp, et al, 'Telomere length predicts replicative capacity of human fibroblasts', Proc. Natl. Acad. Sc, USA, vol.89: pp10114-10118, 1992), Adapted with the permission of the author.

It has been known for a long time by our experience that as we age, our remaining lifetime diminishes. This has been confirmed by actuarial studies and the insurance companies have successfully used the data developed by the actuaries. This relation is shown in Fig. (6-3A). It indicates a finite limit to lifespan.

What is the relation between mean telomere length and aging? In several studies, telomere lengths of older people were found to be shorter than the telomere lengths of younger people from which a definite association between the lengths of these telomeres and the age of cells was suspected. But, in what way the two variables age and mean telomere length are connected? Does shortening of telomeres cause aging or other way round? Is it not possible that age while independently progresses; the shortening of telomeres follows concomitantly? To establish causal relationship one must be able to show that for every change in the independent variable there is a corresponding change in the dependent variable.

To establish that shortening of telomere causes one must be able to prove that a deliberate shortening of telomeres causes aging or vice versa, in repeated trials. This aspect, studied by Richard Allsopp and others (1992) IS displayed in Fig. (6-3).[33]

Fig.6- 3:Relation between telomere length and Aging

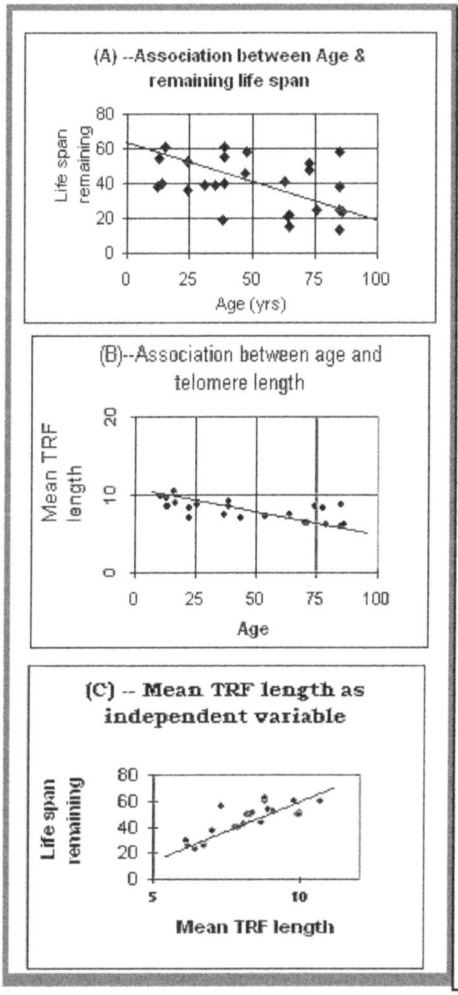

Graph (A) shows the normal actuarial relationship

Graph (B) shows the relation between age and telomere lengths. Using age as the independent variable age and TRF length, as dependent variable. However we cannot jump to the conclusion that age causes shortening of telomeres. This correlation between telomere length and age

Graph (C) shows the relationship between the independent variable TRF length, and replicative age is of limited use. . This shows positive correlation telomere length and remaining lifespan. That means the longer the telomere length the longer is remaining lifespan.

Source: Richard C. Allsopp et al, "Telomere length predicts replicative capability of human fibroblasts." Proc.Natl. Acad. Sc., USA, and vol.89: pp10114-10118 (1992), Adapted with the permission of the author.

The reader should note that because many epidemiological experiments cannot be conducted directly on human beings, they are studied in the laboratories using proxies. In this case, growth of cells in the laboratory jars is taken as age, taking doubling of cell population as one unit of aging.

Statistical Study of the Relation between Telomere lengths and Life span

Then an important question arose. If telomere is a clock, why do we need 92 of them in each cell and millions of them in all? Do we have even hundreds of clocks in our homes to keep track of time? To answer this question some elementary statistical analysis was used.

Look at the two figures (A) & (B) in Fig. (6-3), in which age is taken as the independent variable and telomere length and the number of cell divisions to go, are taken as the dependent variables.

Often, we see some people look much younger than their chronological age suggests. In addition, some people appear surprisingly older than we expect for their age. It means that current age cannot be taken as a good predictor of the remaining divisions of the cell or the length of the life.

In addition, cells drawn from people of identical chronological age were found to show different senescent characteristics not only initially but also in subsequent aging in the laboratory.

This study strengthened the conjecture that age and telomere length were somehow related but was it conclusive proof that age can predict telomere length? What then is a better predictor of the remaining age?

The next experiment confirmed the conjecture. It is suggested that telomeres in our cells are those predictors. Finally, we have one more relation showing the association between telomere lengths and remaining lifespan.

Experiments showed that the life span of cells from unrelated donors decreases with the age of the donor but the correlation was found to be weak. On the other hand, when initial TRF length of these strains was used as independent variable replicative capacity, showed a significant positive correlation. This shows that initial

TRF length is a better predictor of replicative capacity than the donor age. This relation is displayed in graph (C).

Variance: Aging depends not only on the length of the telomere but also on the variance of the lengths. They vary within cells as well as among the cells. One of the reasons for this variance is that every time a chromosome replicates one of the strands shortens. One cell inherits a short strand while the other inherits the long strand. In the next generation, one cell will have a strand of the original length and two of shorter lengths and one of even still shorter length. This process continues until we have a large collection of cells. The statistical distribution of their lengths is found to follow normal distribution, with the familiar bell shaped curve. Under the assumption that the length of telomeres is randomly distributed, the likelihood of one cell keeping it short or long telomere intact over a long period is nil.

The variance of telomere length is even greater among the cells than within cells. Cells found in the same tissue are unlikely to have the same number of telomere bases because of many sources of variation within the cell. Having the same average telomere length does not guarantee the same degree of aging because of the variability in gene expression.

If variability contributes to aging, why do we find that variance decreases as group of cells age? If we consider a large group of cells, the variance is initially high. As the group ages most of the telomeres shorten. The cell in which the telomere shortens to a critical length stops dividing and further shortening stops. Gradually, all the cells with shorter telomeres cluster around a shorter mean.

This variability explains why some people of same chronological age appear to be older than others. Aging is such an irregular, slow, and stealthy process. Aging is like that. That explains why we do not get up on one day to find ourselves suddenly transformed into gray, shriveled, toothless and stooping, old people. Have you ever observed the time in a clock shop? No two clocks ever keep the same time! The hundreds of millions of our cellular clocks do not keep exactly the same time. That explains why aging creeps in irregularly, imperceptibly and stealthily.

Lengthening Telomeres

CAN CLONING RESTORE TELOMERE LENGTH?

When a sheep was cloned about three years ago (called Dolly affectionately) some scientists began wondering whether the cells of Dolly were older than its age because she was cloned from an adult cell. A year later, that question was resolved as they found that the telomeres of the cloned animal were found to be shorter than normal. The implication of that finding is, at least as per the current understanding of telomere research that the cloned animal would age even faster than the normal. That was not encouraging news to scientists who were hoping that someday the cloning technique could be used to produce replacement cells of defective body parts.

In a latest experiment, a group of scientists at the Institute of Advanced Cell Technology Worcester, Massachusetts[34]are reported to have made a breakthrough. They cloned fetuses from adult cells kept in culture until senescence. The cloning was accomplished by transferring the nuclei of senescent somatic cells to the enucleated egg cells and eventually produced calves. In a related experiment, they removed the fetuses of cows at 6 weeks of gestation to compare their cells with those of normal fetal calves. They claim that cloned cells divided on average 93 times as against 61 times for the normal calves giving rise to speculation that cloning might have done the trick of elongating the telomeres. In order to test whether nuclear transfer can rescue aging cells they compared telomere lengths of cloned cells with those of age-matched control animals. They reported that their results indicated a clear increase in lengths of telomeres in cloned cells.

"If this extended life span could be extended to the whole animal," Lanza observes, there is "a real possibility" that cloned animals might live as much as 50% longer than their normal counterparts. If this were translated to human aging, it would mean humans living as long as 150 to 180 years. That prospect should again be taken with a pinch of salt because we don't know first whether the findings on a few cells can be extended to whole human beings.

If telomere shortening is the cause of human aging and if telomerase can repair the eroded telomeres, we are, naturally tempted to try this method for lengthening the human life span.

How do we go about doing this practically? Perhaps we have to adopt some of the clinical methods available in treating cardiovascular diseases. Some of these methods are: (1) Oral administration, (2) Injecting telomerase into blood stream, (3) Administering through a catheter and (4) Introducing genes through viral carriers.

Injecting telomerase directly into blood stream has to be ruled out at the very outset as telomerase is very unstable and cannot reach the targets. As of this date, there is no such oral drug capable of stimulating the telomeres. Angioplasty may be tried for local treatments but not for correction of system. If telomerase genes can be identified and isolated they may be introduced into the blood stream through viral carriers with the hope that they themselves will eventually express and release the required amounts of telomerase. The still undecided point is whether telomeres are the sole cause of aging. If they are not, we are back to the square one.

(B-4). Damage Theories

Damage theories hold that aging is caused by several types of damages acting severally or jointly on the body, wear and tear & accumulation of waste products, cross-linking of connective tissues, free radical damage to DNA and other bodily parts, damage caused by diseases and other factors to bodily parts and damage caused to immune system.

The connective tissues in our body, viz., collagen and elastic fibers deteriorate as they undergo cross-linking. Cross-linking agents are churned out as byproducts of normal metabolism. The cross linking theory was first suggested by Johan Bjorksten, an expert in the chemistry of tanning. This deterioration of connective tissues leads to stiffness of joints, hampers the dynamical actions of the bodily organs and leads to a state of old age.

Aging is caused by damage done to DNA. As age advances, the DNA repair rate falls short of the damage rate. Maximum life spans of various species of animals depend on the respective rates with which they spontaneously repair the damage to their own DNA. Body has the ability to repair this damage only to some extent. Age advances as the DNA repair rate falls short of the damage rate.

DNA can be damaged in several ways. As you know, DNA is a double helix with two strands and from each strand nucleotides stretch out. If genes have to become functional or the DNA has to replicate, the strands and their stretching hands have to be healthy. Any damage to the strands their hands can cause problems. Among the known varieties of damages are:

- Break in one strand
- Break in both strands,
- Linking of adjacent bases,
- Unnatural union of bases compressing the strand
- Formation of what are called dimers
- Debasement of some sites in the strands and Crosslinking.

Defective Mitochondria and Aging.

Damage to DNA is considered of paramount importance in aging. Among the causes of this damage, free radical damage has of late received special attention. If oxidative damage to DNA is involved in aging, longer-living animals should show lower level of markers of this sort of damage to DNA than short-living animals. In a recent study, Gustavo Barja and Asuncion Herrero[35] have measured separately the effects of damage to nuclear DNA and mitochondrial DNA of species ranging in maximum life span from 3 to 46 years. They demonstrated in this important work that the effect of damage to mitochondrial DNA is negatively correlated to maximum life span while the effect on nuclear DNA is not statistically significant. Increasing concentration of oxidative damage to mitochondrial DNA indicates dwindling maximum life span in mammals. Since oxygen radicals are generated by the mitochondria, it means that the maximum life span is inversely related to the rate of free radical generation.

Empirical evidence apart, the authors stopped short of explaining as to why the damage to DNA hinders life span. One theory says that the key seems to be in diminished energy production by the defective mitochondria. Defects in mitochondria may be inherited or may have been caused by accumulated mutations

Sometimes defects in inherited mitochondrial DNA may cause premature disorders that normally afflict many people in their old age, such as deafness, dementia, diabetes, heart disease and muscular problems. The other type of damage occurs through accumulation of mutations in DNA.

It has been proved that certain mutations in mitochondrial DNA caused some cases of neuro-degenerative conditions like Alzheimer's disease

These patterns, combined with the fact that a number of late life degenerative diseases are associated with declines in the activity of protein complexes involved in energy production suggest, that progressive reductions in mitochondrial energy production in muscle, nerve and other tissues could be an important contributor to various age-related degenerative diseases and aging itself.

Damage to DNA is caused both by external agents and by internal agents. External agents like pesticides and herbicides, other food toxins, X-rays, ultra-violet radiation and cosmic rays cause DNA damage. As cosmic rays emanate from the outer space, bombard our planet constantly, and enter all life on earth, we are defenseless against this hazard. Therefore, it is clear that some damage to DNA occurs naturally regardless of any precautions we take.

The internal enemies to DNA are free radicals and strangely our own body heat. This also suggests that some damage to DNA is inevitable and damaged genes are dangerous as they are used to make other molecules that wreck the body. What more scientific evidence is needed to realize that our bodies are not designed for permanent living?

According to the mitochondrial theory of aging, our mitochondria produce oxygen free radicals as a byproduct of ATP production. Even in people who began their lives with healthy mitochondrial genes, these free radicals can wreak havoc by attacking the mitochondrial DNA and reduce the output of energy by interfering with the production of ATP. If the individual lived long enough this process enters into a vicious circle and progressively brings down the energy level in various tissues that results in the symptoms of what we know as normal aging

Damage theory says that this damage is the driving force behind aging. If damage to DNA were such an unmitigated force that whittles down the lives of species, many species including human beings would have become extinct. That has not happened because of the built in ability of many species to repair their DNA damage continually. In that case, we have to take the net repair rate as the determinant of longevity.

Figure (6-4) shows the inverse correlation between oxidative damage and maximum life span of various species.

Fig.6- 4:Relation between Oxidative Damage and Maximum Lifespan of some animals

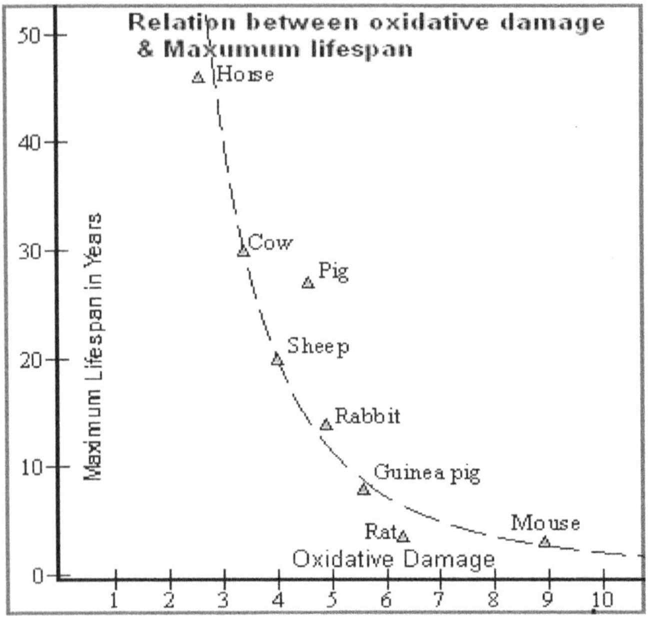

This theory holds the promise of extending max life span if we can find the method of increasing DNA repair rate. This was attempted with some success on one-celled organisms but we are in the dark when it comes to increasing the repair rate in animals closer to humans.

If other molecules are damaged due to any reason, the problem can be remedied by increased turnover of protein molecules though it involves spending of some additional cellular energy, but that cannot be done in the case of DNA molecules. They can only be repaired. So the body takes a great deal of care to repair the damage done to DNA but the rate at which it can repair the damage falls short of the rate at which damage accumulates and the net repair rate becomes negative yielding the body to forces of destruction.

Professors Ron Hart and Richard Setlow[36] conducted in 70s, an experiment to measure the repair rates of DNA in various animals. In that classic experiment, they extracted DNA from the fibroblasts of animals, from a mouse to man and exposed them to artificial damage.

The rates of automatic DNA repair of all these animals were estimated after exposing them to damage causing agents, for several hours. The results obtained showed that there is a remarkable positive correlation between the DNA repair rates and maximum life spans as shown in Fig. (6-5).

These results show that damage to the DNA is universal in all species and its extent varies from species to species. The shortest-lived shrew exhibited the smallest repair rate while the longest-lived man and elephant showed the highest repair rates.

Fig.6- 5: DNA Repair and Maximum Lifespan.

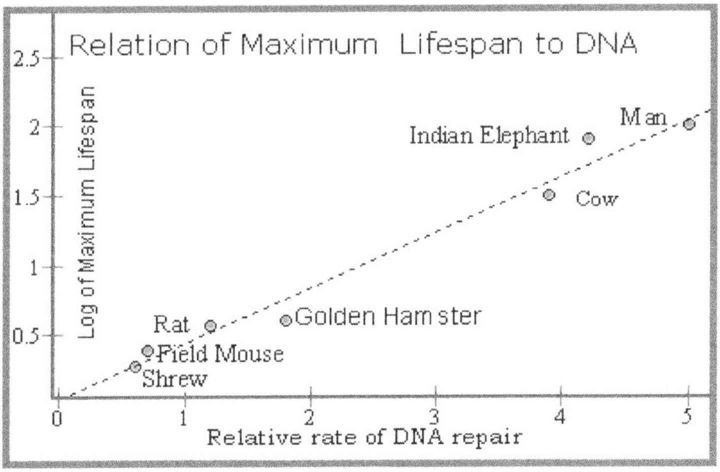

Source: R.W.Hart and Richard Setlow, "Correlation between DNA excision repair and lifespan in a number of mammalian species", Proc. Natl. ACAD. Sc., USA, vol.71: pp2169-2173 (June 1974). Adapted with permission

Damage to the Immune System

The immune system is an amazing feature of our bodies. Like the modern computers, it has a phenomenal ability for creating, storing and using information and for learning. This incredible system can recognize molecules that are alien to the body. It can recognize what belongs to it and what does not. If it doesn't, it is a battle. How does the immune system recognize what belongs there and what does not?

A special protein molecule called Major-Histocompatibility Complex (MHC for short) sits on the surface of almost every cell in our body. It is an identity tag to tell the immune system that this unique cell is a friend, a part of us. The cells on which this identity tag is not found are recognized as foes and expelled. Thus, the immune system differentiates the self from non-self. The idea that immune system must discriminate between the self and the non-self is the corner stone of immunology theory. Organisms like bacteria, fungi, parasites and viruses are quickly recognized as aliens when the immune system is young and efficient.

The skin is not a mere passive cosmetic covering. It is the first line of defense against the foreign invaders. It has sentry cells that detect the microorganisms and warn the immune system of the impending invasion. So

me cells produce acids that hinder the growth of harmful bacteria and fungi.

Joining the skin are membranes that line the internal surface of the body. These membranes secrete mucus that traps microbes. Saliva, nasal secretions and tears contain microbe-killing substances. Hair-like cilia in the air passages leading to the lungs push mucus and debris into the throat from where they are eliminated by sneezing and coughing. If some invaders manage to overcome all these hurdles and reach the stomach they are killed there, broken down by digestive enzymes or trapped in the mucus that lines thee walls of stomach and intestines. Eventually they are eliminated with other body wastes. These are mere skirmishes compared to what happens in the blood stream.

When these outer defenses are breached, the invaders reach the blood stream body fluids and tissues. White blood cells, the big guns of the immune system take over the defense now. About a million of them are born every second in the bone marrow and they emerge to form three distinct divisions- phagocytes and two kinds of lymphocytes, namely T-cells and B- cells. The immune system generates trillions of soldiers each capable of attacking one kind of enemy. During a disease, millions of germs may be generated and each one of them will have the same antigen. Therefore, the T-cells and B-cells must have antigen-specific receptors to fight with the various diseases. The immune system is designed to recognize foreign invaders. To do so it generates millions of different kinds of immunological receptors so that, there will be some complimentary receptor to recognize it and effect its elimination, no matter what the shape or form of the foreign invader is. Thus, immune system can produce groups of T-cells and B-cells that can match any antigen that enters our body.

As age advances this protective response of our bodies diminishes. In addition, the very same immune system, which protects us in our youth, may tend to turn against us as age advances. With age, two things happen to the normal immune system. First, its normal response declines so that it can no longer react effectively against foreign bodies. Secondly, it cannot act against the aberrant cells that may arise in its own body.

In a perfectly working immune system, anything that is different from itself, that does not carry the marks of 'Self', is apt to be isolated and dispatched promptly. The immune system's surveillance against bacteria, viruses and other foreign bodies and aberrant cells protects the integrity of the body. This normal protective response is high in youth and deteriorates with age.

Discrimination between self and non-self is the hallmark of immune system. As age advances it not only fails to act against the invaders but also that discriminatory power is blunted and may begin to act against the very self of which it is a part. It is a double-edged sword. Telomere therapy should be able to lift the sagging immune power of the body by rejuvenating the lymphocytes, by restoring the texture of skin and by fine-tuning its discriminatory power.

It is well known that our bodies protect themselves from outside attacks of disease-causing organisms efficiently in our youth. These organisms are there all around us bent upon getting inside our bodies, craving for the warmth and nutrition. When once they get inside of us they soon multiply in alarming numbers and try to take us over. If these invaders are not expelled our lives hang in balance. It's a question of them or us. Usually we win but not always. The outcome depends upon how quickly and efficiently our immune system gears up for the fight.

Agents of Damage

NORMAL WEAR & TEAR

Wear and tear is natural to everything that is composite. Everything wears out under the influence of elemental forces. Human bodies are not an exception. They last for a long time but they do wear out eventually. At the cellular level, cells too wear out. Unless they are replenished or rejuvenated, body functions decline and old age creeps in.

DISEASES.

In general, chronic diseases gradually destroy or degrade the functioning of the body organs leading to progressive descent into a condition recognized as old age. In particular, certain diseases like Progeria, Werner's syndrome and Down's syndrome present the symptoms of accelerated aging.

Progeria shows symptoms of stunted growth, withered skin, and premature arteriosclerosis. Werner's syndrome is similar to Progeria but onsets later in life. Down's syndrome (otherwise known as Mongolism) is also a premature aging disease.

Diabetes is generally an old age disease but it may strike anyone without regard to age. Diabetes is a disease characterized by excess sugar in blood. An insufficient supply of insulin causes this disease. In the absence of sufficient insulin, glucose cannot be absorbed into the cells where it is metabolized to produce energy. Arthritis, which is characterized by deterioration of connective tissues in the joints, is the most common ailment of old age.

FREE RADICALS

Damage by what are known as free radicals is considered by some biologists to be the primary villain in breaking down the defenses of body and converting us slowly but surely into old hags and delivering us into the jaws of death. What are these free radicals anyway? Free radicals are molecules with a single unpaired electron in their outer shell. The lonely electrons are constantly on the prowl to find a mate. In their anguish to find partners, they roam about carelessly and steal an electron from the neighboring molecules thus damaging them. The damaged molecules divorced of some electrons become crazy and set off a chain reaction of destruction. The chain reaction stops only when the distressed electron finds a mate and settles down with it once again at peace. Why then should we have such destructive agents in our body? Can't we do without them? The answer is 'No.'

Paradoxically the most common free radical in our bodies is oxygen. You know that we can't live without oxygen! It is required for our survival. That is why our cells tolerate the free radicals. When they are confined in their proper places in the body they are safe and serve the vital purpose of maintaining the metabolism. When they are out of bounds they wreak havoc like mad dogs. Similarly, cell division is necessary for life but if the cell division is out of control, the result is cancer.

If free radicals and other Reactive Oxygen Species are considered to influence the etiology of age-related diseases, then the logical answer to that problem is to mitigate their influence in the body. Many free radical scavengers and anti oxidants are suggested in the literature but some are important. One such anti oxidant is Super Oxide Dismutase. (SOD to be short). Simon Melov[37] of Buck Institute of Age Research, Novato, California, and others tested this theory recently. They augmented the natural anti oxidant systems of *C.elegans* with mimetic SOD and found that their life span increased by 44%.

EXTERNAL AGENTS

There are many external agents like cosmic rays, ultra-violet radiation, pollution of atmosphere & waterways and accidents, which cause irreparable damage to the body. Agents like cosmic rays and radiation are beyond our control, while the other like pollution and accidents can be controlled by individual and collective efforts.

(B-5). Rate of Living Theory

Nearly all animals use up approximately the same amount of chemical energy in a lifetime. A little mouse burns the same number of calories per pound in his short two to three year span of life as a mighty elephant in his fifty and odd years of life. So it is the rate of metabolism, the theory goes, that determines the maximum life spans of different species.

The German physiologist Max Rubner first formalized this relationship[38] about a hundred years ago. He measured metabolic rates in five different species of mammals with large differences in their longevity and their metabolic rates. He concluded from this study that, in spite of large differences in their longevity they had approximately an equal amount of mass-specific lifetime energy output. Expanding on this work, another scientist by name Pearl[39] proposed 'Rate of Living Theory' which linked longevity to the metabolic rate of an organism. The exact causal mechanism linking these two was not known but Pearl proposed that longevity was determined by the exhaustion of some vital cellular component. It is supposed that the rate of consumption of this component is proportional to the rate of metabolism. Life is supposed to end when this component is exhausted.

A recent investigator named Denham Harman[40] suggested that the link between metabolism and longevity could be free radicals and other oxidants called Reactive Oxygen Species (ROS to be short) produced in the process of metabolism. Free radicals produced during aerobic respiration cause cumulative oxidative damage that result in eventual aging and death. This is one of the most popular theories of all the theories proposed to explain aging.

Two lines of evidence support this theory. The first is the well-known inverse relationship between mass-specific metabolic rates and longevity found across a wide range of animal species. In general, smaller animals have higher mass-specific metabolic rates and shorter life spans than the larger animals and vice versa. A simple explanation for this phenomenon is that the higher metabolism causes in higher rate of production of harmful ROS. Higher rate of production of ROS results in higher cellular damage and rapid aging. The second line of evidence is based on the biochemistry of aerobic respiration. Although there are many sources of oxidant production, it is believed that most free radicals are produced during oxidative phosphorylation in the mitochondria, which is necessary for production of ATP. It has been shown that virtually all-cellular components are sensitive to the ROS damage.[41]

Slowing Down the Bodily Processes

This theory is based on the idea that slowing down the bodily processes that can be achieved by controlling metabolic rate can retard aging. Various suggestions are made for achieving Control Over metabolic rate. The processes of going into hibernation, deep freeze and caloric restriction have been discussed earlier but caloric restriction alone has been found to be a process that can be easily implemented.

CORE TEMPERATURE REDUCTION

Natural histories of cold-blooded animals and recent laboratory experiments have suggested that by reduction of internal body temperature, maximum life span can be increased. Keeping them in colder environments can lower body temperature of cold-blooded animals; but the ambient temperature cannot influence core temperature of warm-blooded animals as an internal thermostat controls it. Therefore, it is not possible to verify the life span extending effects of lowered temperatures on human beings by manipulating the ambient temperature. As far as this aspect of longevity is concerned, it makes no difference whether you live in Texas or Alaska.

Wherever you live, your internal thermostat keeps your body always at 98.6^{0}F. If you wish to lower the core temperature, this thermostat supposed to be in the hypothalamus, a gland at the base of the brain, has to be reset at a lower level. Some mammals like polar bear are known to do this successfully by going into hibernation in winter to conserve energy. The aim of animal hibernation is survival. They can skip winter with falling temperatures and dwindling food supplies. As winter approaches, the animal's thermostat shuts down and as internal temperature falls, it slips into hibernation. During hibernation, the core-temperature comes close to the level of ambient temperature. Hibernation is distinct from sleep: in hibernation the body temperature drops, heartbeat slows down and breathing rate is greatly reduced. The physiological mechanism by which animals lower their core temperature in hibernation is not exactly known. It is in the stage of science fiction.

HIBERNATION

Arthur C. Clarke describes in his science fiction story 'Child-hood's End', a drug capable of producing prolonged sleep, lowered body temperature, decreased heartbeat, and reduced metabolism.

From such a state a person could wake up rested, renewed and in peak of health. The drug, in other words, could induce a condition that mimicked animal hibernation. Hibernation involves extended absence of food and suspension of some normal bodily functions. It suggests ageing process is slowed down during this time. A hormone like chemical by name Hibernation Induction Trigger is said to induce in humans some properties similar to hibernation but its chemical structure and method of operation are still undetermined.

If it is so, theoretically, we can reduce the metabolism by hibernation or even by suspending animation by deep-freezing the body for as long as we wish and revive it later. It means that your waking life ends when once you go into hibernation or you are frozen and starts again when you will be revived. Naturally, this implies that your waking life docs not increase though the terminal point is extended.

This method is of some practical interest. Metabolic rate of each species was determined by the needs of evolution. Humankind's metabolic rate was determined thousands of years ago when people were leading nomadic lives, hunting for food. Now the pattern of life has changed. Life is now controlled more by the power of mind than by the strength of the muscles. So our metabolic rate as indicated by our core temperature may now be altered, if possible, without coming into conflict with physiological needs.

Humming birds and shrews have high metabolic rate. Humming birds that drop their temperature at night greatly outlive shrews that have not learnt that trick. The humming bird's consumption of oxygen during night falls from a high twenty-four cubic centimeters, per gram of body weight, during day to a bare one cc/gram of body weight, during night. The hummingbird lives for about 8 years whereas the shrew lives for about 6 months on average.

Suspending Animation

Deep-freeze:

Can human beings be frozen alive and brought back to life after some time? Some stray cases where human beings were frozen in natural circumstances and were revived bring us the impression that the method is possible. In 1960, a Russian tractor driver who was lost in a snowstorm was found frozen and stiff but without frostbite and putrefaction after being buried in snow for three hours. He was taken to a hospital, revived and eventually returned to work. A similar case was reported in Tulsa, Oklahoma: A Negro woman found frozen and stiff was revived but not without frostbite of her extremities.[42]

The prospect of preserving bodies by freezing with the hope of being revived later when such techniques will be available is taken seriously by some people. Some associations were formed in New York and California to implement these ideas. A number of people are said to have paid fees to ensure that at death, their bodies would be preserved with the hope of being revived later.

The Cryonics society of California has reported to have actually performed such a cryonic-burial[43]. The essential steps in the cryonics process are as follows. As soon as the person is dead, the body is injected with an anti-clotting agent, heparin. Following that the chest is opened and chest massage applied to keep the brain supplied with blood. Then the body is transferred to a heart-lung machine and the body temperature is lowered to 8^0 C, by packing it with ice. At this point, the blood is withdrawn and replaced with a solution of salts and the wonder solvent DMSO (Dimethyl Sulfoxide). The temperature of the body is then lowered to minus (-) 790 C and transferred to liquid nitrogen storage at minus (-) 1900 C, in liquid nitrogen.

Experts in this area are divided about the technical possibility of such a procedure. Storing certain inert organs is a different matter from storing actively metabolizing cells, as the cells may disintegrate when brought to a standstill. The current situation is that certain mammals were revived after less than an hour of cold storage without adverse effects. At present individual cells, for example, sperms and eggs and a number of isolated cells from various tissues, and eukaryotic cells like yeast are routinely frozen in liquid nitrogen for long-term future use.

Caloric Under-Nutrition: Study of the effects of Under-Nutrition on Model Animals

Studies on the relationship between energy consumption and disease can be traced back to the beginning of 20th century.

Moreschi[44] demonstrated that underfeeding could impede the growth of tumors. Five years later Rous[45] confirmed and expanded those findings. No further progress was made until 1935 when Clive McCay[46] demonstrated broad disease preventive effects as well as extension of life span, because of caloric restriction. In 1940s, Albert Tannenbaum[47,48,49,50,51] demonstrated that energy restriction *per se* in rodents could inhibit tumor initiation and growth. He also showed that, in many circumstances, fats could promote the growth of already initiated tumors. Jose et al later extended these findings to humans, in a particular case[52]. They reported in 1969 that Australian aborigines who were usually unable to develop antibodies, showed after mal-nourishment, unexpectedly increased proliferative responses of T. lymphocytes upon stimulation with certain phytomitogens. The same kind of increased T. Lymphocytes was observed in experiments with animals. Further, they reported having observed the same or enhanced immune response even at very low level of protein intake. Good et al[53] and West et al[54] reported that caloric restriction produced some additional good results among which are delayed allergic reactions, increased capacity of lymphoid cells. They reported to initiate graft-vs. -host reactions, increase in immune response to certain types of tumor cells and increased capacity to resist certain types of viral infections.

The age-old practice of dietary control has at last gained some respectability in scientific circles due to sacrifices made by rodents. Sixty years ago, some scientists at Cornell University studied the effect of food intake on the life span of rats and found that caloric restriction extended the maximum life span by 33%. They found that the rats which were fed low calorie diet retained their youth longer and suffered fewer late-life diseases than their counterparts which were fully fed.

Several recent accounts of laboratory experiments with rats have reported surprising results following two different methods of dietary control, viz., strict regime of caloric under nutrition and intermittent fasting. The group kept under strict regime of under nutrition lived 60% longer than the fully fed rats. The intermittently fasting group lived 30% to 48% longer than the uncontrolled group.

Richard Weindruch of the University of Wisconsin and Roy Walford of the University of California at Los Angeles in their experiments with rats rediscovered in their laboratories that diet restricted rats outlived the gluttonous rats. Dietary control in this context means limiting total intake of calories without loss of critical nutrients. Now scientists have started realizing the value of under nutrition as a viable method of life extension.

Among the several procedures scientifically investigated caloric restriction has been found to be the only procedure found to be effective in slowing the aging process; though the investigation has so far been limited to non-human creatures like single celled protozoa, round worms, fruit flies, fish and rats. Interim results of experiments on monkeys confirm that finding.

In all these investigations, the animals were restricted to diets with 30% to 50% fewer calories than the diets given to control group. However, care was taken to ensure that the test subjects ingested enough proteins, fat, vitamins and other nutrients. If the nutrient intake is taken care of, caloric restriction was found to increase not only the average life span but also the maximum life span of a population.

In addition to extending the average and maximum life spans, low calorie diets were found to postpone many diseases like cancer, cataracts, hypertension and kidney failure encountered in late life; slow down undesirable increases in cross-linking rate, free-radical production and oxidative damage to tissues.

Further, the procedure was found to be responsible for postponing the declines in protein synthesis, muscle mass, control of blood- glucose, female reproductive capacity, and DNA repair rate and learning ability.

Almost in every measure of aging studied about 90% of the sample remained younger. In normal aging, ability to synthesize proteins declines leading to decline in muscle mass; ability to clear glucose from blood declines, which may progress into diabetes; and alertness decreases. In calorie restricted rodents all these changes were delayed keeping them youthful much longer.

Did caloric restriction per-se yield these benefits or reduction in some other component of restricted diet did the trick? It turned out that the first possibility was right because it was found that reduction of fat or carbohydrate or protein, while keeping the caloric intake high, did not yield the desired result.

Some results of experiments with other model organisms also are given in Table (6-1).

Table 6- 1: Effect of Caloric restriction on the life span of model organisms

Name of Creature	Range of lifespan	Diet	
		Unrestricted	Restricted
Protozoa	Max life span	13 days	25 days
	Ave life span	7 "	13 "
Water flea	Max life span	42 days	60 days
	Ave life span	30 "	51 "
Bowl & doily	Max life span	100 days	139 days
spider	Ave life span	50 "	90 "
Guppy	Max life span	54 months	59 months
	Ave life span	33 "	46 "
Rats	Max life span	40 months	60 months
	Ave life span	30 "	45 "

Table compiled with data collected from various sources

In all these experiments, the diet-restricted rats were found to be more alert, sexually active to a more advanced age and lived longer than the fully fed. A number of biochemical measurements indicated that diet-restricted animals were found to be bio chemically younger than their chronological age indicated:

- Blood cholesterol increased with age in fully fed rats but not in restricted ones.
- Change in collagen was less severe in restricted rats.
- Restricted rats retained the catalyzing enzymes three times longer than the fully fed ones.
- Fat-energy conversion rate was higher in restricted animals.

Diet restriction rejuvenated the immune response system of the controlled animals. Diet restriction was found to have counteracted the usual decline of ability of bodies to eliminate toxic materials and their increasing trend of auto-destruction. Finally, it reduced the frequency and number of diseases.

The effects of caloric intake on body weight, caloric restriction on percent survival and on average and maximum life spans are illustrated in three diagrams of Figure (6-6).

Fig.6-6: Effect of caloric intake on (a) Body weight (b) Percent survival and (c) Maximum Lifespan

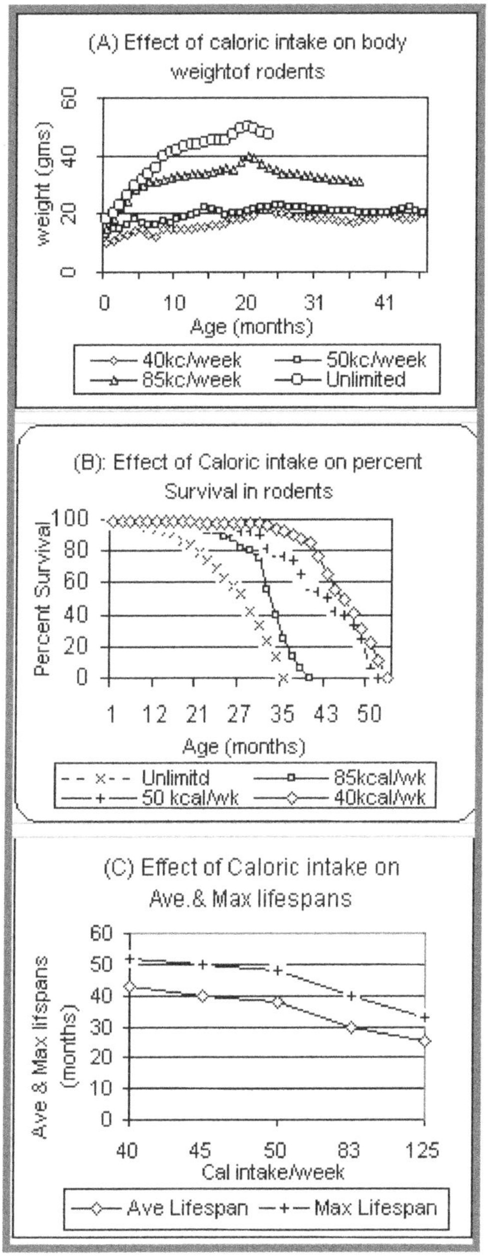

Diagram (A) shows body weight increased as caloric intake increased.

Diagram (B) shows percent of surviving rodents decreased as caloric intake increased.

Diagram (C) shows the longevity of rodents decreased as caloric intake increased.

Source: Richard Weindruch, Roy Walford, et al, "Retardation of aging in mice by dietary restriction: Longevity, Cancer, Immunity and lifetime intake", Journal. of Nutrition, vol.116, pp 641-654, (April 1986)- Adapted with permission.

The dramatic reversal in positions of all the curves shown in the diagram (A) and (B) may be noticed. The lowest curve (representing lowest caloric intake) in the diagram (A) corresponds to the highest curve in diagram (B) representing the highest percent survival. Similarly, the highest curve in diagram (A) (representing highest caloric intake) corresponds to the lowest percent survival in diagram (B). Diagram (C) shows directly how average and maximum life pans of rodents decreased as caloric intake increased. This graph uses caloric intake as the independent variable and measures average life span and maximum life span as dependent variables. The decline of both average life span and the maximum life span as caloric intake increases can be clearly seen in this graph.

Any procedure that results in inadequate development and function of brain is not recommendable. Fortunately, in all these experiments brain weight increased at the same rate in the controlled group as in the uncontrolled group.

Diet restriction not only increased the average life of these animals but also retarded the aging of brain cells. Even in intelligence tests, restricted mice were as good as the fully fed mice, or even better. Restricted mice that are 50% older displayed more alertness and intelligence than the fully fed mice did. Adult initiated diet restriction regime was found to keep the brain clearer and to retard its aging.

Evidence from Gene Expression Analysis

Now computer technology has made its foray into biology with its venture of probing the activity of the genes. In a recent study commenced by Dr. Richard Weindruch and Thomas A. Prolla, at the University of Wisconsin, a gene chip was used to detect the activity of genes. They used a technique in which a mouse muscle tissue was used for gene expression analysis. The chip made by Affymetrix corp. of Santa Clara, CA, was programmed to recognize the activity of some 6,000 of mouse genes whose DNA sequence had been decoded and filed in public data banks.

As usual, they divided their study subjects into two groups, rats normally fed and those carefully underfed; and performed the experiment of injecting the calf muscles of both groups into the gene chips. They report that the results obtained confirmed some of their previous findings. One percent of those genes, that were involved in stress response activity were found to be much more active in normally fed mice than their counterparts in underfed mice are. On the contrary, genes involved in generating energy from glucose became less active in the former group than in the latter group. To be short, the gene expression chip gave them a profile of younger age in underfed rats. They expressed hope that they would eventually be able to pinpoint the genetic changes that underlie the aging process.

EXTRAPOLATION OF RESULTS

Can these results be extrapolated to human beings? Perhaps, yes. Some historical evidence supports this confidence. One of the Greek writers[55], referred to earlier, reported that they found in Pandora valley of India, people living up to 170 and even 200 years and opined that the reason for their longevity was their moderation in eating habits! However, such claims have to be confirmed by systematic data about human response to reduced caloric intake. Data taken from certain geographical areas may give clues but cannot be taken as scientific evidence as people there are forced to live in poverty without essential nutrients in their diets and such data might contain the influence of other factors.

Why can't we conduct these experiments on human beings and the results confirmed? Of course, they can be done but we have to wait for a long time, a few decades after the experiments are begun, because, as you know, normal human life span, which is the subject of our study is quite longer and the experiments have to be co-extensive with it. So far, nobody seems to have started carefully controlled systematic experiment of long-term caloric restriction in average weight humans although we get indirect pieces of information about the effects of low calorie diet on their longevity and health. Interesting results were obtained when eight people living in a self contained enclosure called 'Biosphere-2' near Tucson, Arizona, were forced to curtail their food intake for two years. Roy Walford of the University of California, an expert on aging worked as the team's physician and monitored various aspects of the group's physiology. As per his observations, caloric restriction led to lowered blood pressure and glucose levels and lowered serum cholesterol level.

UP-GRADATION OF EXPERIMENTAL SUBJECTS

However, scientists have recently upgraded their experimental subjects to primates that are closer to man in evolutionary scale. In a study initiated at the National Institute of Aging in 1987, Rhesus monkeys and squirrel monkeys have been under observation. Another project involving only Rhesus monkeys started at the University of Wisconsin. Rhesus monkeys live typically 30 to 40 years and squirrel monkeys live about 20 years. We are not yet in possession of the complete results of these experiments although some intermediate results have been trickling in. The results reported so far are encouraging. Blood pressure and glucose and insulin levels in blood are lower in dieting animals than in control group. The dieting animals appeared healthier and happy. Table (6-2), illustrates these details.

In animals on Caloric restriction, these changes occur partially due to amelioration of oxidative stress and due to a fundamental metabolic shift from growth and reproduction strategy to one of life maintenance.

Table 6- 2:Comparison of Health Statistics of Dieting and fully-fed Monkeys

Compared Variables	Unrestricted diet	Restricted diet
Food Intake	588 cal / day	477 cal / day
Body weight	31pounds	21 pounds
% wt from fat	25	10
Resulting measures of health		
Blood pressure-sys/dias	129 / 60	121 / 51
Glucose level (mil.gms/deci litre of blood)	71	56
Insulin level(microunits/ deci. litre of blood	93	29
Triglycerides (mil. gms/deci litre of blood	169	67

Molecular Basis of Caloric Restriction's Ability to Delay Aging

Free Radical theory

After knowing how it happens, the intriguing question, in many scientific phenomena is why it happens so. Why does caloric restriction yield these benefits, through what mechanism? There may be several reasons like reduction in body metabolism, internal temperature and reduction in the production of free radicals. A dominant section among biologists thinks that caloric restriction extends survival and vitality by limiting the injury to mitochondria by the free radicals. Mitochondria are the tiny intra cellular structures that act as power plants of cells by producing ATP (Adenosine triphosphate). Unfortunately, the same machinery, which produces ATP, also produces free radicals. Once produced, the free radicals do not care whether certain components are useful and attack everything in their stride - proteins, lipids and DNA anywhere in the cells. They deal more deadly blows to the DNA in mitochondria as they are in the proximity of production. The theory suggests that, over time, the accumulated damage to mitochondria will result in decline in production of ATP and increase in production of free radicals, a double deal that destroys the tissues.

Nevertheless, what is the mechanism through which caloric restriction helps in achieving the results we have been seeking? Some biologists are of the view that caloric restriction somehow strengthens the antioxidant enzymes in their fight against the free radicals. Richard Weindruch of the University of Wisconsin thinks that CR cuts short the production of free radicals themselves. The argument goes like this.

If free radicals are the by-product of oxidative process in the body there must be something wrong in the oxidative process just like the production of cinders and smoke in the incomplete combustion of damp wood. Since oxidative process is by itself a life phenomenon it is a necessary process, but there must be deficiency in the oxidative process, which leaves behind these unwanted free radicals. Pumping more food into the body demands increased oxidation for digestion and assimilation. At higher levels of food intake one has to use more efficiently the oxygen needed for digestion and assimilation of the increased food. Added to that, our normal involuntary respiration is not quite efficient to meet the oxygen demands of the huge quantities of food that is not really required by the body for sustenance. The consequent inefficient digestion and assimilation encumbers the body with residual waste products. The gross waste is eliminated partially through the excretory organs while the finer waste as free radicals remain inside and attack the cells unless and until they are scavenged out of the body. Therefore, reduced food intake demands less of oxidative activity. Lesser activity of free radicals will leave mitochondria healthier produce ATP more efficiently. The fact that yogis consume much lesser food than us, the ordinary people and yet manage to enjoy better health than us, is a fitting testimony to the efficacy of caloric restriction.

One important question that arises in this context is whether a lifetime of abstemiousness is necessary to receive the benefits of caloric restriction. William Mair et al report in the columns of Science[56] that it is indeed not necessary but this finding is questionable, as the damage already done to the body cannot be completely eliminated by late resorting to remedial measures.

After all this, scientists are not content to leave it there, as the remedy seemed to them more painful than the disease. Who, in this age of primacy of sensual pleasures, is willing to control the palate even if it confers longevity? Therefore, the search is on for finding the biochemical pathways that confer the benefit of longevity by caloric reduction. If they understand the biochemical pathway through which caloric restriction achieves extension of longevity, they hope to do the same by tinkering with those mechanisms without reducing a morsel of food! Therefore, the research was on for what they call C.R. mimetics, or those substances or methods that mimic the effects of Caloric Restriction without reducing a morsel of food.

Mimicking Caloric Restriction

One among such methods attempted by Su-Ju-Lin[57] and her colleagues at the Massachusetts Institute of Technology is notable. The researchers claim that their investigations led them to the conclusion that caloric restriction extends life span in yeast mechanism but it is not the proximate cause. It redirects the biochemical pathway that catalyses the activity of a gene called SIR2. They demonstrated how the shunting of carbon metabolism towards mitochondrial TCA cycle and concomitant increase in the efficiency of oxygen used played the central role in this process. They induced caloric restriction in yeast by limiting glucose availability or by genetically crippling their ability to sense and respond to glucose. They found that this genetic crippling of glucose sensing ability increased yeast longevity by 20 to 40%.

This gene responsible for this result is named SIR2 to mean Silent Information Regulator No.2. rDNA genes produce small circles that are harmful. The SIR2 protein silences or suppresses the expression of deleterious genes, in this case rDNA. This is accomplished by making the material that envelopes the DNA, wrap more tightly around the histones thus denying the cell access to the underlying genes. They claim that yeast cells which had been denied the influence of SIR2 lived shorter lives while those supplied with an extra copy of this gene lived longer. Let us see now the connection between under-nutrition and the silencing activity of this gene.

Yeast cells grown with very little of their food, namely sugar in the form of glucose lived longer than normal. This did not happen when the action of the SIR2 gene was disrupted. This leads to the inference that excess metabolism disrupts the activity of SIR2 gene and diminished metabolism assists the silencing activity of this gene. From this, it can be deduced further that the metabolic activity and the silencing activity of this gene compete with each other for some chemical. They identified this as a chemical called NAD. SIR-2 cannot perform its silencing activity unless adequate NAD is available to it. When there is high metabolic activity the cell is busy converting glucose to energy and deprives SIR2 of adequate NAD. The resulting inefficiency in silencing the expression of rDNA genes causes the infirmities of old age and aging as observed by them in their lab. This means that the rDNA circles are the culprits we have been looking for. If we can find some intervention that effectively silences the expression of rDNA genes, it would amount to mimicking the caloric restriction regime. If this biochemical mechanism found in yeast can be extended to human beings, our decks would be cleared and we will be able to give better direction to our fight against aging. Our ideas about hibernation, lowering the core temperature and diminishing the metabolic activity coalesce here.

How does CR spur SIR2 activity?

Caloric restriction spurs SIR2 activity through a process well known as aerobic process. Metabolism can take in two different paths. First glucose is metabolized to pyruvate. Then the pathway bifurcates into fermentation and respiration.

Respiration oxidizes glucose to carbon dioxide generating 28 ATP molecules per each glucose molecule, whereas fermentation to ethanol generates only two ATP molecules for each glucose molecule. When glucose level is high metabolism takes the route of fermentation, when it is low, metabolism prefers the respiratory route. When metabolism takes the respiratory route, it consumes more oxygen and shunts carbon to mitochondrial TCA cycle, thereby increasing electron transport and production of more ATP molecules. Does this metabolic shift towards aerobic process occur under conditions of caloric restriction? To decide this the MIT team conducted experiments to compare oxygen consumption by yeast cells grown in glucose solutions of 2% and 0.5% strengths, and found that the metabolic shift did indeed occur.

In order to give credit to CR, we have to answer another question. Under the regime of CR, is this metabolic shift necessary? The answer is 'Yes.' To put the question in another way, if the metabolic shift is somehow prevented artificially, can CR regime still lead to life extension? The answer is 'No'. That clinches the issue. It is the metabolic shift towards the aerobic process that is the proximate cause of life extension and Caloric Restriction's role is to facilitate this metabolic shift towards aerobic process.

(C) Replacement of Body Parts

(C-1):History of Transplant Operations

The history of body part replacements goes back to the 4th century BC when a Hindu Ayurvedic physician & surgeon named Susruta successfully replaced a lost nose and ear lobes from tissues taken from other parts of the body.

Towards the end of nineteenth century, Alexis Carrel who perfected joining blood vessels tried to graft organs, but they were later rejected. During the 1950's, French surgeons attempted removing kidneys of guillotine victims and implanting them in uremic patients but the kidneys failed after a few days due to rejection crisis. In 1954, the first successful transplant of kidney was made between identical twins, in a Boston hospital. The recipient of kidney lived for eight years and the operation was considered historic as it opened up a new era in transplant surgery. Successful grafting between non-twins followed this in 1964.

The next part to be tried for implantation was liver but liver is a more sensitive organ and the recipients did not survive for more than a few days. Lung implantation took place in 1969 in a Jackson, Mississippi hospital but the patient died of kidney trouble after eighteen days. The operation that would have the most widespread effect on human condition is heart transplant not only because of its importance in the body machine but also because of the widespread damages to hearts of modern people.

To extricate from the ethical problems involved in using animal organs, one has to think of mechanically made artificial organs. It is doubtful whether human bodies accept artificial organs.

(C-2): Limb Grafting

It would be of immense benefit if techniques could be perfected to attach limbs to people maimed in accidents but again it is not always feasible due to the problem of procuring proper limbs. This has been tried successfully in some cases where sharp instruments severed fingers, hands and legs without disturbing the internal architecture of the limbs and the severed limbs were speedily brought to the operating room along with the patient. In many such cases, partial restoration of the functions of the limbs was reported. Beyond such reattachment of ones own limbs there does not seem to be much scope in this area for, who would cut ones own healthy limbs and donate them! It is not known whether dead person' limbs can be usefully grafted. The alternative is attaching mechanical limbs.

The foremost problem in making mechanical limbs attached to human body functional is creating communication between residual anatomy of the maimed body and the attached part. In the chain of components required to restore function to mechanical limbs that include materials, mechanisms, actuators, electronics and microcomputers, communication is the weakest link. Available prosthetic technology can control only one joint at a time. This limitation highlights the mismatch between capabilities of natural human body and the prosthetic parts. Biomedical researchers are opening new vistas as they are steadily bridging the gap between men–machine interactions. For example, National Aeronautics and Space Administration have developed a robotic hand that comes close to human dexterity in its ability to move joints independently.[58]

Biomedical technology can restore lost function of amputees if they can:

(1) Express cognitive control over relevant motor functions in their residual anatomy to send signals and (2) can receive and decipher the inward coming signals.

Individuals who lost function of limbs due to either amputation or paralysis can satisfy the first requirement as they can sense and imagine manipulating their non-functional joints. They can express motor control over their lost parts through nervous or muscular activity directed to their residual limbs and these signals can be registered by appropriate technology. The other requirement is more challenging due to the complexity involved in human movement control. Action originates with a few neurons in the motor cortex. This triggers a large neural network that coordinates the activities of several effector muscles after receiving and processing feedback information from tactile, visual and positional senses. Transforming this tangled mess of electrical pulses into graceful movements is an incredible but routine accomplishment of healthy human body.

It is well nigh impossible to duplicate exactly the human sensory-motor system, nevertheless steady progress is being made as seen by the many robotic devices that are being developed. Biomedical technology can be used to restore at least some degree of almost any function. Among the techniques are:

Brain Machine Interface and Peripheral Machine Interface.

In the former, brain is in control of movements, in the latter, the brain is bypassed and communication is established with peripheral muscles and nerves. For individuals with missing or paralyzed limbs, BMI can be used for achieving some degree of cognitive control of artificial limbs at least for some primitive motions such as grasping. Peripheral-machine-Interfaces operate by functional electrical stimulation of muscles and peripheral nerves. However, they can be programmed for specific pre designed movements, because volition cannot be expressed without brain. Scientists are trying to overcome this difficulty by creating a record of specific movements and corresponding Electromyographic (EMG) patterns recognizable by decoders. Therefore bionic restoration of movements have to be designed depending on user characteristics. Different approaches are now under study.

CAN HEADS BE TRANSPLANTED?

As observed earlier, the present quest for immortality is not concerned with perpetuation of species or even family line, it is rather aimed at perpetuating individuality, continuity of individual identity. As the head is the repository of individual consciousness this intriguing question arises: Is it possible to sever a head from it's body and keep it alive without brain damage?

Transposing heads is a bizarre scenario. Transposition involves severing two heads simultaneously from their bodies; keep them alive for a while and attaching them to exchanged torsos. Can heads be kept alive without their torsos? Experiments conducted at the Case Western Reserve University School of medicine in the seventies, suggest that it is not impossible. Heads of monkeys were surgically removed, mounted separately and kept alive by connecting them to life support system. Dr. Robert White, a neurosurgeon who conducted these experiments said, "The monkey [head] became awake and acted like a monkey, they followed you around [the room with their eyes]. They would try to bite your finger off. There was every indication that brain was functioning as it was when it was on its own body"[59].

Transposing heads seems to belong to the realm of science fiction but if one considers the experimentation already done in this area, one wonders whether it will become real life science in course of time. It is reported that Dr. Robert J. White and his associates at Cleveland Metropolitan General Hospital removed a monkey's brain from its skull and kept it alive on artificial circulation for seven hours in 1963. The fact that it was still consuming oxygen and releasing carbon dioxide into circulating blood seemed to prove that it was still alive. By 1966, this team of surgeons went ahead a little further, connected the removed brain of a dog to the circulatory system of another dog, and found that the brain was alive for two days. The astonishing event occurred in the next stage of their experiments they were able to severe whole heads and maintained them alive for six hours by connecting them to the circulatory system of another dog.[60] This suggests the novel idea that heads can be transposed. In fact, even the idea is not new as a story in ancient Indian literature reveals. This story appears in a renowned ancient Indian work called Katha Saritsaagaram (Ocean of tales) and retold by the Nobel Prize winning German author, Thomas Mann, in his small book 'The Transposed Heads'.

Problem of Individuality

There is an important philosophical problem to reckon with in all cases of body part replacements. Does the individual who has his body parts replaced feel the dilution of his individuality? Perhaps it may not be so in every case. In case of transplanted heads, there is no need for such a feeling as the old personality acquires a new torso. One may argue that consciousness cannot be totally ascribed to brain alone as it has to receive signals from other parts of body to be conscious; but this can be answered by saying that the determinative faculty resides in the head and therefore the individuality is determined by the head and not the torso. This aspect is dramatically portrayed in the story referred to above.

Replacement of diseased or worn out body parts has been a recognized method of extending the functional life of people. The current practice of replacement of body-parts is by procuring useful organs from willing donors and transplanting them into the patients needing them. This procedure suffers from two limitations. The first is availability of required organs in good condition. The second is matching problem. Biomedical engineers try to solve these problems by manufacturing artificial organs. This endeavor has now developed into a special branch of technology known as Bionics.

(C-3): Tools for Grafting-

Bionics- current developments

Bionics is the science of designing instruments or systems modeled after living organisms. This theory follows the analogy of car part replacement. When the car malfunctions or refuses to move, we try to identify the malfunctioning part or non-functioning part and replace it and get the car going. Conceptually it is easier to understand this than others but its practical application to human bodies is far more difficult and risky.

Bionics has a fascinating history. The earliest cases of bionics can be traced back to human efforts to replace severed hands and legs. Progress in this field has come slowly and gradually since the beginning of 16th century. In other fields like organ transplantation progress has been slower with decades of gaps between the first attempts at a new procedure and routine success. The reason for the slow progress of this technology is its interdisciplinary nature. The progress in this field has been contingent on the insights gained from other disciplines like physics, chemistry, and immunology among others.

A chronology of development of this technology in three different groups furnished below shows the punctuated progress made in the field of biomedical engineering.

Can we have body parts off the shelves? Recent developments and current research in bio-medical engineering have opened up such possibilities and other new vistas in bionics.

Spurred by a severe shortage of donor organs, biologists, biomedical-engineers, and physicians have been combining their talents for designing artificial replacements for the human body. Here are highlights of some of the current scenarios of biomedical research.

MECHANICAL HEARTS

Yes, You can have a heart and other limbs taken from the shelf and attached to you when you need, if the current research succeeds.

Heart transplants are considered as miracles because of their impact on the quality and length of life, but like all miracles, they are rare. It is estimated that in United States alone about four hundred thousands of people develop heart failures every year. Of these about thirty thousand to hundred thousand with end-stage heart condition could benefit from heart transplantation, but only about two thousand people receive transplants because of donor shortage. This has inevitably led to a system of rationing. Committees determine whether a candidate is eligible based on many factors as age, habits that effect health, conditions that limit quality of life and life expectancy. They strive to give the available hearts to those who have the promise of making the best use of this gift.

This system of rationing leaves thousands of other patients and their kin in an uncomfortable position. To avoid this situation physicians and biomedical engineers have been working for decades to develop artificial hearts. A daring goal that has captured public imagination is the implantation of a completely artificial heart into patients with end-stage heart failure. Barney Clark, a Seattle dentist was the first to receive the totally artificial heart Jarvik-7 (named after the designer) in 1982. Clark went through a great ordeal because the he was tethered to an external power unit and he was never able to leave the hospital room during the 112 days he lived thereafter. Though another patient, who later received a Jarvik-7, lived for 620 days, artificial heart in the public mind was associated with ordeal of Clark.

This concern was answered by the venture of another company called Abiomed. They designed a system in which the artificial heart-pump, a miniature computer, power converter and cables are all implanted into the body and it is supplied energy by radio frequency transmission through the skin. This device was tested first on a patient named Robert Tools. He was able to venture outdoors a couple of times after the implantation. He lived for 151 days carrying the one- Kg- weight heart-pump inside. He died possibly of a blood clot in the machine.

Artificial Blood

Donor Blood used for life saving transfusions has always been in scarce supply. Nearly for three centuries, physicians have been looking for blood substitutes. In 17th century, Christopher

Wren thought that ale and wine could be used as substitutes for human blood. Not until the beginning of the 20th century, could blood transfusions be made scientific and safe. Modern era of blood transfusions began after the discovery of blood grouping method by Karl Landsteiner in 1901. Ottenburg, in 1913 was the first to apply blood group serology to transfusion practice. Even after Ottenburg's description of blood compatibility testing, transfusions were limited due to lack of proper anti-coagulants and storage methods. This situation was remedied by the discoveries made by the necessities of the world wars and blood transfusion has become the standard tool in the surgeon's kit.

The principal constituents of blood are plasma, red blood corpuscles, white blood cells and platelets. Hemoglobin, a conjugated protein in red blood corpuscles transports oxygen to tissues of the body and picks up carbon dioxide from them. White blood cells perform the function of immune response. Cellular fragments called platelets perform the function of blood clotting and wound healing.

Transfusion of donor blood has become a standard and safe procedure now; but due to various reasons like donor blood shortage, stringent storage requirements of human blood and difficulty of sterilizing it, have led to the search for blood substitutes. However, they cannot succeed unless they satisfy stringent conditions like efficiency of carrying oxygen, being free from pathogens, storability, and overcoming blood group matching criteria.

ARTIFICIAL LIVER

Liver is a sturdy organ that can regenerate itself after acute injuries, but it cannot do so if it is damaged by cirrhosis or fibrosis caused by alcohol abuse or hepatitis infections. In advanced state of liver disease, the only hope to the patient is transplantation of a healthy liver. In acute cases, the patient may die before a donated liver could be procured. In chronic cases the liver may have been so damaged that even transplantation of a healthy liver may not work. This situation has prompted the scientists to invent an artificial device that could be used as a stop-gap arrangement until a donor liver could be procured for transplantation or the patient's own liver recovers health.

The solution suggested is to come up with some machine which works extra corporeally like the kidney dialysis machine. This device called Bio Artificial Liver, houses metabolically active hepatocytes (liver cells) sandwiched between artificial capillaries and plates. Patient's plasma is circulated through it. The problem faced by the designers of BAL reactor is to decide which of the many functions liver performs should be given importance in the design. Normal liver cells produce many kinds of proteins, they produce bile and regulate fat, carbohydrate and protein metabolism, they detoxify the ammonia products of nitrogen metabolism, and they breakdown alcohol and drugs.

Restoring Vision

Is it possible to restore vision through bionic eyes? Vision is a complex form of information processing that depends on a remarkable neuro-processor at the back of the eye called retina. Light passing through the pupil of the eye is focused onto the retina's sensory neuro epithelium. This results in the formation of an image on the millions of photoreceptor cells in the outermost layer of the retina. These cells transform the luminescence and color pattern of the projected light into electrical and chemical signals. Visual information thus transformed into electrical signals is then carried by the ganglion neurons of the optic nerve to the visual cortex of the brain. If this mechanism breaks down anywhere in the path, blindness can result.

The transformation of visual scenes into electrical images (a group of signals that can create a visual image) carried by the optic nerve to the brain requires that sensory neurons are stimulated simultaneously and in correct order to enable three dimensional objects to be correctly encoded. Researchers propose to do this by retinal implants.

There are two kinds of implants- sub retinal and epiretinal, under development. Both methods have their pros and cons. Experiments in animal models where recordings were made with electrodes attached to the inner or outer retina reveal that injecting a tiny current is sufficient to excite retinal neurons. It is encouraging to note the feasibility of eliciting action potentials in the visual cortex using electrical impulses generated by the implants but a number of obstacles remain. In spite of promising leads and a lot of publicity, restoring vision through retinal implants remains a distant dream.

RESTORING HEARING (COCHLEAR AND BRAIN-STEM IMPLANTS)

Hearing loss- partial or total has a number of reasons. Defects either in the outer ear or middle ear can result in conductive hearing loss, i.e. the ability of the hearing mechanism in the ear to conduct sound waves to the auditory nerve is either dampened or lost. Conductive hearing loss can usually be remedied by scraping the stapes bone in the ear to enable it to vibrate better or by surgically replacing it with a teflon prosthesis. Using hearing aids can also alleviate this condition. The other variety is sensory neural deafness, which is harder to cure. This is caused by the loss of sensory hair cells in the fluid filled, snail-shaped organ of the inner ear, called cochlea. This organ converts the sound waves into electrical signals and sends them to the brain via the auditory nerve.

They started doing this by cochlear implants. Implanting an electronic chip to mimic the functions of cochlea benefit those individuals in whom the auditory nerve is in satisfactory condition. Cochlear implant is a microelectrode array implanted in the cochlea that directly stimulates the auditory nerve. Early CI's were designed primarily to replace the lost function without regard to the way the brain processes and adapts to auditory information. For some profoundly deaf individuals, this device is not useful. For them, a device, which bypasses the cochlea and the auditory nerve and directly stimulates the brain stem, is required. Such Auditory- Brain-Stem implants have been under development since 1970's but they have had only limited success. The electrode array in the new generation of these implants penetrates the depths of the ventral cochlear nucleus instead of the surface of the nucleus. This facilitates stimulating the neurons directly.

REPAIRING SPINAL CORD

A few years ago, the prognosis for recovery of a patient after spinal injury was bleak and research in the field of restoring spinal function was almost labeled as a lost cause. Let us see how the situation is changing now.

Spinal cord is a strand of nervous tissue that is tightly enclosed in the bony vertebrae of the spinal column. The spinal column receives sensory information from the skin, the muscles, the joints and other tissues of the body. It transmits the information in the form of electrical signals to the brain along millions of nerve fibers. The motor commands generated in the brain are sent to the spinal cord along nerve fibers, which terminate in the local spinal motor circuits. From here, electrical impulses reach the muscles via peripheral nerves and coordinate their movement.

When the spinal cord is injured, the first stage of injury involves mechanical tissue destruction and in the second stage, local blood supply is disturbed. There have been attempts to minimize this damage with neuron-protective agents but the results have not been quite satisfactory.

There are four principal strategies for repairing the damaged spinal cord.

1. Promoting the re-growth of interrupted nerve fiber tracts.
2. Bridging the spinal cord lesions with scaffolds that are impregnated with nerve growth factors.
3. Repairing the damaged myelin (insulation surrounding the axons) and restoring the conductivity of nerve fiber in the lesion area and
4. Enhancing the plasticity of Central Nervous System by promoting the growth of the undamaged nerve fibers surrounding the lesion.

Damaged nerve fibers attempt at regenerative sprouting but the inhibitory proteins produced by the central nervous system thwart that attempt. Attempts are now being made to neutralize the inhibitory molecules by application of antibodies against them and other methods.

Lesions caused by the damage act as gulfs that interrupt the communications from either side. In anatomically incomplete spinal cord lesions, nerve fibers can bypass the injury site if we can build persistent tissue bridges but sprouting nerve fibers are hesitant to cross these bridges. Bridges that form a growth-permissive scaffold within the lesion may facilitate the growth of regenerative axons to fill the communication gap. Several laboratories are currently exploring the use of various kinds of cells as bridging material. The principal problem in these efforts is to find a method of integrating the bridges into the spinal cord tissue without inducing scar formation.

- A sheath made of insulating material called myelin surrounds the nerve fibers. When this material at the site of damage is lost, electrical communication is interrupted. Current efforts are aimed at implanting the lesion sites with cultured neural stem cells in order to promote myelin production.
- It is common knowledge that if a child breaks it's bones the injury is easily repaired. The reason is that, in the young, the nerve fibers remaining in normal condition after the injury react to compensate the neural imbalance by sprouting new fibers quickly. When an adult sustains an injury, this compensatory action is very slow and sluggish. In animal experiments it was found that by using antibodies against the factors inhibiting the growth of nerve fibers in the spinal cord, compensatory growth could be enhanced[61, 62]

TISSUE ENGINEERING

A brand new field of engineering called Tissue Engineering has been emerging. This field relates to manipulating living cells to restore, maintain or enhance tissues and organs. Tissue engineering conjures up visions of organs built in the laboratory being transplanted into desperately ill patients. The potential impact of the success of this new branch of engineering is quite broad. If successful, the engineered tissues could reduce the need for organ transplantation. Engineered tissues may accelerate the development of new drugs that could cure diseases and eliminate the need for transplantation.

The challenging question before the researcher in this field is how to engineer the tissues to form organs of various complicated shapes, sizes and functions? The challenge involving size and shape is answered by suggesting that biodegradable scaffolds resembling the desired organ could be built and cultured cells could be coaxed to grow in them. These scaffolds provide the physical and chemical cues to guide their differentiation and assembly into 3-dimensional tissues. Coaxing cells to form tissues in a reliable manner is an engineering problem that must be accomplished within the constraints of reliability, cost, social acceptability and government regulation. We can adopt three different strategies to treat injured or diseased tissues.

1. Freshly cultured cells may be implanted at the site and coaxed to grow.

2. The second method is implantation of tissues assembled in the laboratory from cells and scaffolds. Scaffolds are temporary supporting structures. Biological scaffolds are similar in function to the engineering scaffolds we find in high-rise construction sites. The difference between the two is, whereas the construction scaffolds in engineering sites are manually removed after serving their purpose of supporting the men and materials, the biological scaffolds are allowed to degrade in the body. Biological scaffolds are porous degradable structures fabricated from natural materials like collagen or synthetic polymers. They can take the shape of sheets and gels or they may be complex structures

3. The third method envisages implantation of a scaffold at the site of damage and stimulate the body's own cells to populate it and repair the tissues. Three dimensional polymer structures can be built from a series of very thin layers. Structures have been built by seeding different cell types for each layer sequentially on biodegradable scaffolds made from synthetic fibers. Such a structure was sculpted in the shape of bladder and seeded with urinary epithelial cells and smooth muscle cells, was implanted into dogs. This artificial bladder is reported to have attained near normal function[63].

New methods for fabricating more demanding applications scaffolds are under way. For example, 3-dimensional polymer structures are being developed from thin 2-D layers based on computer models of desired tissues derived from MRI or tomography image of patient's original tissue.[64]

The next stage after readying the required scaffolds is to populate them with cells. The cells from the patient (known as autologous cells), the cells from donors (known as allogenic cells) and cells from other species (known as xenogenic cells) are three different sources that can supply the cells needed for transplantation. Each category can further be differentiated into stem cells, adult cells or a mixture of both. Allogenic cells have been used successfully to treat skin problems, diabetes and liver disease. The use of xenogenic cells evoked mixed reaction and it remains to be seen whether they will gain social acceptance.

Engineering Challenges

An important problem to be solved in grafting the engineered tissues on to the patient is, providing blood supply to the grafted cells because the cells engineered outside the body do not have their own blood supply. After implanting in the patient, the new cells consume very fast the oxygen available in their neighborhood and it will take several days for the tissues to grow their own new blood vessels. They are trying to solve this problem by implanting cultured cells directly into the existing vascular beds, at least in the case of liver and spleen. Hepatocytes injected into human liver showed engraftment and sufficient biochemical activity to give some relief from liver disease[65]. Similarly, diabetic patients implanted with pancreatic islet cells exhibited normal glucose tolerance for several months after the procedure66. However, this procedure of exploiting the existing vascular beds cannot be exploited in the case of bones or tendons. Other methods have to be devised.

Tissue engineering can be applied even to the development of drugs. The greatest impact of tissue engineering in the coming decade will perhaps be in designing in physiological models in laboratory for developing molecular therapeutics. However, we cannot deny that most of these procedures are no better than dressing up the body with borrowed robes.

(C-4) Stem Cells.

What are stem cells? The name is again metaphorical, arising out of comparison with a plant stem. Just as the stem of a plant grows branches, shoots and blossoms, the embryonic stem cells, beginning from a tiny state are supposed to blossom spectacularly into various organs of the body.

The origin of stem cells is the zygote. We start our lives as zygotes or single fertilized eggs. In time, this divides into two cells, four, and so on in geometric progression to become the embryo. At this early stage of development, the embryo has no resemblance at all to what we are, but the DNA in the nuclei of these cells contains the programmed information for developing gradually into coordinated limbs and organs of the body, in course of time. These cells are called stem cells because out of these emerge other cells which go to make other organs. No one had ever seen stem cells, though physicians have been doing bone-morrow transplants. It is just assumed that such cells existed, but that situation changed after two sets of researchers – John Gearhart's group at Johns Hopkins University and James Thomson's team at the University of Wisconsin, isolated human stem cells.

GROWING BODY PARTS WITH STEM CELLS

The problems of procuring human parts in time for transplantation and the possible rejection by the recipient body can be solved if the research going on for regenerating body parts using one's own stem cells bears fruit. Some reports indicate that this dream has been coming closer to realization. It has been reported that Dr. James Thomson of the U of Wisconsin has been able to direct stem cells to make heart tissues. Dr. John Gearheart of the Johns Hopkins University has crafted human stem cells to make neurons. The biotechnology laboratory at Hyderabad, India, is reported to be coaxing the stem cells taken from an embryo to develop into the retina of a recipient subject.

These results raise hopes of using stem cells for therapeutic uses of stem cells but at the same time, they open the Pandora's box of ethical problems.

REJUVENATION BY REPLENISHMENT OF WORN OUT CELLS

Another area of interest in stem cell research is rejuvenation by replacing worn-out cells with fresh cells. Some companies[67] started contemplating about using stem cells while some have already started experiments. A sample of bone marrow containing stem cells could be extracted from the pelvis and the cells could be grown in a culture. They could then be mixed with blood and the mixture injected into the body. The mixture containing the stem cells and the products of stem cells is expected to home in to the locations of damaged organs (say a damaged liver or kidney) and revive them. If these experiments succeed, stem cells can be used as natural defenses against ageing. Our bodies can then be tuned up with stem cells and perhaps we can lead healthier lives longer. At this early stage in research, we cannot say whether the stem-cell techniques could be turned into human therapies.

(D) Esoteric Phenomena

Scientific Case for Life after Death

Is there life after death? This old question is the main theme of the famous Katha Upanishad that is translated into several languages. Does the question appear antithetical? No, what this question really means is, the enquiry whether consciousness survives death of physical body. According to present state of medical theory, physical death ends the personality but metaphysics always held that possibility of something surviving physical death, while psychologists never ruled out the possibility of paranormal experiences. Several instances of paranormal experiences were recorded in the past, but what lends currency and importance to this topic is a recent incident in which a cardiologist shared that experience in a hospital and he is reluctant to dismiss it as hallucination.

Paranormal Experiences

This incident is reported in the October 2003 issue of Readers Digest under the title 'The Scientific Case for the Human Soul'.

In the summer of 1991, one Pam Reynolds learned that she had a life-threatening bulge in an artery in her brain. Neurosurgeon Robert Spetzler told her that in order to operate he would have to stop her heart. By all clinical measures, she would be dead for up to an hour.

> While Reynolds was put under anesthesia, leads from a machine that emitted a clicking sound were plugged into her ears to gauge her brainstem function (the brainstem plays a part in controlling hearing as well as other involuntary actions). Additional instruments tracked her heartbeat, breathing, temperature and other vital signs. Her limbs were restrained. Her eyes were lubricated and then closed.
>
> As the surgeon powered the surgical saw to open the patient's skull, something occurred that sophisticated monitoring devices never registered. Reynolds told later, that she felt herself "pop" out of her body and she looked down on the operation from a vantage point just above the surgeon's shoulders. She "saw" the surgeon holding something that resembled an

electric toothbrush. It appeared as if they were about to operate on her groin.

She thought '*That couldn't be right*' because this is brain surgery. Reynolds then assumed that whatever they were doing inside her skull had triggered a hallucination.

However, even though her eyes and ears were effectively sealed shut, she told what she perceived was actually happening. The surgical saw resembles an electric toothbrush. Surgeons were, working on her groin to insert catheters and thread up to her heart in order to connect her to a heart-lung machine.

The surgeon gave the order to bring Reynolds to "standstill"-draining the blood from her body. By every reading of monitoring instruments, life was supposed to have left Reynolds' body.

Surprisingly, Reynolds reported that she found herself traveling down a tunnel towards a light. At this end, she saw her dead grandmother, relatives and friends. Time seemed to stop. Then an uncle led her back to her body and instructed her to return. It felt like plunging into a pool of ice water. Later, after regaining consciousness, she told Spetzler all that she had seen and experienced.

The surgeon is reported to have remarked "You are way out of my area of expertise." Several years later, Spetzler is reported to have said that still he didn't know what to make of it.

The kind of experience described above is known as Near Death Experience (NDE). Doctors are prone to dismiss such experiences as hallucinations brought about by changes in the dying brain. The trouble with that explanation is that hallucinations could occur only if brain maintained some function. Once it becomes dysfunctional, it is analogous to a computer unplugged from power source. It can't do anything let alone hallucinate.

Another cardiologist and researcher Michael Sobom compared what Reynolds said she saw and heard, with Spetzler's surgical transcript and found that during the period, Reynolds experienced the tunnel, she had no brain activity at all. "She met all clinical criteria for death," said Sobom, "she had no blood in her body, she had no vital signs at all. So, was this death? And, if it was death, what was this experience that she had while in that state?"

Near Death Experience Episodes

- In a study published in the British medical journal *The Lancet*, Dutch cardiologist Pim Van Lommel recounts the NDE of a 44-year-old cardiac-arrest victim who was clinically supposed to be dead. He was rushed to the hospital where the doctors restarted his heart with defibrillators. A nurse removed the man's dentures so that a breathing tube could be inserted in his throat. Once stable, the man was moved to intensive care. A week later the man saw the nurse who removed his dentures and recognized her, though in the only prior encounter he was in a state of coma or clinically dead. "You took my dentures out of my mouth," he told the nurse and went on to describe the other details his disembodied self had viewed.
- In an attempt to gauge the frequency of NDE's, Van Lommel and his fellow researchers interviewed 343 others who suffered cardiac arrest and survived. "Eighteen percent have a story of very clear consciousness," says Van Lommel. These stories comprised of a general feeling of peacefulness to full-fledged NDEs
- Another study of British researchers published in the journal *Resuscitation*, found that 11% of the resuscitated subjects had memory recall of the unconscious period. Six percent of those after cardiac arrest reported NDEs.
- Both Van Lommel and the British researchers believe that these findings suggest consciousness could exist in the absence of a functioning brain. " You can compare the brain to a T.V. Set, says van Lommel. "The T.V. program is not in your set."

It is understandable that Spetzler and his ilk were baffled by this phenomenon because their training was governed by the positivist philosophy that does not admit of any paranormal experiences. However, these recent events are neither the beginning of such experiences nor the bewilderment of medical professionals the dead end of knowledge. Our ancient thinkers knew of similar and more pronounced of such experiences. They not only understood what happens behind the veil of death but also seem to know how to gain control of such a situation.

There are accounts of some ancient Indians performing the wonder trick of migrating into lifeless other bodies and animating them. This is known by the Sanskrit name '*Para-Kaya-Pravesa Vidya*' literally meaning the art of entry into other bodies. Ancient literature on the topic says that this task is accomplished by controlling the life breaths that animate the body.

Five life breaths called *prāṇa, apāna, vyāna, udāna* and *samāna* performing different functions permeate our bodies. Corresponding to these five principal breaths there are five auxiliary life breaths called *Nāga, Kūrma, Krikara, Devadatta,* and *Dhanamjaya*. When a person dies, these breaths exit the body one by one, the last being *Dhanamjaya*. The Yogi who is adept in contemporary transmigration performs his feat by exercising control over the last named breath.

Where is the consciousness then? Spetzler thinks it is in every cell of the body. Let us examine the validity of this presumption.

We know that each day about fifty billion cells in a body die. Fresh cells are then regenerated to replace the lost cells. There is intensive turnover of cells every day. If that is the case, your body now is not the same it was when you were a child. That can be verified from the difference in structure and resemblance in your photographs at these different times. Yet, you feel that you are the same person as when you were a child and claim the same identity. This phenomenon leads us to make two different hypotheses about this continuity of consciousness.

The first hypothesis: Even as some cells die continuously, the rest of the cells pass on the consciousness to the new comers like inheritance and continuity is maintained.

However, this hypothesis runs into difficulties when we consider the death of brain cells. When the brain cells die, continuity of consciousness is lost even though the rest of the cells are living. This suggests that the continuity of consciousness is not the handiwork of cells. Therefore, we are not inclined to subscribe to this hypothesis.

The second hypothesis: Some entity oversees all the changes occurring through time, while it remains invariant and independent. As long as this independent and effulgent entity animates the entire body, consciousness is felt. When it begins to withdraw from the brain, the brain cells die first and the body survives for some more time as in Alzheimer's disease. When this entity withdraws entirely from the body, the body dies. This is a more sensible hypothesis.

Even though most of the medical professionals are reluctant

to utter the word "Soul", there are many who are not satisfied with the explanation that NDEs are triggered by the failing brain. Continuity of consciousness is maintained, not because of the continuous turnover of cells but because the turnover of the cells themselves is engineered by the independent entity called soul. When it withdraws, it takes with it all the functions of the body lock, stock and barrel and the body falls dead.

CHAPTER SEVEN

Appraisal of Anti-aging Theories

> Life's but a walking shadow, a poor player
> That struts and frets his hour upon the stage
> And then is heard no more. It is a tale
> Told by an idiot, full of sound and fury,
> Signifying nothing.

---- Shakespeare (Macbeth, Act 5, Scene 5)

That is how Shakespeare appraised the worth of our lives most of which signify nothing. Instead of making our lives more significant, we make endless noises about our ability to hang on forever to our insignificant lives. Are they just sound and fury or do they lead us to the enchanted land? Just read on and decide for yourself.

Can you conjure up the vision of old people over sixty being transformed into bouncing youths and run their lives all over again? That is the idea of reversing age. Such an event seems to be still shrouded in mystery though some scientists are claiming that it will soon become a reality. Certain procedures are already applied to treat age-related afflictions. If heart is not functioning well it is replaced with a new heart; if kidneys are not doing their job it is entrusted to dialysis machines; even diseased lungs are transplanted with new ones. Nevertheless, these procedures that replace part for part of body are all expensive, superficial and uncertain of success. Even if successful, they may not be trusted to work for long and bring the pep and vigor of the body it used to possess in the bygone days. We may go deeper and remove the free radicals, repair the damaged DNA and strengthen the immune system and yet these are just patchwork and not renewal of the system. Here comes the telomere therapy, which offers hope of overall renewal of the system, to some extent.

1.Appraisal of Telomere Therapy

Frantic efforts are now going on in various biological laboratories around the world to take control of the mitotic clock mentioned earlier. Theoretically, if this clock can either be reversed or reset, the processes under the control of this clock start over. If the aging process is like the unwinding of the mainspring of a clock, what should we do to reverse aging? What should we do to renew the aging body? Guess. Rewind the mainspring of the clock before it stops. As we said earlier if telomere shortening is the analog clock in our bodies, we should arrest it's shortening to stop aging and re-lengthen it by some method to renew the body. This is easier said than done at least for the present because of tough problems in its implementation.

In order to lengthen the telomere its proliferative potential has to be augmented. This has to be done through supplying, from outside, an enzyme called telomerase to the shortening telomeres, as telomerase is not naturally found in the cells. Once this rejuvenating enzyme is supplied, the telomeres will then start to regrow the base pairs. As the telomeres increase with addition of base pairs, the body is supposed to be rejuvenated; but supplying the required telomerase to the cells in required quantities seems to be a formidable problem though not insurmountable. An obvious method that suggests at first is to inject telomerase into the blood stream but unfortunately, this method does not work because telomerase is very unstable and is easily destroyed by the normal enzymes in the blood. When it is injected, it does not stay long enough in the blood to reach the chromosomes. It is not that easy as one would wish.

At present three methods are identified for resetting the telomere clock.

- Introduce a telomerase analog.
- Introduce a gene that can express telomerase
- Introduce an inducer that would unlock your own telomerase gene.

All these methods face some counter arguments.

1. Telomerase analog is a synthetic, artificial enzyme that is more stable than the natural version. Synthetic telomerase

molecules may evade destruction for long enough time during which they can reach the cells to lengthen the telomeres before they are broken down.

2. Since telomere genes are normally repressed, constructing genes that would manufacture telomerase in normal somatic cells is another method. This would avoid forcing telomerase through blood stream but in practice, this approach bristles with many difficulties. First, It involves knowledge of DNA sequence, which is not yet fully known. Once the complete sequence is known an artificial gene has to be built to produce telomerase. Then the sequences have to be inserted into a viral carrier and injected into body to have them infect every one of the trillions of cells in the body and expect telomerase expression. We cannot be over- optimistic about insertion of telomerase gene into human body due to these unresolved problems although research in this direction is progressing fast. Even if the full DNA sequence is known, the problem of finding a suitable carrier has to be solved. Even then, there is the very important problem of producing the exact amount of the needed telomerase.

If the new gene doesn't manufacture enough telomerase, the cell may fail to divide, if more, it may result in too long telomeres causing cancerous tissues. Therefore, it is not enough if you are able to supply telomerase genes to cells; you should also be able to deliver the optimum quantity. To accomplish this we should know the optimum quantity required and having known that we should be able to produce that optimum quantity, nether less nor more. That is, when the new gene is built we must also be able to devise a self-regulating mechanism to accomplish that task automatically.

3. The third alternative, viz. discovering inducers, which would unlock your own telomerase genes is in experimental stage. Scientists of Geron Corporation in Menlo Park, California, are reported to have succeeded in doing exactly that. Their research claims to have discovered a human telomerase gene that can lengthen the shortened telomeres by knocking down the upper limit on the number of cell divisions. The method is to turn on the switch of the gene that makes the enzyme component known as hTRT (human Telomerase Reverse Transcriptase). When they inserted this gene into human cells along with a signal to turn on its switch, they found that the cells

continued dividing past the Hayflick limit and quickly rebuilt their shrinking telomeres back to their original lengths. It remains to be seen whether this will prove to be the fountain of youth.

PREMATURE AGING:

The diseases that should claim our first attention in the context of telomerase therapy are those that exhibit early old age symptoms. Everyone dreads the physical decline associated with aging but people afflicted with an inherited disease known as Werner's syndrome and Hutchinson-Gilford's Syndrome and progeria have to face that dread much sooner than others. People afflicted with Werner's syndrome suddenly exhibit signs of premature aging, while they are still in their twenties. Their hair grays, their skin loses suppleness, and cataracts cloud their vision. They get cancer, heart disease and host of other diseases, which normally don't strike until late in life, and most of them die before they are fifty years old. Progeria is also premature aging disease. Children afflicted with this disease suffer from strokes, heart disease, baldness, thinning skin, lack of fat and age spots. Some have arthritis and osteoporosis but the most prominent is the vascular disease caused by aging cells.

If Werner's syndrome and progeria are diseases in which old age symptoms appear in children, then we should look into their symptoms to assist our guess about the potential benefits of telomere therapy. In a number of studies, telomere shortening was found to be associated with progeria, Hutchinson-Gilford's syndrome, and Alzheimer's disease. Therefore, these should be our targets for application of telomerase therapy. In this case, the exact clinical method to be used is not yet clear, as this is not a local area therapy.

It is very tempting to imagine that telomere lengthening could eliminate all these diseases but perhaps it won't happen so. From our present state of knowledge what we can infer is that telomere therapy will restore you to your earlier original status that existed when you were younger, it wont make you strong if you were a weakling when you were young; It won't make you a genius now if you were a commoner then. If you had any congenital diseases, they do not vanish now; if you had any missing organs, they cannot be regenerated now. If you had genetic predisposition to strokes and heart afflictions, which would have appeared even while you were young, they will continue to be there but perhaps they can be ameliorated after this therapy. In short, telomere therapy won't make you a new person.

What telomere therapy does is, it may help you to start over a part of your life and feel younger, but are all young people the same? There are innumerable variations among the young; lean and stout, pretty and ugly, strong and weak, smart and dumb etc. Such idiosyncrasies will continue to be there. It does not transform you into a new person but you may feel restored. What you can hope for from telomere therapy is younger cells and healthier vessels, healthier skin, and perhaps postponement of some cardio-vascular diseases.

Applicability of Telomere Therapy

We cannot claim that telomere therapy has universal applicability; its usefulness is limited. Some of the areas of its usefulness are considered now.

STROKE & HEART DISEASE:

The chances of occurrence of strokes and developing heart diseases may be diminished. However, disease of the heart itself is not reversible since the heart muscle, once destroyed is gone forever and cannot be regenerated. If the therapy is applied before the heart attack ever comes, it can perhaps be prevented but nothing can be done after the attack to replace the lost heart muscle. Heart failure, as far as it relates to inelastic arteries and constricted capillaries, may be prevented and cured by telomere therapy.

BLOOD PRESSURE

Blood pressure is a complex problem. High blood pressure is the body's attempt to compensate the diminished flow of blood through constrictions in capillaries caused by plaque deposition. Several factors such as diminution of arterial and capillary volume caused by plaque deposition, elasticity of blood vessel walls and changes in the function of heart muscle itself contributing to high blood pressure may be treated by telomere lengthening as all these adverse changes are known to result from telomere shortening.

ANEURYSMS

Aneurysms are caused by aging cells in which telomere is almost exhausted. Therefore, we can expect that re-lengthening the telomeres can prevent aneurysms. Vascular diseases like heart disease, high blood pressure, stroke, heart failure and aneurysms will be among the first targets of telomere therapy.

SKIN:

The first thing that comes into our minds, on hearing about the rejuvenation of cells, is skin. If aging cells can be enlivened in test tube by applying telomerase, we can successfully use this as a therapeutic technique on burn victims. The aging skin cells on the victim's body can be taken out, enlivened with telomerase therapy and put back on the victim's body along the periphery of burn area where the skin cells died. That would allow the enlivened and replanted cells to proliferate and cover up the burn area with new skin. As this is a local area therapy, it is done easier than those involving all the cells of the body.

The cosmetic aspect of skin is of paramount individual and social importance because color of the skin, its texture, elasticity and suppleness sway the emotions of people in love and hatred. It is here that the effects of telomere therapy will be strikingly visible. However, the deeper significance of skin is that it forms the outermost defensive barrier of the immune system. The skin of the older people is so thin that it is prone to be lacerated easily and prone to contract infections that may not heal easily. Telomere therapy may be able to restore the skin to its former health.

LIVER

Another area in which telomerase therapy might be successful is liver. Cirrhosis of liver is the seventh leading cause of death in the world. Cirrhosis is the scarring of liver. Scarring hinders its ability to regenerate itself, by blocking blood flow and creates symptoms, which will inevitably lead to death. Toxins like alcohol and hepatitis viruses are known to cause this condition. In a recent experiment, Japanese scientists found that the telomeres in cirrhosis patients were unusually short. Linhard et al[68] and his colleagues at the Dana-Farber cancer institute and Harvard medical school in Boston tested this finding in their laboratory mice into which cirrhosis was artificially induced by injecting a liver-poisoning chemical. Select groups in them were injected with a virus carrying a copy of mouse telomerase gene. In this group, it was observed the development of cirrhosis was blocked. They concluded from this that it is reasonable to expect that telomerase could inhibit development of liver cirrhosis in humans also.

THE IMMUNE SYSTEM

Immune system is the general guardian of the body but guardians too age like their wards. Aging brings in its wake debility of immune system making it sloppy and inaccurate. It cannot mobilize its troops as efficiently as it used to do. Some of the lymphocytes do not divide as rapidly and reliably as they used to do. Worse still, some of the cells are unable to kill the invaders as efficiently as they used to do. Some of the lymphocyte chromosomes shorten because of their previous battles in which they were obliged to work harder expending their telomeres. Shortened telomeres in turn result in slower rate of division, which will eventually cause breaches in the defense.

Immune system has the ability to memorize its experiences. This memory keeps it battle ready in case old offenders dare to visit. As we grow up, we acquire immunity to a variety of antigens, which might have attacked us earlier. As we age, this memory fades; as it becomes fuzzy, there is inaccurate discrimination between self and non-self and the defenses are breached. Telomere therapy can improve the immune system by augmenting the proliferative capacity of lymphocytes and restore its former accuracy by improving the retentive capacity of its memory.

ARTHRITIS

Is there any hope of curing arthritis with this therapy? The answer is yes and no. A disease of the joints, this is of two varieties known by the names Rheumatoid arthritis and Osteo-arthritis. Immune system is among the causes blamed for arthritis. It is said that rheumatism is an autoimmune disease. Due to faulty discrimination, the immune system over-reacts and attacks its own body resulting in inflammation and destruction of joints. Therefore, telomere therapy should be able to relieve arthritis, not because the therapy acts directly on the joints but because it improves the immune system by fine-tuning it. Osteo-arthritis is a different matter. It's entirely a creature of aging. Most probably, shortening telomeres are at the root of this disease. Cells in the joints used very frequently lose their ability to divide and replace themselves. The more a joint is used the more the cells are called upon to divide and replace themselves. As some cells are lost, the remaining put up a losing battle until the cartilage of the joint is destroyed. Therefore, this disease also may yield to telomere lengthening.

NERVOUS SYSTEM

Nervous system is similar to immune system. The difference between the two systems is that the neural cells of the nervous system do not age except those in our nose. The cells in the nose divide continuously and replace the lost olfactory neurons giving us the smell sense. As the division continues for a long time, their telomeres shorten and the sense of smell is blunted. Lengthening the telomere of these cells will start the division of residual cells and the acuteness of the sense of smell returns, though with some unpredictable changes in the smell perception.

CANCER

Telomeres have diametrically opposite clinical roles. If they are too short a plethora of age related diseases set in and if too long, the dreaded cancer will be the result. Parasites also will thrive with longer telomeres. The curative technique here is shortening the telomeres and stopping the runaway growth of the malignant cells. The method of shortening is using telomerase inhibitors. This inhibitor is designed to block or slow down the action of human telomerase. This inhibitor forces the malignant cells to rapid senescence and death. Telomerase inhibitors are already reported to have been isolated and further work is in progress at Geron Corporation in California.

There are a few questions to be answered before settling on the telomere therapy to cure cancer. The first question is whether the inhibitors would work fast enough to keep pace with the growth of tumors. If the rate of growth of tumor were more than the rate of inhibition, the resultant would be some malignant growth however small it may be. In critical places of the body like the vicinity of heart or brain, even a small growth would be fatal. While we are targeting the malignant cancer cells with our telomerase inhibitors, they may also produce the undesirable effect of inhibiting the growth of cells at the other end of the scale. Suppose the vascular cells have telomeres, which are shorter than cancer cells and divide faster than the latter, what would be the effect of telomerase inhibitors? While the inhibitor is working at both ends of the scale, the patient may succumb to some vascular disease due to shortened telomeres even before the procedure had a chance to cure the cancer. Perhaps this kind of catastrophe could be avoided if a method could be developed to localize the work of the inhibitors to desired locations. Although telomere therapy can cure cancer theoretically, these questions and associated risks have to be studied. Finally, nobody knows yet whether telomere therapy alone can cure cancer.

Parasitic Diseases

These are perhaps special targets for telomere therapy because of their genetic structure. Parasites cannot survive unless the telomeres are re-lengthened. Human telomere has six-letter sequence of nucleotides while parasites have seven-letter sequences one such being that of the malarial parasite with sequence of TTTAAGG. The extra 'T' in the sequence can be used to develop a specific telomerase to force the parasite to age and die. How wonderful it would be to eradicate this disease, which has been killing tens of millions of people around the globe and inflicting huge economic losses! There are some other potentially curable parasitic diseases if this technique can be perfected.

Doubtful Conditions:

We cannot yet arrive at any conclusion about an extremely important condition. That is menopause of women. Perhaps it is not related to cellular senescence. Available medical literature indicates that each woman has a stock of eggs that are depleted month after month until menopause. This process is controlled not by the mitotic clock we have been discussing, but an independent timing mechanism that counts down the estrogen levels in the body. Therefore, it is presumed that since telomeres are not involved average age of menopause in women will remain unaffected by telomere therapy revolution.

DISEASES FOR WHICH TELOMERE THERAPY IS IRRELEVANT.

There are a number of diseases, which do not involve telomere shortening and hence such diseases are not candidates for telomere therapy. We know that many of our diseases are not related to aging. Congenital malformations, genetic diseases, infections, organ losses and mental diseases are not age-related diseases. Diseases related to malnutrition, be it due to lack of food or overeating are not related to aging. If somebody is born with cleft palate, or has sickle-cell anemia or muscular dystrophy, telomere therapy cannot help them. There are numerous diseases, which are independent of telomeres. As such, millions of people in poor countries who suffer from kwashiorkor cannot be helped by telomere therapy nor can millions of bald-headed males look at it with any hope.

It is very tempting to imagine that telomere lengthening could eliminate all these diseases but perhaps it won't happen so. From our present state of knowledge what we can infer is that telomere therapy will restore you to your earlier original status that existed when you were younger. It won't make you strong if you were a weakling when you were young. It won't make you a genius now if you were a commoner then. If you had any congenital diseases, they do not vanish now. If you had any missing organs, they cannot be regenerated now. If you had genetic predisposition to strokes and heart afflictions, which would have appeared even while you were young, they will continue to be there but perhaps they can be ameliorated after telomere therapy. In short, it won't make you a new person.

What telomere therapy does is, it helps you to start over a part of your life and feel younger, but are all young people the same? There are innumerable variations among the young; lean and stout, pretty and ugly, strong and weak, smart and dumb etc. Such idiosyncrasies will continue to be there. It does not transform you into a new person but you will feel restored.

What you can hope for from telomere therapy is younger cells and healthier vessels, healthier skin, and postponement of strokes and heart related diseases. The chances of occurrence of strokes and developing heart diseases will perhaps be diminished. However, disease of the heart itself is not reversible since the heart muscle once destroyed is gone forever and cannot be regenerated. If the therapy is applied before the heart attack ever comes, it can be prevented but nothing can be done after the attack to replace the lost heart muscle. Heart failure, as far as it relates to inelastic arteries and constricted capillaries, can be prevented and cured by telomere therapy.

While telomere therapy emits many rays of hope, it is not the panacea for all the ills of human body; certain maladies will continue to be there.

Even if a perfect telomerase enzyme is found and precise clinical methods are developed, it is a moot question whether telomere therapy is panacea. Obviously, it is not a cure-all therapy and not a panacea for all the ills of human body as there are definitely some problems with longer telomeres and in some cases, telomere therapy is not relevant and hence it is not indicated.

For the time being we can only speculate the answer to this question by observing what kind of symptoms develop when the telomeres are abnormally short and then make reasoned guesses in different situations from our knowledge of etiology. The basic principle, however, is that diseases which are supposed to result from telomere shortening can be cured by re-lengthening it. Others are not the candidates for telomere therapy. Telomerase therapy is not the panacea for everything.

2.Appraisal of Gene Therapy.

The foundations of gene therapy were rooted in the advances made by molecular genetics, cell biology and virology during 1950's and 60's. A few scientists suggested in 1960's the possibility of obtaining therapeutic effects by introduction of genes into mammalian cells. Early in the 60's, investigators suggested that it might be possible to introduce genes into mammalian cells to obtain therapeutic effects. The first clinical trial was conducted in 1970's but ended without demonstrable efficiency. Thereafter, there was little further progress until techniques for gene cloning and viral vectors were developed in the mid 1980's. Meanwhile public enthusiasm was roused by premature claims of biologists and publicity on gene therapy. Consequently, when biologists failed one after another to demonstrate by experiment the safety and efficacy of gene therapy, disillusion and widespread criticism followed.

Problems of Gene Therapy

Identification of Specific Genes

The first problem is identification of specific relationship between diseases and the genes that cause them either by their presence or absence. The prerequisite for obtaining the desired effect is to know what gene causes what effect. As already mentioned, a single gene may not cause a single effect. A group of genes may cause a single effect or a single gene may cause several effects. Unfortunately, there is no one to one relationship between genes and their effects. There may be one to many or many to one relationship. That complicates the matter and makes it hard to identify the relationship between genes and ageing process. In addition, a host of practical difficulties often plagued the scientists.

Finding a Safe Method of Introducing Genes into Human Body

The second important problem is finding a safe viral vector to do the job of spreading the required gene clones in the target body. It is this second problem that has plagued the biologists from the beginning and thrown them into public focus. With the deaths of a couple of therapy related deaths the criticism against gene therapy has acquired sharp teeth and led to regulatory investigations. The question to be asked now is: Do a couple of deaths linked to the therapy justify abandoning the promise and potential of gene therapy? None who is aware that a few failures including some fatalities are inevitable in most of the clinical trials would venture to answer this question in the affirmative. That is why the investigators have not trashed the whole field of gene therapy but only recommended that fundamental endeavors should be made to improve the methods of gene delivery.

Gene therapy was first conceived as an approach to deal with inherited genetic disorders for which alternative therapies did not exist. Therefore, criteria that were adopted in the beginning for using gene therapy were imminent threat to life and non-availability of alternative therapies to meet the situation. In this background, it is hard to appreciate why experimentation on human beings started hastily. Scientists were aware that rare diseases of immune system like SCID XI would be perfect candidates for gene therapy. They used this therapy successfully on a young girl named Ashanti De Silva in September 1990, at the National Institute of Health. Buoyed by this success and hoisting Ashanti as their showpiece some other scientists began experimenting gene therapy on human beings with a dozen other diseases. If you can cure SCID XI with gene therapy, it does not mean you can cure any other disease with gene therapy! Apparently, they seem to have overlooked this fact when they began applying this therapy on human beings in other cases. When one after another began failing, euphoria subsided and disillusionment followed.

The irony of this situation is that gene therapy is in the dock even though it is one of the most closely scrutinized fields of medical research. Perhaps the scrutiny was imperfect due to conflicts of interests; perhaps the experimenters were a little hasty. Be it whatever, it does not appear proper to victimize a whole field of science because a couple of deaths occurred. Initial failures, accidents and some fatalities are inevitable to any developing field of endeavor. Shouldn't we be more concerned about other medical fields in which hundreds of patients die every year due to the incompetence of either the physicians or nurses or negligence of hospitals? At a time when the concerned scientists should be rejoicing for their achievements, their achievements are obscured by the recent events and they stand in the dock of public opinion receiving withering censures.

After all this despair, the path ahead of the concerned scientists is now clear. To bolster the sagging reputation of gene therapy, they should stop grandstanding saying or "Oh, I am sure it is going to cure a number of diseases." or, "Ah, I have found it, this is going to work", at the slightest indication of some new result. They should wait patiently until the theoretical basis of a proposed procedure is well established, tested on non-human subjects, and the results are proved safely on human beings before they make any claim of success. It goes without saying that circumspection and great care should be exercised before extending their success with non-human subjects to human beings.

Safe methods have to be explored and protocols established for every new therapy aimed at a different disease.

Though the recent events may make the working lives of gene therapists a little more difficult, public criticism should be taken as a guidepost from which they should turn toward safer methods. On the part of scientists they can take, the data derived from the trials conducted so far try to design safer delivery vectors. On the part of regulators, they should ensure that overseeing staff is not loaded with people having conflict of interest. Time and a few successes will provide the perspective for the public to make fresh judgment. Once this happens acclaim will not be far off.

3.Appraisal of Damage Repair Methods

This is a promising area in longevity research as plenty of evidence has accumulated about the detrimental effects of damage to DNA. Available evidence suggests that longevity is prolonged roughly in direct proportion to the ability of the organism to repair the damage done to DNA. Species have varying damage repairing capability. Except using some anti-oxidants to fight free radicals, we do not yet have any other method of remedying the damage done to DNA.

4.Appraisal of Replacements

Extension of life or some functions of life by replacement of certain worn out or diseased body parts has been in vogue for some decades. Perhaps, it is the only successful method that has attained some practical usefulness.

Apart from the medical problems involved in implanting, these operations have other problems like availability of healthy replacement organs at proper time and place, and moral considerations. If a particular organ transplant is medically feasible and healthy human organs are not available, then one is tempted to go kill an animal and bring its organ for replacement, with the hope that it helps. If human bodies accept non-human organs and people live by such transplants, woe to the animals, they would be slaughtered in tens of thousands. If science succeeds in making such organs acceptable by human bodies, a market will develop in trading human parts with all its inherent evils. Such an event is not far off with the enormous possibilities spawned by genetic engineering.

In his science fiction thriller Chromosome 6, Robin Cook[69] suggests how creating a new breed of transgenic animals can solve the rejection problem. Certain human genes are transferred into the fertilized eggs of animals and the organs from the resulting new generation of animals are used in transplant operations. He describes how the animals can even be custom made (so they can serve as 'immunological doubles') to suit each individual, by transferring the genes from chromosome 6 of the individual, which contains the histo-compatibility complex. The monetary rewards involved in such operations inevitably attract business minded people to enter the field. You have only to read the novel to see how the greed of medical professionals attracts organized crime into the picture and to what extent profit motive drives corporations.

The other alternative is manufacturing synthetic body parts using the bionic technology discussed earlier but most of it is still in experimental stage. However, we can look upon it with some hope.

USING STEM CELLS FOR GENERAL BODY TUNE-UP AND FOR GROWING BODY-PARTS

Rejuvenation of the whole body system by using elixirs, potions, extracts and the like has been in vogue for a long time. The latest method under scrutiny is using stem cells to do that work.

From the potential of these cells, two questions arise. Can we take control of these cells before they are committed to develop into particular organs say, heart, kidneys or blood and direct them to develop into the organs of our choice? The second question is: Can we grow these cells in culture and send them to places in the body either to assist the deteriorating organs or to regenerate the decayed parts? If the first question can be answered in positive, we need not worry any more about procuring scarce organs for transplants, because they can be grown in the laboratories. Although some research has been going on in this area fruitful human applications seem to a distant vision if not a utopia.

In answer to the second question, some biotechnology companies have already begun thinking about clinical methods that would use stem cells to reverse the effects of aging on skin, flesh, blood, and other organs. It is reported that a company named Aastrom Bio-sciences has been developing a machine to culture stem cells. Concerned scientists are predicting that stem-cell treatments may be available in about five to ten years.

5. Appraisal of Rate of Living Theory

Rate of living theory is based on the major premise that there is a fixed reserve that we may call 'Life Maintenance Reserve' for each species. Minor premise is that the maximum life span of each species is a function of this reserve. From this follows the conclusion. The sooner an organism exhausts this reserve, the sooner it reaches the end of life. Life in the slow lane as a method of prolonging life span has great appeal due to various reasons; but this theory has its difficulties too.

The first difficulty is that even within a species, all members may not have the same initial fund of 'Life Maintenance Reserve'; it may be genetically determined. Secondly, this exhaustion of reserve may not be completely under individual control. Exhaustion of this reserve is partly a random phenomenon determined by some endogenous factors like genetic stability, metabolic capacity and stress response ability of the organism and exogenous factors like environmental impact.

However, the redeeming feature of this theory is the realization that at least two endogenous factors, namely metabolic activity and stress response are within the ability of individual to control. Now there is enough evidence gathered from experiments with small animals and primates to prove that metabolic activity can be whittled down by caloric restriction. Until recently, this method has been enjoying only limited respectability as it is based on empirical finding without a theoretical basis. With the recent researches opening up some clues to the biochemical pathways through which caloric restriction works to bring down the metabolic activity, it has been acquiring the respectability of a theory.

Caloric Restriction

A reliable theory could be built up if research findings of the

M.I.T scientists are corroborated by others. If the idea that inefficient silencing of certain genes causes aging is accepted, this theory fits in nicely with the evolutionary theory. When food is scarce, an organism chooses to postpone reproduction. As postponement of reproduction is closely linked to extension of longevity, it is reasonable to assume that natural selection favors genes like SIR2 which are linked to lower caloric intake and higher longevity via reproductive delay.

Pranayama & Caloric Restriction

Caloric Restriction is the only method, which has consistently shown to prolong lifespan not only in round worms, and fruit flies but also in monkeys. We have referred earlier to MIT experiments that demonstrated caloric restriction works by diminishing metabolism, which in turn provides adequate NAD chemical to SIR2 protein that curtails the deleterious activity of rDNA genes. Any process that mimics this biochemical process should then be able to do the same trick.

Prāṇayāma or regulation of *prāṇa* has long been known to confer the benefit of extending lifespan of humans but the scientific explanation behind it has been limited to the process of better oxygenation of alveoli in lungs. We think that this research has established reasonable links between the *prāṇāyāma* technique and modern science.

That being the state of science, with great potential to herald a new era for better or worse, it is time to examine the consequences of success in our attempts to prolong longevity. Although the hectic efforts of molecular biologists around the world are motivated by pure academic interest, it will be pushed aside in the wake of success. Once success comes, they will be relegated to the background and political and commercial interests will dominate. Scientists will lose control of their brainchildren. If the past developments in the fields of physics and chemistry are any guide to us, politicians wait around the corner to take control of the achievements of scientists.

The consequences of prolonged longevity are not the concern of individual alone. They permeate the whole society. The consequences will be neither wholly good nor wholly bad but overall we can predict that the net result may be prejudicial to the societies and life on earth. Forestalling death will inevitably deepen many of the social crises looming large. As these aspects bear heavily on the welfare of our future generations, we have to wake up and scrutinize them, and then take appropriate actions to regulate the biological research setting aside the virtues of cloistered academic freedom. Now is the time to scrutinize them.

CHAPTER EIGHT

Consequences of Longer Life

Promised Land or Purgatory

> That is the greatest torture souls fed
> In hell: In hell they must live and cannot die.

---Daniel Webster (Duchess of Malfi, Act IV)

Success in our efforts to reset the age-clock and achieving the ability to make it a routine operation is not an unmixed blessing. It has its problems too, perhaps of the magnitude never imagined before. The first problem is personal. As poet Byron expostulated, 'what is the use of garlands to the wrinkled brows'! The second is social. As the hurdles to telomere therapy are overcome one by one, do you expect to approach the frontier bustling with healthy bi-centenarians living in peace and plenty looking forward to another century? It is a dream! Hordes of unexpected problems will come to greet us. As menopause of women is not affected, so are the fertility rates while the death rates will decline following the longevity therapies. The combined effect of stationary or increasing birth rates and declining death rates will be explosion of population with bias towards older cohorts, with all the attendant problems.

Demographic Projections for USA

What is the present estimate of longevity? The longevity forecast by the Social Security Administration is conservative. Their actuaries assume that gains due to advances in sanitation and conquest of many infectious diseases, have already been made and that future gains, if any, due to the efforts at curing diseases like heart attacks, cancer, strokes etc. will either be marginal or slow to accrue. Without taking into account speculations about future gains they, project average life expectancy in the United States will be about 83 years by 2050. Some demographers put it at 85.

Fig.8-1:Growth of centenarians in U S A

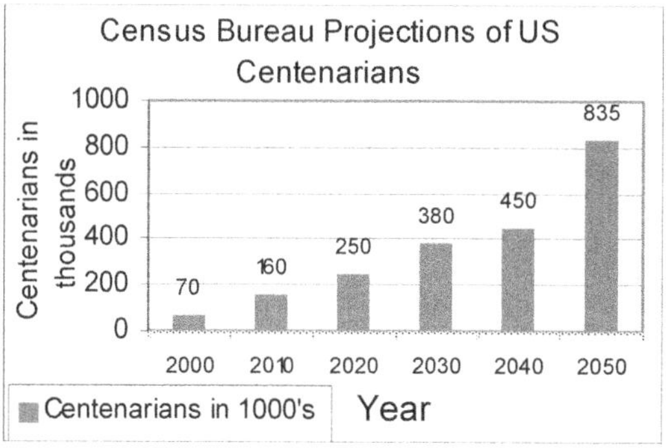

The US Census bureau predicts that the ranks of elderly people will swell and their number will be in the range of 60 to 78 millions by the year 2030. Social Security Administration's projections are only slightly lower. The 76 millions of baby boomers born between 1946 and 1964 are not only the largest American generation but are also set to become the longest-lived in American history. This upheaval did not start with them alone. In 1900, the average life expectancy of an American was a mere 47 years. It climbed to about 71 in 1970 and went up further to 76. Even if the upward movement stops now most of the 70 year olds living now can expect to live into their 80's because the higher your age now, the better are your odds of living longer. Growth of centenarians in USA is displayed by the bars given in Fig. (8-1). If there is a real breakthrough in telomere therapy or some related genetic therapy or changes in lifestyle, the scenario will magnify many times with centenarians bustling around everywhere.

Just to get a sampling of what will be in store for such a society, let us consider the present demographic situation in which telomeres do not figure at all. To get a perspective of how such societies will be, what kind of problems they encounter, how they are being tackled it is instructive to study the society where the life expectancy has already reached the highest and remained there for some years.

OLD AGE SERVICES--THE JAPANESE EXAMPLE

Since 1980's, average life expectancy of the Japanese has surpassed Sweden and remained highest in the world, being 76.6

for males and 83 for females. The proportion of people 65 and over is 14%, which is double of what it was 25 years ago.

As in many eastern countries, family support system was prevalent in Japan too, before the advent of state supported social security and health care system; but it has been breaking under the pressures of modern civilization and increasing population.

As more and more Japanese enter the ranks of the old, less and less proportion of them has been living under the care of their children. In 1957, their children cared for 80% of the old Japanese. It was down to 55% in 1994. It is not hard to imagine the reasons. From one side higher life expectancy that entails higher incidence of disability, swelled the ranks of the disabled-old, and from the other side family support structures have been breaking under the strain of the modern life styles involving high mobility. The situation is aggravated by the fact that two successive generations have become old and are surviving together. How can 60 to 70 year old children who have themselves become weak be expected to support their 80 to 90 year old parents?

U S SOCIAL SECURITY SYSTEM:

Services like old age pension system and health care services have not been able to cope up with the increased demand caused by burgeoning ranks of the seniors. They have not been able to fill even the gap left by the decline in family support. According to SSA, there were 20 net consumers in 1900 for every net economic producer.

The need for services will continue to grow. It is estimated that by 2025 the number of elderly who need care will grow from 2 million in 1993 to 5 million.

Some of these thorny problems are gaping at us even without improvements in life extension techniques.

Aggravation of the Problems by Longevity Recipes.

Consequences of any successful immortality recipe, which can be administered on a mass scale, whether it is telomere therapy or something else, are the same. In an ideal situation the proportion of young adulthood and middle age will expand and the proportion of old age can be expected to diminish as schematically shown in Fig. (8-2).

The economic consequences of such a demographic change and redistribution of age cohorts will be very severe. According to Social Security Administration, there were twenty net consumers in 1900 for every hundred net economic producers. Even now, the ratio of net consumers to economic producers is precariously balanced. Increase in lifespan is likely to disturb that delicate balance and lead to the collapse of the Social Security System.

Fig.8-2:Hypothetical age distribution for a life span of 130 years

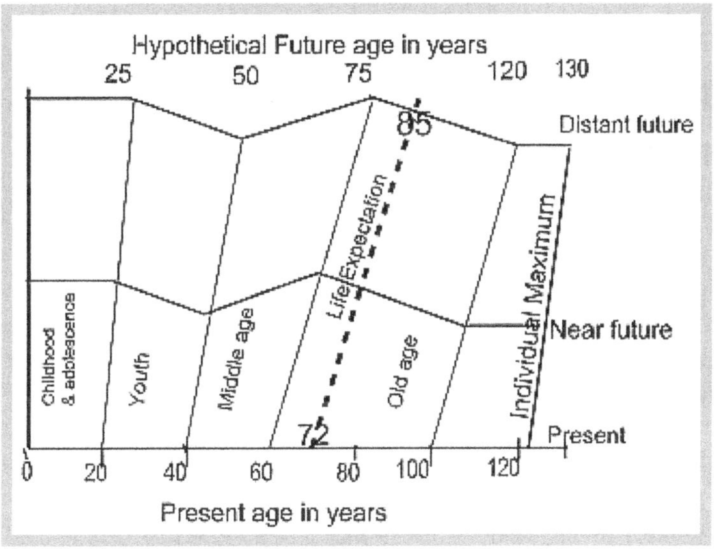

Many social institutions like social security and old age pensions were based on the premise of limited life expectation. Actuarial calculations, on which life insurance policies and annuities are based, will go awry. If that happens, social security system will have to be redesigned. When Social Security was setup the retirement age was fixed at sixty-five. It has to be moved up as people begin living longer and claim their entitlements for a longer period. It has to be restructured with lengthening periods of work, shortening periods of entitlements or a suitable combination of both. After the limited entitlements are exhausted the elderly will have to be left on their own as filial emotions cannot be counted to come to the rescue of the old if, as contended by economists, altruism has no place in economic calculations.

Because of this intractable problem economists whose science is built upon the hedonistic principle that self-interest sorts out everything, are now rethinking in terms of altruism that has all along been an anathema with them. Once they dismissed altruism as not worthy of consideration in economic analysis but they are busy now developing theories suggesting that the families should fill the gap generated in transfer payments (to the elderly) by governments' inability to meet the expenses, as they love their seniors. Economists have at last found a use for altruism!

That is only about the transfer payments to the elderly. There are other problems like medical care of the elderly, tenure of employment, contracts, life insurance and annuities and legal problems. If longevity increases without corresponding betterment of health in the older cohort, it will not be a welcome change, as the cost of providing medical care to the burgeoning old people will skyrocket. Employers have to begin rethinking about retirement age. For every given percent increase in longevity, they have to increase the retirement age by a certain percent. If retirement age moved upwards, there will be lesser opportunities for the younger people to enter into labor force unless additional opportunities are created and there will be conflict of generations. Annuities have to be redefined to cope up with payments over an extended period. Inheritances will be smaller as people who amass wealth in their heyday live longer into their ripe old age and spend it. We are already facing some of these thorny problems even without any further improvements in life extension techniques.

If on top of this situation, everybody can buy the means to extend life span, over the counter like a commercial product, the consequences that will follow will be frightening. If such means to immortality are used with success on a mass scale, great upheavals in societies can be expected because of adverse economic and environmental consequences.

Apart from the problem, of old age pensions and medical care of the elderly, there are other problems like tenure of employment, contracts, life insurance and annuities, legal and social problems.

EMPLOYMENT PROBLEMS

The terms of service contracts have to change extending the period of active service that entitles one for pension. Employers have to begin rethinking about retirement age. For every given percent increase in longevity, they have to increase the retirement age by a certain percent.

PROBLEMS OF GENERATION GAP

If retirement age is moved upwards, there will be lesser opportunities for the younger people to enter into labor force unless additional opportunities are created. Otherwise, there will be conflict of generations. Inheritances will be smaller as people who amass wealth in their heyday live longer into their ripe old age and spend it. Heirs will perhaps resent it.

LEGAL AND FINANCE PROBLEMS

How do courts define life sentence then? Jurists have to revise the penal code redefining life sentence to avoid overcrowding of jails. They may also have to redefine the word juvenile in order to classify offenses as juvenile and adult.

Annuities have to be redefined to cope up with payments over an extended period. Life annuities will perhaps disappear, as insurance companies can't afford to pay over indefinite periods. Life insurance premiums are fixed based on life expectancies of different people. If people live longer than now, perhaps the premiums can be smaller than now.

TERMINAL ILLNESS PROBLEMS

In the vast majority of cases, death does not come as a discrete jump from a state of perfect health to cessation of life; rather it comes as the culmination of a continuous decline and long suffering due to one or more diseases. The intensity and length of suffering make many a patient think of putting an end to their pain and suffering by terminating their lives. If one's termination of life is voluntary and is done by one, it is known as suicide. In modern times a variation of suicide has come into vogue to describe the termination of the life of a person desiring to die painlessly but does not either know how to accomplish it or does not have the courage to do it alone by oneself. That is known as assisted suicide.

EUTHANASIA:

If someone like a medical practitioner puts the terminally ill patient to painless permanent sleep, it is known as euthanasia or mercy killing. Euthanasia is a Greek word meaning 'good death' used generally in the context of mercy killing of terminally ill patients who have been undergoing a lot of suffering.

As life spans expand, so will the number of terminally ill patients and their problems, giving a fillip to the controversy over euthanasia. Though mercy killing was not altogether unknown earlier, euthanasia has acquired more importance now a days because of the advancements in medical technology enabling doctors to prolong the life of patients by connecting them to what are known as life support systems.

In many cases life support systems serve to keep the patients alive but in a vegetative state. One of the arguments in favor of euthanasia is that it is useless to keep a patient in vegetative state for long without any hope of recovery. True, but why then should you put them on mechanical life support systems at all? It is better one does not go on life support systems in the first place. If due to some reason it becomes necessary to go on life support system, it is better one gets out of it when one is still conscious. If the life support machines are disconnected later due to lack of hope of recovery, it is unnatural intervention because the patient has no part in that decision. Sometimes, in some emergency circumstances, the patient may not even know that he/she is on life support machines and the physicians, by law, are not allowed to disconnect the life support system. How long then should the comatose patient continue in the vegetative state as nobody including the physician has the right to pull the plug? Only courts can decide it. That is expensive and time consuming.

To overcome this legal tangle a new instrument called "Living Will' has come into vogue. Every individual can draw a living will in which he specifies how he should be treated in his terminal illness - whether he wants to be put on a life support system in case of need, if so how long, whether he wants to die in hospital or at home and so on. Those who think that it is neither necessary nor desirable to prolong the life of an incapacitated body may choose never to go through the trauma of life support systems. Those who think that it is better to accept the inevitable death with dignity and poise than to go through the trauma in hospital where pipes go into his cavities, needles pierce his body, movements are restricted and where eventually he may die on a lonely hospital bed, should make the necessary provisions in their living will. Those who wish to avoid dying in hospital surrounded by machines without the soothing presence of their loving families that softens the final departure should rely upon living will as the legal instrument to fulfill that desire. Without that living will, doctors in America fasten their terminally ill patients to life support machines and try to maintain them even in a vegetative state.

Even if it is mercy killing, euthanasia it is not without legal and ethical complications. At present in USA, a wafer-thin barrier divides the legal and illegal domains of euthanasia. For example, if a doctor willfully or negligently increases a patient's morphine drip to the point of lethal dose he is considered to have committed a murder, even though morphine is used to alleviate pain. On the other hand, if a doctor does not perform the cardiopulmonary resuscitation procedure on a heart attack patient, he cannot be blamed. Dr.Kevorkian wanted to make euthanasia legal and doctor assisted. Probably many cases of euthanasia occur quietly in many hospitals but Dr.Kevorkian had to pay the penalty for being too open and bold in his campaign for legalization of euthanasia.

The objections to euthanasia are mostly ethical and transcendental. Euthanasia goes against the principles of a physician, expounded in the Hippocratic oath by which new doctors swear before entering into their profession. One argument against euthanasia is that the physician who pronounces that a disease is incurable and decides upon euthanasia thinks his knowledge is infallible and excludes the possibility of saving the life by some method not known to him. Another strong argument against suicide is that it does not solve the problems to escape, which the individual has chosen the path of suicide. Following suicide, the soul which lost it's body prematurely, finds itself in a limbo, a state of temporary damnation, wanders here and there for a while and enters into another body suitable for continuing the same experiences that were interrupted by the suicide. The soul remains embodied as long as the karma is not exhausted. Untimely separation of the body results only in delaying the soul fulfilling its destiny and attaining freedom. Suicide therefore does not solve one's problems.

To assess the magnitude of the terminal illness problems, let us examine some statistics of deaths in USA during 1998. Of the 2.3 million people who died in 1998, 33% succumbed to diseases or accidents without having the need for any terminal care, 44% were admitted to hospitals and died there in the usual hospital environment without the comfort and solace of the family environment. That is the tragedy of living in a technologically advanced country devoid of family values. The remaining 23% were admitted to hospice programs, which made no attempts to extend life by artificial means but provided comfortable environment and eased the pain of dying. These figures show that almost none among these 2.3 millions died at home in the soothing family environment. The Hospice movement that started in 1960's gained some popularity in these four decades perhaps because people consider the hospice environment better than hospital environment.

Some Fringe Benefits

However not all is gloomy. A few fringe benefits accrue out of extended life spans. A desirable consequence will perhaps be multiple careers. With extended life spans, an engineer can become a doctor or doctor can become an artist or a businessman and still have enough time to retire. Dedicated people who were pursuing some project that could not be completed in a short life will find fulfillment in extended lives provided the infirmities of old age do not overwhelm them.

Can Our Ecological-System Sustain Long-Living Societies?

If telomere technology or some other technology which produces the benefit of extending the maximum life span on a mass scale, while there is no decline in birth rate to compensate for the decline in death rate, there will be an inevitable explosion of population with swelling of upper age cohorts and addition of new cohorts aged over hundred years. With the change in population patterns and the consequent additional demands on the resources, can the earth's ecological-system sustain long-living societies with their avowed desire for limitless economic growth?

Filial emotions may not work because economists have been crowing in our ears that altruism has no place in economic calculations. Even supposing that all these problems will somehow be sorted out by our ingenuity, do we have enough resources to support the change in population patterns and the consequent additional demands? Can the earth's ecological-system sustain long-living societies with limitless economic growth? This question cannot be answered definitely until we examine the resource use pattern of peoples.

ECONOMICS OF FOOD PRODUCTION & ECOSYSTEM RESILIENCE

There is abundant evidence that we do not prudently use the resources of our small planet. In the name of food production alone, incredible amounts of resources are being squandered away to support an indefensible eating habit. The force with which meat-eating habit exerts its destructive influence on the ecology of the earth cannot be over-emphasized. Let us examine a few facts before we search for answers.

The unnatural meat-eating habit of people has far-reaching ramifications. You have to understand how this habit has been contributing to the distortion of economic use of resources. It is said that the grain fed to the livestock in U.S. can easily support five times the entire human population in the country.[70] It is estimated that by resorting to the indirect method of obtaining food by feeding grain to the cattle and taking their bloody flesh, we get back only 10% of calories and 6% of protein of what we would have got by consuming grains directly.[71] What a waste machine have they invented! For every 16 pounds of grain and soybeans we feed to beef cattle, we get back barely one pound of meat; the rest is turned into manure. We lose 88% of protein by feeding grain to pigs and 83% by feeding to poultry.[72]

Translating these figures into land use patterns; we need 3¼ acres to feed one mouth with beef per year. To feed the same mouth with lacto-vegetarian food we need only 1/6 of an acre. In other words, one acre can support twenty times as many vegetarians as those eating modern American style meat centered diet.[73]

It is estimated that the world's cattle alone consume a quantity of food (grains) equal to the caloric needs of 8.7 billion people, nearly double that of the entire human population. The lion's share of this consumption is attributed to the cattle in USA alone.[74] According to the USDA statistics, one acre of land can grow 20,000 pounds of potatoes. If used for meat production, the same land can yield less than 165 pounds of beef.[75]

What are the consequences of this beef mania? Its consequences are spreading everywhere like wild fires.

Deforestation

To maintain this wasteful diet style of Americans, increasing tracts of land are required to grow livestock feed and to create grazing grounds. Where from does this additional land come? Obviously, by leveling down timberlands and forests.

We have been witnessing a lot of deforestation in our generation but as this has been taking place at a time when rapid urbanization also has been taking place in many places, uninformed people carry the wrong notion that urbanization is the cause of deforestation. It is true that rapid rise in population and urbanization exert some pressure for new land but the lion's share of the blame for deforestation should be laid at the door of the meat industry in America. For every acre of land used for new urbanization programs, 7 acres are used for cattle grazing and raising cattle feed.[76] Therefore, it is clear that the root cause of the spate of deforestation occurring for the past few decades has been the organized meat industry. The bureau of land management who are supposed to be the guardians of our lands help the cattle ranchers by leasing them huge chunks of forest lands at cheaper than market rates. The forest service which is supposed to be the guardian our forests help them by allowing them to clear cut the forests for use in livestock production which is the major cause of top soil erosion.

It is astonishing that not only the great timberlands in America are being leveled down but also the great rainforests of some tropics are being systematically destroyed to support the voracious American meat habit. At the present rate of deforestation occurring in America, it appears; that the country will be stripped bare of all its forests in about fifty years.[77]

Tropical rain forests are the habitats for about half the species of life on earth. In order to make room for cattle feed lots to manufacture meat demanded by the American food industry, these forests are being denuded wiping out the habitats of thousands of species of life. Every year, hundreds of species are becoming extinct. It is believed that precious life forms are becoming extinct at the rate of about a thousand a year[78]. Many unsuspecting catastrophes occur due to the eradication of some life forms. For example, efforts that have been going on in the USA to devise methods of controlling agricultural pests without using (or at least by drastically reducing) chemical pesticides will receive a setback. Being personally involved in these efforts the author is aware of the tremendous importance of what is called "Integrated Pest Management" that envisages reducing the usage of chemical pesticides to the maximum possible extent and supplementing that effort with natural control of various kinds of pests. Natural control involves study and identification of predator-prey relationships and promoting some predators for every kind of identified harmful agricultural pest. It is known that ecological diversity is a factor for success of the integrated pest management. Some of the migratory birds of the Central American rain forests visit the USA and take their prey of some harmful pests in every season. Disappearance of these birds is a set back to the IPM (Integrated Pest Management) efforts.

Furthermore, there are vast resources of unexplored medicinal herbs concealed among the flora of these forests, The plant kingdom still holds the promise of supplying the herbs necessary for our pharmacopoeias and helping to produce drugs to cure some of the obstinate diseases bothering mankind now. With the decimation of the forests, the home of flora, that promise and opportunity will be gone forever.

Soil Erosion

Experts in archeology tell us that soil erosion is one of the principal causes of the decline and fall of civilizations. America's farmlands which used to have about twenty one inches of top soil about two centuries ago are now left with barely 6 inches of top soil and its rate of loss is accelerating[79, 80, 81.] We have been using chemical fix-ups to compensate for the loss of fertility due to topsoil erosion, but continued use of chemical fertilizers worsens the situation. It takes about five hundred years to build an inch of topsoil[82]. We are losing it at the rate of one inch every sixteen years.[83]

Depletion of Water Resources

Most of us are used to take for granted the availability of water, but this precious resource has been becoming scarce year after year. Good water has been disappearing fast and if this trend continues disaster will strike. The root cause of this situation is again the voracious meat-eating habit and the methods used to produce that substance. About half the water used in US goes to irrigate the land used for growing livestock feed and fodder for production of meat[84]. Enormous amounts of water are required for washing away the cattle excrements.

It takes about 100 times more water to produce a pound of meat to than to produce a pound of wheat. It takes less water to producc vegetarian's food for a whole year than to produce the food of an omnivore for a month.[85] This unquenchable thirst of the meat industry has not only been distorting the efficient use of resources but also depleting the ground water reservoirs in many places.

This utterly disproportionate use of groundwater by the meat industry is intolerable waste of a precious resource and weakens the economy by transferring resources from their more efficient use to inefficient uses.

Fossil fuel consumption

Fossil fuels are non-renewable sources of energy. An enormous amount of fossil fuel is wasted to produce meat products. Meat products are inefficient in producing energy with a given amount input. To produce one K.cal of energy from plant protein sources, 2.2 K.cal of fossil fuel is input into plant protein production systems. To produce the same amount of energy from meat production systems, about twenty-five K.cal of fossil fuel, on the average, has to be input. That is about 11 ½ times more than the former. Though some margin can be allowed as animal protein has more biological value, the difference is enormous and cannot be justified on grounds of economy. As production of lacto-ovo-vegetarian diet requires significantly lesser amounts of fossil fuel than meat-based diets, it is clear that the former is more sustainable than the average meat-based American diet requires.

FEEDLOTS & POLLUTION

The current American style of meat production in feedlots not only wastes prodigious amounts of water but also produces in its wake horrible pollution in the remaining waters.

A few decades ago, before the land grant institutions invented the feedlots, cattle used to live in their natural surroundings like farmers' ranches. In eastern countries, they live in the backyards of family homes in villages. Their excrements used to be returned to earth scattered everywhere over the croplands and used to serve as organic manure for the crops of next season. The feedlot concept has changed all that. It has put an end to the natural ecological cycle. Tens of thousands of animals are now crammed into feedlots with limited space, for forced-feeding. The quantity of the excrements of animals so concentrated is so immense that it staggers one's imagination. It is estimated that, in twenty-four hours of time, the animals destined to end up on America's dinner tables generate an incredible 20 billion pounds of waste.[86] This excrement is said to be 20 times as much as the excrement of the entire human population in the country.[87, 88, 89.] As all this waste is generated in a few concentrated places, it cannot be returned to the croplands where the cattle originally belonged. It is all washed away into the nearest waterways.

It is reported that the run-offs from these feedlots are: *"ten to several hundred times more concentrated than raw domestic sewage. When the highly concentrated wastes in a run-off flow into a stream or river, the results can be, and frequently are catastrophic."*[90] It is mind boggling to think of the pollution generated by beef production in feedlots. Animal wastes account for more than ten times as much water pollution as the total amount attributable to the entire human population in the country.[91] Meat industry alone accounts for more than three times as much organic wastewater pollution as the rest of the industries combined."[92]

This is a cursory review of how a nation's avoidable diet-style has been wreaking havoc with the ecological-system of the earth. By now, you, as a thoughtful reader must have got a fair understanding of how the meat habit and the pernicious methods of its production have been causing horrible pollution of our waterways, and depleting the aquifers in several regions of the country. You must have realized how meat-eating habit led inevitably to the leveling down of the great timberlands in our country exposing the denuded lands to topsoil erosion. Several tropical rain forests, the habitats of thousands of rare species of flora and fauna are destroyed to support the eating habit of Americans and other beef lovers. Slavery to the meat habit has not only been contributing to the distortion of economic use of resources but also been causing all-round destruction of our ecosystem.

Sustainability of Ecosystem

Now we get back to the question of sustainability of long-living societies and limitless economic growth. On the face of it, it appears absurd to expect everybody to live for several centuries eating flesh foods obtained by making imprudent use of resources, destroying non-renewable resources and raising prices of other commodities by distortion of resource use. However, some people bring in technology argument to counter every argument decrying misuse of resources, as if technology is a license to misuse available natural resources. Therefore, we have to examine whether technology can save the human beings while they degrade the ecological-system relentlessly and whether there are no limits to the carrying capacity of the earth's ecological system.

Traditionally, policy economists have never considered environmental factors in policy making, presumably under the belief that competitive economy will be able to take care of those distortions. Unfortunately, such competitiveness is not prevailing exactly in such areas where environmental factors impinge. Instead of making the meat industry realize the enormous social costs they are imposing on the society and taxing them to compensate for the social costs imposed by them, they are given forestlands at below market prices and allowed to denude those lands. Instead of forcing them to clean up the waterways, which they have been polluting with enormous amounts of fluid waste concentrated with highly poisonous pollutants, they are given water subsidies. Environmental factors are thus neither automatically taken care of by the economic system nor taken care of by the regulators.

In addition to the foregoing situation, there is a general belief among economists that environmental degradation increases for some time as economy grows but then comes down, as people become richer, the pattern resembling an inverted U-shape. This again is a fallacious belief because as people have become more affluent, they have been substituting traditional staple food with more meat and fast foods made of meat and sugar, thus giving a fillip to the meat industry that pollutes increasingly. There is no guarantee that more money generates better habits. At best, that belief may apply to some selected items of hygiene like sanitation and water supply and some selected pollutants only, like carbon monoxide and sulfur dioxide. On the other hand carbon dioxide is shown to be an increasing function of per capita income[93, 94]. The inverted U-shape is a far cry from reality.

There is a direct link between agricultural production, dietary preference and environmental degradation. In an interesting study, Lucas Reijnders and Sam Soret estimated the relative environmental burden of agricultural production and meat production in the industrial countries[95], by comparing the input usage in producing equal amounts of vegetable protein and meat protein.

Relative differences in input usage for production of meat-based protein are significantly higher than for same amount of processed protein based on Soybeans. In addition, environmentally injurious emissions are also significantly higher for production of meat protein. For producing one unit of meat protein as against producing one unit of soybean protein, the following multiples of quantities of inputs have to be used.

Land use	6-17 times
Water use	4 to 26 times
Fossil fuel use	6 to 20 times
Phosphate rock	7 times

The resulting polluting emissions released by meat production are several times more than the emissions released by the production of soybean protein. The multiples are:

Biocides	6 times
Acidifying chemicals	7 times
Copper	100 times

Similarly they weighed a number of environmental factors that cause damage to ecosystem and human health and assigned what they call eco-points, based on the extent of damage to different agricultural products and practices. In this assignment, meat and integrated agriculture got the highest score and organically produced vegetables the least score.

Unlimited Economic Growth cannot be sustained

The present scale of economic activity can't be sustained for long if we fail to maintain the ecosystem resilience. If we aim at enlarged and long-living societies, that will be a utopia as long as our current habit of giving precedence to our palates and life styles over sustainability of the ecosystem in which human race is only a part, continues. Our biosphere has limits and any attempts to encroach upon them will defeat the very purpose of such attempts.

Ideally, people should have the freedom to choose the food they want to eat but under current pressure of burgeoning population, scarcity of resources and over-consumption, dietary freedom may soon become a luxury.

Possible Solutions

We explained earlier the fallacy of composition. What is good for an individual or a few of them is not necessarily good for a huge mass of individuals. Prolongation of life should therefore be a selective option, which should depend on individual effort and should not be an assembly line technique.

Any research efforts aimed at discovering practical methods of leading to life extension and lasting health because of individual effort are welcome and should be encouraged. Then, only those who deserve them as fruits of their sincere effort will get them. To put it in economist's parlance extra longevity must have a high price like that of a rare painting because the product is so very precious and has severe consequences to the society. If it is mass-produced and priced in dollars and cents, a huge market in trading longevity will emerge with all the attendant complex consequences discussed above. Moreover, the question arises what is the good of prolonging the lifespan of entire communities. Therefore the price of such a valuable prize should be high in terms of discipline, self sacrifice and dedication to preserving life–life in general, not one's own life. Only those who adhere to such discipline of body and mind deserve that prize and not all and sundry. Fortunately, modern scientific research has unwittingly turned in that direction after many detours.

Long living societies living on beef is an antithetical concept. Is it not ridiculous to aspire to live on forever by putting an end to some other lives, just to satisfy the palate? Even at the existing rate of increase of population, ecosystem cannot sustain the human society unless we change our food habits. It is therefore imperative that, as far as research is concerned with collective benefits, it should be focused on making people healthier and attaining the full length of the allotted life span. Efforts of individuals to live any length of time they can, are of course always welcome.

One such method which is non-invasive, safe and which requires individual aspirant to strive for it with discipline is now attracting the attention of the scientific community. That is caloric under-nutrition, otherwise called Caloric Restriction.

Barring certain biological recipes, which are still in experimental stage, nothing else seems to have garnered as much support as caloric restriction to be one of the few ingredients of the elixir of life. This will be elaborated in a subsequent chapter.

Be Friendly With Nature.

Beef acts like a double-edged sword, cutting the individual life span from one side and cutting the humankind from the other by destroying the ecosystem of this small planet. It is up to us to decide whether we want to promote health, wealth and longevity by being friendly with Nature and promoting its diversity or we want to invite disaster by our relentless attacks on Nature. Now humanity is like the proverbial impatient chef who cut open the goose, which used to lay a golden egg a day, with the hope of grabbing all the golden eggs at once, but, found none in its womb. The sooner we realize this and are friendly with Nature the better for all of us.

For what good is this entire struggle? For whom? In what way is society benefited by this exercise on a mass scale? Will the longer life bring any sanity to people? Will the societies be any more peaceful? Perhaps they will not be. Recriminations and exploitations will continue, and wars will continue to be fought and the world will be not a whit wiser. It may be worse because some modern Hiranyakasipus may arise and hold the world at ransom. The world had been experiencing such situations off and on. It is preposterous to expect all of us to become immortal with drugs bought over the counter. Nothing is more dangerous to the world than the availability of such made-easy methods that do not involve any individual effort, self-discipline and sacrifice. Therefore, it is in the interests of humanity, such therapies based on genetic engineering be limited to ameliorate the sufferings of people without ever becoming the key to immortality.

For Whom The Bell Tolls.

Prolongation of lifespan should be a selective option, which should depend on individual effort and not be an assembly line technique. We must repeat that life is a trust, which should be earned by every individual who deserves it with disciplines of body and mind. Then, only those who deserve them as fruits of their sincere effort will get them. It is for them the bell tolls. Fortunately, modern scientific research has turned in that direction after many detours.

It is a welcome sign that such a method is now attracting the attention of the scientific community. That is known as Caloric Restriction. Are all the calories we have been pumping into our body necessary to maintain it? The answer is emphatic 'No'. Part of that pumping is due to our cravings and part of it is due to our misplaced protein mania. Before we see what this protein mania is and how it can be tackled, we need to know how nutrition standards are fixed by the governmental agencies. As soon as we raise the point of reduction of food intake, vested interests raise a big clamor about the dangers of malnutrition as perceived by them. Therefore, it is worth our while to make a brief examination of nutrition theory and related facts.

CHAPTER NINE

The Role of Diet in Longevity Planning

To lengthen thy life, lessen thy meals

---Benjamin Franklin (Poor Richard's almanac)

Large groups of people in affluent societies are under the spell of an obsession concerning meat as food. They use the idea of meat as a supplier of protein as a handy argument to cover up their cravings of palate. The argument goes like this. The more protein we ingest the better we are; therefore, we must eat meat to get enough protein. Let us verify whether the claims that more protein is better and that meat is the exclusive supplier of protein are justified by nutrition theory. Although the word "nutrition" appears to be commonplace, most people are ignorant of the facts of nutrition theory. Therefore, elements of nutrition theory are presented here.

Elements of Nutrition Theory

THE NEED FOR BETTER NUTRITION EDUCATION

Even medical professionals do not seem to have good knowledge of it, as it is not taught in medical schools. As of 1987, only 30 out of the 125 medical schools in USA required their students to take a course in nutrition. The nutrition training received by an average U S physician during four years of his medical training is a mere 2½ days. Thus, physicians in whom American public keep a lot of trust are unable to help the latter in this important aspect of health. Of late, some efforts have started to remedy this lacuna in American public life, with the birth of Physicians Committee for Responsible Medicine (PCRM.), a Washington D.C. based non-profit organization. It consists of about three thousand physicians who are members of American Medical Association, and a few thousands of lay members. They hope to change the American consciousness on human nutrition, in particular among the medical community. It would therefore be appropriate and instructive to make brief review of nutrition theory here.

The three most important actors, among others in the Nutrition Act in the drama of life, are carbohydrates, fat and the proteins. It is the proper balance and harmony among these three actors that makes the drama of life possible.

Carbohydrates:

There are many things in the world which can make you happy- riches, good food, nice buildings, sex, nice companionship, music, beautiful scenes - any of them can make you feel good. However, if you are not in a state of good nutrition nothing but good food can make you happy. Your feeling of happiness even with the best of external objects is limited by the state of your health and your internal feeling of wellness. That state of well being i.e. the state of feeling good and buoyant comes only when you are generally healthy and when the sugar in your blood is in right concentration. If that condition is not met, none of the above material conditions can bring you happiness.

The functioning of almost all the cells in your body depends on blood sugar as these cells draw their energy from this sugar. When the blood sugar level falls below the optimum level, the healthy body sends signals in the form of hunger and lassitude. (Exception to this rule is the case of obese people). Then your natural response would be to eat some food. If this food contains adequate amount of nutrient known as Carbohydrate, soon your blood glucose level be restored and with it the state of your well being.

Did you ever wonder what that feeling of wellness is? Whither does it come? It is the consciousness of energy released into the body by the nutrients of the food we have eaten. When a log of wood is burnt, the resulting energy is radiated into the surrounding atmosphere in the form of heat and light. The very same energy gives locomotion to the living things. It creates thoughts and feelings. It is the stuff without any form, yet it resides in many forms. It is the stuff which you cannot see but that enables you to see. It is apparent sometimes as light and sometimes as heat, sometimes as both and sometimes as that invisible form electricity. Living things can capture energy from the nutrients of the food they eat and harness it for the body's work now or in the future. This wonderful phenomenon is what distinguishes the animate from the inanimate.

The food energy with which we are concerned now comes from the three nutrients known as carbohydrates, fat and proteins. These are the principal actors in the drama, 'The Battle of Life'. Most of the foods we eat contain mixtures of these nutrients in varying proportions though nutritionists classify them according to the predominant nutrient in each of them.

Most of the energy we get is derived from carbohydrates and the rest from fat and proteins. But, how much energy is to be demanded of each of the nutrient? How do we go about getting them from various foods in order to fulfill body's needs? That is the central idea of nutrition theory.

FAT-The Reservist:

While carbohydrates and fat, both provide the ammunition to the individual fighting the battle of life there is an important difference in their method of supply. The energy emanating from carbohydrates can be stored in limited quantities for brief intervals of time and is available for body's use almost immediately. Fat cells are like reserve soldiers commissioned into activity when the first line of defense is either weakening or absent. All the surplus energy not used by body is stored as fat in the body. In times of starvation or substandard level of nutrition, fat comes to the rescue of body and supplies the energy needed by it.

A wonderful example of the rescue mechanism of the body-fat is found in the polar bear. As the bear limbers about on an iceberg, great masses of fat ripple beneath his thick fur-coat. When he hibernates, his body oxidizes the fat extracting thousands of calories of heat to maintain his body temperature during sleep. He wakes up thinner by hundreds of pounds when spring comes. Fat not only supplies emergency time energy but also protects the body from mechanical shocks and extreme temperatures and protects the skin health.

Protein: The Hero

Protein is the hero of the drama of life. We have been using the word "Protein" as if its meaning and import are well known. Indeed, many of you are aware of their role in our lives at least vaguely. Now we'll pursue this topic a little more discussing their composition, their role in human nutrition and the technique of extracting them and using them to our best advantage.

The word "protein" calls to mind life processes in general and as proteins are essential to all living things. Wherever there is life, there are proteins. We are 20% protein. Unlike the other two energy nutrients carbohydrates and fat, the molecules called "proteins" are a class by themselves. While carbohydrates and fat are interchangeable depending on the situation in the body, the role of proteins is unique. They are indispensable and their rolc is versatile. Body parts like muscles, skin, hair, and hemoglobin of red blood cells, insulin and other hormones are protein molecules. Like a general participating in a combat in an emergency, protein molecules step in and supply energy when other nutrients of the body are in short supply. These molecules play a preeminent role in the vital processes of the body. That is why they are called "proteins", a Greek word meaning "pre eminent."

The Creative Principle

Where do we find proteins and in what form do they appear? Proteins are ubiquitous in the world of living matter and they greet us everywhere in our daily life, like bones and cartilage that form the structural framework of animals, their flesh and their blood. Even their hair contains protein. The objects of our constant attention for manicure are protein. The silk in our apparel, the wool in our winter garments and the leather in our shoes is all protein. How can we tell which, among the hundreds of objects, is protein? Are there any physical characteristics like size, shape, or color by which we can tell whether an object is protein? None of these physical features can give us the slightest idea of whether an object is protein or not. Only chemical analysis can tell us whether an object is protein.

Anyway, why do we need these proteins at all? Isn't it possible to get along without these deriving the necessary energy from carbohydrates and fat? Unfortunately the answer is 'No' because carbohydrates and fat do not contain nitrogen necessary for the sustenance and growth of organisms. Proteins alone can supply nitrogen; Proteins perform several important functions in the growth and maintenance of our bodies. Their versatility is astonishing. The basic chemical reactions called metabolism are dependent on proteins. As catalysts of metabolic reactions, proteins are known as enzymes; as transport vehicles of oxygen to several parts of the body, they are known as hemoglobin; as regulators, they are known as hormones; and as protective agents, they are called antibodies. It is obvious then that these lively things are indispensable to life whether one is male or female, young or old. Children need protein for growth and adults need them for repair and maintenance of tissues.

Therefore, we have to eat proteins but none of the objects mentioned above are fit for eating. We don't eat wool or silk or nails (When you see somebody biting their nails don't think they are eating protein, they are simply biting dust). We have to look for edible protein. There are two methods of determining the protein content of foods. One is the nitrogen content method and the other is the amino-acid content method. Proteins are known to contain about 16% of nitrogen on the average. Using this information, the first method roughly estimates the protein content of a food by multiplying the nitrogen available in it by a factor of 6.25 (100/16). Since this is only a crude method, this is generally verified by experimental studies. The other method that is more reliable arrives at the protein content by adding the weights of their chemical constituents known as amino acids. In spite of the limitations of the existing methods, approximate estimates of protein contents of various foods are made available to us by the biochemists.

The original source of all proteins is plants. Plant proteins ingested by animals are transformed into milk, eggs and meat but one should note that consumption of this transformed proteins, as food by humans is a very inefficient way of using proteins.

How Much Protein? The RDA Story[96]

Ordinarily a person can judge the sufficiency of food intake by the sense of gradual fullness he or she achieves eating food; but that feeling of fullness does not tell us anything about the nutrient content of foods eaten and hence not much about their sufficiency or otherwise for maintenance of health. The widely adopted method for estimating the protein requirements is based on the concept of nitrogen balance. Of the three energy nutrients, since protein alone contains nitrogen, it should be possible to determine the intake of protein by following the path of nitrogen in the body and determining how much of it is lost per day through urine and feces by an average person. Adults in normal health are supposed to be in neutral nitrogen equilibrium; that is nitrogen lost by a healthy adult through bodily excretions is exactly replaced by the dietary intake. Normally children and pregnant women are kept in positive nitrogen equilibrium to meet their special needs.

Adopting this concept, National Research Council (NRC) of the National Academy of Sciences (NAS) developed recommendations for dietary allowances of protein. Note that these are allowances, not requirements. The allowances are developed in three steps starting from the concept of nitrogen equilibrium of a healthy standard person. Let us follow these steps and see the limitations of RDA

Estimates of the minimum amount of dietary protein required have usually been based on the quantity of nitrogen needed to replace the nitrogen lost by body in metabolic process (FAO/WHO, 1965,73). The less efficient a body is in utilization of nitrogen from dietary nutrients, the more will be the nitrogen output measured from its excretions and vice versa.

Depending on some experimental studies, NRC/NAS reported (RDA, 9th ed., 1980) that the average quantity of body protein lost in metabolic processes for every Kg of body weight is 0.34gm, with a standard deviation of 0.05. This 0.34 gm. represents the nitrogen utilization efficiency of an average subject. That means nitrogen lost by the most efficient subject can be set at 0.24 (0.34 – 2 standard deviations) and the least efficient subject at 0.44 (0.34 +2 standard deviations) The NRC / NAS started with a little above the upper limit of 0.45 with the plea that they want to cover the 97.5% of the population or almost all.

In the next step, they added 30% to compensate for the loss of efficiency 'around the required level'. That comes to 59gm/kg.

Finally, it was further raised by a factor of 100/75 for meat eaters and 100/55 for vegetarians for the ostensible reason that ordinary foods are not as good as lives about to emerge, i.e. eggs. That comes to 0.79gms for meat eaters and 1.07gms/kg of bodyweight for vegetarian groups. Remember this factor is based on the group average as the committee had to make general recommendation encompassing a whole range of foods.

The unreasonableness of this method is obvious. Recommendations to benefit almost all people can be made in two different ways.

- Recommend a standard corresponding to the highest fraction of population with a proviso for making upward corrections for safety by the rest of the people who fall short of the standard.
- Set the standard at so high a value that corresponds to the nitrogen output of a very small fraction whose nitrogen utilization is least efficient, and ask the rest of the whole majority to follow suit.

Which of the two methods is better? The method that sets the standard for the preponderate majority or the method sets the standard for a small minority. Which method wastes resources and imposes unnecessary burden on individuals with the only result of damaged health?

The method adopted by NRC/NAS follows an elegant statistical theory, but the applicability of the theory to a particular situation needs to be carefully judged. Who is to decide whether 97.5% of population should be covered or 99.9% should be covered?

If you increase the number of persons to be covered just by 2 more percent, you have to set the basic minimum at 3 standard deviations above the mean, according to this theory and that would be [0.34 +(3 x .05)] = .49gms. If you want to be fair to the remaining 1/2% of population, according to the same theory, you have to set the minimum at infinity! Following their reasoning, if it is raised by 30% and then raised again by factors suggested by them, it would total to 0.79gm for omnivores and 1.07gm, for vegetarians, per kilogram of body weight.

Throughout this procedure, qualified statements and equivocation in language can be found. All this manipulation is clearly arbitrary.

Using an extreme value as a guide can be appreciated in circumstances where a slight departure from it may result in a disaster as in the case of fording a stream. Where such a danger is not imminent as in day to day dietary practice, the procedure which takes the least efficient nitrogen conservator as the standard for the rest, results in avoidable waste of national resources and causes drain on your purse as you have to shell down almost 60% more on your grocery bill. Above all, it is spending your money for ruining your health.

In the second step they added 30% more to the previous figure of 0.45 (0.45+0.14=0.59 rounded to 0.60), to compensate for the loss of efficiency suffered at higher levels of protein intake and rounded it upward to 0.6gms. (Note that they increased it first and then claim some more addition because protein usage is inefficient at higher levels!). These kind of inflated figures have resulted in the current obesity epidemic sweeping across the country. In my calculations shown below, I have omitted this 30% increase as redundant as I consider it more injurious to health than helpful, and certainly contra indicated to people aspiring for longer life spans.

In the final step, the value is raised by a factor of 100/75 (0.6x 100/75 =.8) for omnivorous groups, and by 100/55 (0.6 x 100/55 = 1.09) for vegetarian groups to account for the supposed sub-substandard efficiency of mixed diet relative to egg-protein. These operations bring the total to 0.8gms/Kg of body wt for omnivorous groups and 1.1gms for vegetarian groups. These factors are based on the Net Protein Utilization (NPU) concept. I have raised the NPU of vegetarians from 55% to 65%, as vegetarians do not survive on single foods, they eat complex mixtures of foods whose complementarity raises the NPU at least by ten percent.

Therefore, one has to take these results with a pinch of salt. Take the experimental data presented by them and make your own decisions. For example if you are an average individual, you may take 0.213gm of nitrogen output per pound of bodyweight and work out the requirements as follows. --------------------------------

Step 1: Basic amount of usable protein required for

Nitrogen Balance = 0.21g/lb of body wt. (i.e.0.45gm /Kg)

Step 2: At this rate total amount of **usable** protein required by a median ht Male of weight 150 lbs,

= 0.21 x150 =**31.50gm**/day and by a median ht female of wt 125 lbs = 0.21 x125 =**26.3** gm/day

To cover inability of body to make full use of protein, inflate the usable protein by the reciprocal Of NPU percentage in order to get the values of crude protein which when ingested and assimilated, would yield the above values. It is known that the NPU of NV foods is 100/75 on the average and the NPU of plant proteins, is about 100/65.

Step 3: Now on the basis of the above dichotomy we can compute the total protein requirements of average vegetarian and non-vegetarian males and females as follows.

Total (crude) protein requirements

(1) Vegetarian male: 31.5x100/65 =**48.5** gm/day

(2) Vegetarian female: 26.3 x 100/65 =**40..50** gm/day

(3) Non-vegetarian male 31.5x100/75 = **41.3** gm/day

(4) Non-vegetarian Female 26.3 x 100/75 =**35.0** gm/day

Note: 1 Kilogram=2.2046 pounds or 1 Pound=0.4536kilograms.

For the convenience of the readers, the above results are tabulated in Table 9-1.

Table 9-1: Protein Requirements of Vegetarians & Non-Vegetarians (Longevity seekers)

Std man: 150 lbs Std woman: 125 lbs	Vegetarians		Non-Vegetarians	
Males: (1566 cal) Females: (1318 cal)	Ave Male	Ave Female	Ave Male	Ave. Female
Usable protein / lb of body weight.	0.21gm	0.21gm	0.21 gm	0.21 gm
Usable protein per person (rounded)	32 gm /day	26 gm /day	32gm /day	26.gm /day.
Total protein/ lb of body weight.	0.328gm	0.328gm	0.284gm	0.284 gm
Total protein per person (rounded)	49 gm /day	41. gm /day	41 gm /day	35 gm /day
Calories / gm of usable protein	49	49	49	49

This represents nearly 3/4 of the RDA for high quality of homogeneous food. In the RDA calculations discussed above, a standard quality protein was taken as datum and recommendations were developed making allowances for mixed diets, as not all foods contain the standard quality of protein. Raising it further on the ground that it is not equal to egg protein is not warranted as the **concentration of nutrients in egg is meant for production of life and not meant for eating by human beings**. Such concentration is not necessary in our foods meant for routine eating and survival. No more bloating of this figure is necessary unless one's aim is satisfaction of one's gluttonous nature.

How Do Proteins Do Their Work?

PROTEIN QUALITY AND AMINO ACIDS.

Now it is necessary to answer the question: how is the protein content in various foods determined? The one common factor beneath all the physical differences of color, shape, etc of proteins is their chemistry. The protein in your kitchen garden is, the same as that of Tuna in the Pacific Ocean or the protein in your body; but from the point of view of nutritive value, they are not the same as they differ in quality.

Food proteins provide building blocks called amino acids for synthesis of body proteins and other tissues. When we speak of the need for proteins, it can be translated to mean the need for amino acids. The body degrades the food proteins and uses the amino acids released for reconstructing body proteins. In the absence of dietary proteins, proteins of the body tissue degrade to maintain metabolism. During the process, although some of the amino acids are recycled, nitrogenous byproducts of metabolism are excreted in urine and feces. Therefore, dietary amino acids are required continuously to replace these losses, even after the body growth has ceased.

There are 20 amino acids. Most of them can be synthesized chemically. Some of these amino acids can be synthesized by biological systems from other constituents while some of them cannot be synthesized. Those that cannot be synthesized by body must be supplied to it from outside. These are called Essential Amino Acids (EAA). Among the smaller living organisms, there is a wide and unpredictable range of essentiality for amino acids. Among the higher organisms, disparities narrow down and similarities become clearer.

Mammals cannot synthesize nine amino acids, viz., Histidine, Isoleucine, Lysine, Leucine, Methionine, Phenylalanine, Threonine, Tryptophane, and Valine. Therefore, they have to be ingested in their diets. The first in the list is essential for infants, but its need is not demonstrated for adult humans. The remaining eight EAA are indispensable for maintaining our normal nutrition.

UNION BEHAVIOR OF AMINO ACIDS:

One of the most intriguing problems is the cooperation of the EAA. Like trade union members, they hang together. If any one of them is excluded from the job for any reason, the rest strike work or dodge and slow down.

In order for the amino acids to work full steam and carry out protein synthesis all the eight must be present in food simultaneously and in required minimum quantities. If per chance one of them is deficient, then the efficiency of the rest of the members is proportionately reduced. Similarly, if seven of the amino acids are in right proportion and the eighth is above required level the extra becomes redundant. It means that the least available amino acid in food is called the limiting amino acid. If the limiting member satisfies its own minimum requirement, no problem arises because that condition implies that all the rest satisfy their requirements. To make things a little more complicated, our body needs each EAA in a different quantity. If the EAA pattern we eat matches exactly with the EAA pattern required by our bodies then the food is said to have full biological value (B.V), otherwise the biological value of the food is lowered. For example the required minimum of lysine per day for an adult weighing 70kg is 840 mg. Suppose the food you eat on a day contains only 420gm. of lysine, while the rest of the EAA are all present in their full quotas. Then this 50% shortage of lysine can thwart the work of others; the body utilizes only half of each available EAA. Remember that the chain is as strong as the weakest link.

Therefore, biological value is the proportion of protein assimilated by the body from the quantity absorbed by the digestive system. The amount actually digested by the digestive system from the ingested protein is another variable called digestibility. The product of these two is known as Net Protein Usability (NPU)

Usability of crude protein depends on two factors - digestibility and the biological value. Since either can vary either way, their products taken as the Net Protein Usage. Thus we can say that NPU = BV x digestibility.

Using NPU as coefficient, we can compute the Net Protein Value (NPV) of various foods from their crude protein values. For example, we know that NPU of egg protein is about 92%. A whole egg weighing 50gm has a crude protein content of about 6gm. Multiplying 0.92 by 6 we get 5.5gms as the NPV of a 50gm. egg.

NPU is the concept used in raising the values of RDA in the final step of calculations shown in previous pages.

QUALITY VS. QUANTITY

Quality of nutrients is as important as their quantity. A related concept is Protein Efficiency Ratio (PER). This is a measure used to assess the quality of protein. It is measured as the ratio of weight gained by experimental subject in a given time to the weight of protein consumed by it.

Quantity is not the only criterion for protein ingestion. For example an infant weighing about 5kgs needs about 10gms of protein a day for its healthy growth, Surely, this much of protein can be easily obtained from a 0.5 oz hamburger, but do you ever think of feeding your infant with hamburger or ham? The toothless infant with tender digestive system needs a little nutritive and easily digestible food which when gulped can support its life and promote its healthy growth. No other food available in nature satisfies all these demands as the breast milk of mother. It is the same with the egg of the bird. Like the fetus of the mammal, it has to be self-supporting from within. That is why the nutrients in eggs and milk are of the highest variety.

As animals grow, be they humans, cows or cats, their growing bodies need more nutrients. Though milk has high quality nutrients, it has not enough of them to promote growth. Therefore, milk alone cannot supply the growing needs as the infant grows into a toddler, into a child and an adult. As body weight increases, quantity of nutrients needed in food attains significance along with their quality. Then you have to look for proteins with high quantity too.

If you eat foods containing protein of high quality but in low quantities, you need large quantities of them to supply the recommended daily allowances. If you consume large quantities of such foods, your stomach and digestive organs refuse to admit quantities beyond a limit. How then can we manage to resolve this quality-quantity dilemma? We have to choose reasonably good quality foods and try to attain our requirements without overloading the system.

In order to accomplish that goal, we need knowledge of quality-quantity relationship of a good number of foods. Fig (9-1) shows the protein quality-quantity continuum of several available foods.

Fig. 9-1: Quantity-Quality Continuum of some Foods

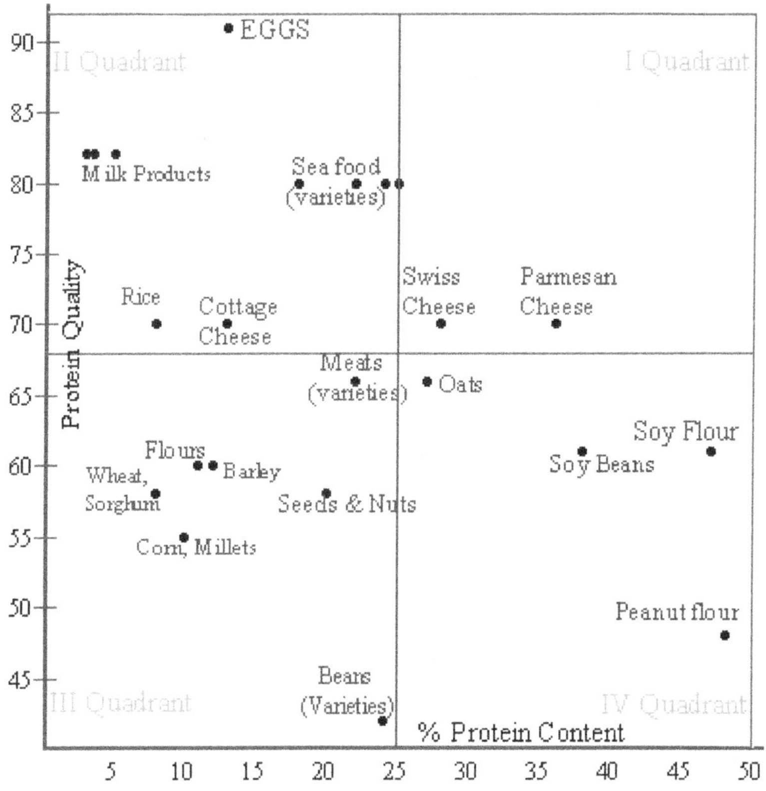

A look at the diagram shows that all most all the plant foods are clustered in the III quadrant (anti clockwise) and almost all animal foods are clustered in the middle with the exception of sea foods and eggs which are way high. This division indicates that the primary sources of all foods, viz. plant products are also the sources of high yielding proteins though the quality varies from low to medium when considered individually. Animal products, as we might expect from their positions in the food chains contain the high quality proteins

QUANTITY SCALE:

A perusal of the diagram reveals that soybeans, especially in their processed form rank the highest among all known foods in terms of quantity. They rank higher than lamb, chicken, steak and even higher than turkey known to contain large concentrations of protein. Certain kinds of cheeses like Parmesan and Swiss cheeses, which rank with pork, follow next. Coming next in the order are various meats and various kinds of pulses (legumes) and beans. Scattered low in the scale are both animal and plant products. Eggs and milk rank below average in quantity scale but they reach the top in quality scale. Grains like wheat, rice and oats also rank low in quantity but their quality is average or above average

QUALITY SCALE

The quality scale of proteins ranges from 40 to 94 in terms of NPU scores. Animal foods occupy the high places in this scale but, surely, they are not the highest. Eggs alone occupy the highest place on the scale (NPU=94%) yielding the next place to milk (NPU=82%). The next place is shared by a variety of foods. Sharing the next rank are brown rice and a variety of cheeses. In the descending order of quality, meats do not come even in the second or third rank.

In case of plant foods, while the upper limit of quality reaches that of meats at 70%, the lower limit falls as low as 30% to 35%. With an NPU ranging from 65% to 70%, meats rank next to brown rice.

Having seen the placement of meats in the middle of the quantity-quality continuum, don't jump to the conclusion that they are the best foods for you. If, after having read what has been explained at length in the previous pages, you jump to the conclusion without reflecting, you read in vain. From the point of view of body, all that is required is to supply it with adequate quantities of EAA in a definite proportion. If this condition is met, the body doesn't ask whether the proteins supplied to it are first class or second class nor does it ask whether they have come from veal or beans. Whether you supply the EAA in the form of Hamburgers or Frankfurters or in the form of rice and beans, the body is indifferent to the source. Why then should you burden it with things unnecessary and suspected to be obstructing its normal functioning? That's what happens when you jump on foods associated with high risk factor. Notice that high fat foods in the IV quadrant are associated with high risk factors.

Substitutability and Complimentarity

Substitutability and complimentarity are terms that occur frequently in economics textbooks. Economics textbooks define two goods as substitutes if one can replace the use of the other without loss of satisfaction to the consumer. Two goods are said to be complimentary if their joint use enhances the satisfaction of the consumer over the level obtained when either one of them is consumed separately. Thus, one can view coffee and tea as substitutes, coffee, and sugar as compliments. Here we define two diets as substitutable if one can replace the other without loss of protein value. Two foods are complimentary if their combined intake enhances the sum of their separate N P U's.

What is the significance of substitutability and complimentarity in the kitchen? There is a lot of it. Go back to Fig (9-1) and look at the diagram. Think of pushing a dot to the right. What does it mean? Since the horizontal scale represents quantity, pushing a dot in the diagram to the right means increasing the quantity. Suppose now you have doubled the quantity of oatmeal eaten. In terms of the diagram, it means shifting the dot representing oatmeal to a position that is double the present horizontal distance. This position represents higher quantity and quality of protein than lamb and chicken and is close to pork and beef. This example makes a point clear. Meat is certainly substitutable and all excuses regarding its inevitability are flimsy. There are different ways of doing it.

Perhaps no other subject is riddled with so many misconceptions, fallacies and prejudices, as 'food and fitness.' Most of the ideas examined in the previous pages were with a view to separate fact from fiction. Table (9- 2) summarizes that separation.

PROTEIN MANIA & CONSEQUENCES

The phrase "under-nutrition" used in previous chapter refers to any level of protein intake less than that recommended by the NRC/NAS. They recommended a whopping quantity 60 gm per Kg of body weight as the minimum. Several experts consider this a bloated figure and disagree with the recommendation. Unfortunately, the official scientists have adopted dubious criteria in arriving at their conclusions. They used the criteria of body weight and the ultimate body size of experimental animals. The regimen used for growing big rats quickly was recommended for human nutrition. This diet is right if we want big kids sexually ready at a very young age.

Table 9- 2: Facts & Fiction of Nutrition Theory

Fiction	Fact
1. Meat gives more strength than other foods	Food chemistry has helped us dispel this myth. Body tissues are built after the food ingested is broken down into amino acids Amino acids of meat and other foods are indistinguishable. The body needs EAA in the right proportion and the source of EAA is irrelevant to the body.
2. Meats provide the highest quality proteins	An understandable misconception in meat centered societies, but facts are otherwise. Protein quality is interpreted as NPU. Eggs and milk have concentrated protein.
3.Plant foods are deficient in some EAA. Hence they cannot match meat	The fallacy in this argument lies in assuming that each plant source is eaten separately. Composite diets cover the deficiencies of individual foods are covered by their mutual complimentarity. The EAA pattern of composite diet can match that of meat protein.

Fiction	Fact
4. Plant foods lead to obesity	This is again a fallacy. The excess intake of calories is the cause of obesity rather than carbohydrates. Calories come from carbohydrates, fat and unutilized protein. Plant foods, by themselves, do not contain any more calories than meat measure for measure. The cause of obesity is gluttony, not plant foods.
5. Meat is healthier overall	This is pure fiction and needs to be cleared up. If you compare the vegetarians and non-vegetarians in homogeneous social groups in equal income bracket, you will find plenty of evidence to show that vegetarians enjoy better overall health than non-vegetarians do.

At this stage of civilization, is it necessary for us to be as big and strong as in Stone Age, when survival of human beings depended upon their success in running fast and hunting for food?

Based on these questionable recommendations of protein requirement, people were indoctrinated by vested interests to believe that more protein is better for health and touted their meat products for sale. Another salient feature of this indoctrination was the false idea that meat alone can supply the needed proteins. What we want now is to maximize intellectual power, youthful life and longer life span. Therefore, you do not need a diet that makes you a mastodon.

You need a diet that keeps you healthy, and alert, a diet that prolongs your youthfulness and longevity. "So what? ', you may say, "we can ingest all the required protein and a little more, what is the harm in having a little more than necessary?" That is the point. In matters of diet, more is certainly not better as we learned in the previous chapters. If we cannot strike the optimum, a little less is better. That is the essence of caloric under-nutrition theory, which is the only theory, discovered so far that promises to decelerate the aging process.

Therefore, in order to achieve your objective of prolonged youthfulness and longer life, it is imperative that you should chop off all excess protein from the bloated RDA figures. No more than 0.34 to 0.40 gm's per Kg of body weight is necessary to keep you healthy and efficient. In order to reap the benefits of under nutrition, it is necessary to follow deliberate diet restriction program whether you are naturally lean or fat. It is not only the natural leanness but also the deliberate effort to restrict diet that counts.

When to Eat?

When should we eat? What a question! We can eat whenever we want. Is that Right? Wrong, because the answer is intimately connected with the program of caloric under-nutrition. By eating at all odd times, while watching T.V, while watching a movie in a theatre or just when you do not have anything to do; militates against all canons of 'Right food' because it pumps in more food than is necessary. Let alone the recent revelations of molecular biologists, even the traditional wisdom of human physiology dictates against eating whenever you please. How do we determine the intervals at which food should be eaten? Normally hunger is the guide, but to see how eating at all odd times upsets the digestive, assimilative and eliminative rhymes of the body, consider these physiological processes.

For any single meal to be digested and assimilated, the normal sequence of events beginning with salivary action in the mouth and the processes performed by the stomach, liver, pancreas and ending finally with the eliminative action of the intestines, takes about 12 to 14 hours, in a normal individual. In a subnormal individual whose digestive and metabolic processes are below par, these digestive events will be delayed and a single meal requires more time to be digested and assimilated. A normal person takes about 28 hours for digesting, assimilating and to eliminate the residue of two meals. As we have only twenty-four hours a day, it is clear that no more than two meals are called for in a day.

The stomach is constructed of several thousand glandular structures whose function is to secrete a digestive fluid suitable for splitting up complex foods, and passing them on to the small intestine for further treatment by digestive enzymes. All this takes time. The stomach phase alone of the digestive process requires four to five hours under normal circumstances.

Functional activities of the body and normal work cause some wear and tear and fatigue produces toxins within the tissues. To keep the body healthy, it should be recouped and the toxins have to be removed by blood circulation. This recuperation, repair and detoxification take time. A period of rest and recuperation should follow the hard work done by the digestive system to digest the protein content of a meal. While the stomach muscles require rest for reconstruction and detoxification, the digestive glands too require enough time for replenishing their store of essential chemicals derived from the passing blood stream. What has been said about stomach is true of other digestive organs also. There should be a long gap between meal and meal. Those of you who are used to take three meals a day should seriously consider reducing to two. Those of you who are used to taking two meals plus a breakfast should carefully time them so as not to leave too much residue in the digestive system. A long rest period between meals allows for functional resting during which not only the digestive system but also other organs of the body are recouped.

When more than two meals are ingested within a twenty-four hour period there is overcrowding of the digestive system. With more than three meals a day, the digestive canal is overloaded right up to the top without any gap. In addition to that, when there is constipation, there is not only an absence of rest period but also the intestinal canal is filled up with undigested and imperfectly digested food residues. That toxic condition is the root cause of many chronic diseases.

How Much To Eat?

That is the biggest conundrum faced by the public, scientists and the policy makers in the public health field, with no end for the debate in site. No issue in public health has been more controversial than this as the food industry interests are at odds with the interests of the public and public policy makers.

In order to answer this question we have to take into consideration a related problem also known as obesity that has been recognized as the mother of many diseases.

FAILURE TO ADAPT TO EVOLUTIONARY CHANGES

Energy Dynamics: Relevance of many chronic diseases like hypertension, coronary heart disease and diabetes in modern society can be traced to the mismatch between the energy balances of modern consumers and the energy balances of our prehistoric ancestors who depended on hunting and gathering for their daily needs of food.

Evolutionary changes are driven by the need of the animals to efficiently allocate their energy budgets between their maintenance and reproductive needs. Much of the hominid evolution took place in open woodlands and grasslands where sustenance is harder to come by than in the dense forests where chimps, gorillas and orangutans evolved and continue to live. In the dense forests, wandering a short distance was enough to find the food required for the day whereas the hominids had to trek many miles a day to find food. The potential benefits of cost effective movements were inconsequential to apes, which moved short distances whereas the hominids stood to gain a lot by cost effective movements.

Non-human primates moved on all the four when they moved on ground. Anthropologists assign varied reasons as to why bipedalism evolved but the most important is perhaps the need for economy in usage of energy by the hominids in their quest for food. Bipedalism is economical form of locomotion as two limbs do the work of four, and they do it more efficiently. The energy thus saved can go into reproduction and the freed forelimbs can help to do many chores like catching food and rearing the children better. Climatic changes spurred further this morphological revolution occurring in the hominids living two million years ago. As the African continent became drier, forests gave way to grasslands leaving food resources distributed more sparsely farther and wider.

Thus, evolution has adapted to the need of our prehistoric-ancestors for foraging far and wide, which activity involves consumption of a fair proportion of the food energy acquired by foraging. Thus, evolution helped the species to survive and perpetuate, by enabling it to exert more with hands and feet. Unfortunately, we have become the victims of our own evolutionary success as our brains developed and invented systems that enabled us to get all the food we want without much exertion to our limbs. Our species evolved first to make use of our hands and feet to procure food. Most of us, no longer do that. We depend on our brains to earn our living. That ability has become the bane of our civilization because we do not exercise our limbs enough to justify the amount of food we ingest. There is a wide gap between the energy we acquire by eating and the energy we expended to acquire it. That is why we see many apple-shaped and pear-shaped people in urban areas struggling to carry their weights. Obesity has thus become the central problem in the current health studies.

OBESITY

Fatness exceeding certain limits goes by the name of obesity. People slip into obesity. Most of them dislike that condition and want to get rid of it. Why? Just for the sake appearance. --They like to appear trim and handsome, more so in America where the culture worships slenderness. More than color of skin, the structure makes the body look handsome. By structure, we mean size, shape and weight of body. Few of us are endowed with body structures that approach the ideal, but even a girl whose body proportions are a far cry from the ideal, looks better if she maintains the normal weight. There is a more important reason for people wanting to get slender – that is the desire to improve health and live longer. Obesity is the mother of many diseases. When scientists speak of obesity raising the risk of mortality, they mean the association between obesity and heart disease. Obesity is known to cause hypertension, high blood cholesterol levels and diabetes. All these three are risk factors to heart disease.

HOW BIG A PROBLEM IS OBESITY?

Millions of people all over the world are obese. Obese people are found mostly in the urban rich. In villages where people toil, the percentage of obese people is remarkably less. It seems this is a malady of affluence

In 1985, a panel convened by National Institute of Health declared that one out of every five adults in America is obese. That is about 34 millions. Over the past decade, it has increased further and now about half the American adults are obese. They warned that the increasing incidence of obesity in childhood and adolescence would aggravate the obesity problem in future.

Recent studies from many countries suggest that where you are fat is more important than how fat you are. Male-type of obesity with bulging bellies is more harmful than the female type with bulging hips and buttocks. While women may not face as much risk of heart disease as men, they face more risk of cancer than men.

EXTENT OF OBESITY

Since 1980, weights of people in USA have been increasing alarmingly. National Health and Nutrition Examination Survey (NHANES) found that people considered clinically obese shot up from 14.5% in 1980 to 22.5% in 1994. They warned that the increasing incidence of obesity in childhood and adolescence would aggravate the obesity problem in future. The fatness problem has been reaching epidemic proportion in USA. The rapid rate at which percentage of obese people has been increasing can be seen from a perusal of Table 9-3

Table 9- 3:Increase in Obesity of Americans.

Age group	Percent increase in obese population during	
	1976-80	2003-2004
2-5	5	13.9
6-11	6.5	18.8
12-19	5	17.4
20-74	15	32.9

Source: National Health and Nutrition Survey

The situation in other parts of the world cannot be assessed reliably as the availability of reliable data is spotty, but the International Obesity Task force of WHO declared a worldwide epidemic of obesity.

Next to USA, this epidemic is showing virulent increase in England. Their National Health Survey reports that obesity has increased from 6% to 15% in men and from 8% to 16.5% in women between 1980 and 1994.

In Canada, obesity went up from 6.8% to 12% in males and from 9.6% to 14% in females between 1978 and 1992.

In Europe, unpublished data from former East Germany suggest that between 1985 and 1992 prevalence of obesity went up from 6.8% to 12% in men and 9.6% to 14% in women. Data from

Finland, the Netherlands and Sweden suggest that obesity has been increasing slightly among men but none in women.

In South America, data from Brazil show that obesity increased from 3.1% to 5.9% in men and from 8.2% to 13.3% in women between1976 and 1989.

In Western Pacific region, obesity in Australia increased from 9.3% to 11.5% in men and from 8% to 13.2% in women between 1980 and 1989.

In S. E. Asia, some meaningful data came from two studies in Thailand suggesting that obesity increased from 2.2% to 3.0% in men and from 3% to 3.8% in women between 1985 and 1991.The Nutrition Foundation of India estimates that about 100 million Indians are overweight.

In Africa, Mauritius reported trend data, which suggests obesity increased there from 3.2 to 5.6 in men and from 10.4% to 15.2% in women between 1987 and 1992.

Anyway, how to find whether one has slipped into the risky zone, because different weights are desirable to different heights? Using a formula, which yields an index number that incorporates into it the values of both height and weight, solves this problem. The resulting value is known as Body-mass Index or BMI for short. This index is obtained by dividing the weight in kilograms by the square of the height in meters of an individual. (BMI=Wt. in kg / height in m^2). Generally, a BMI of 25 is considered as the warning limit. (See Table A9-4 in appendix)

Unethical business people look for an opportunity in every trend whether health, sickness, or death to make money. As fast food chains multiplied and their sales skyrocketed, all sorts of people profited. Those who contributed to the expansion of the waistline of Americans by catering to them greasy fast foods and those who peddled potions and programs for weight reduction are well known. Now it is the turn of a new breed, those who cater to the needs of the families of the dead, to make more money from the fatty dead.

OPPORTUNITY TO FUNERAL BUSINESS

Now it has become more difficult for families of the departed to give final send-off, as they have to bargain for larger caskets and for larger cemetery plots. Undertakers have to devise euphemisms to tell the families politely that their loved ones won't fit in a standard casket.

Therefore, fatness is not a problem to everyone; it is an excellent opportunity to manufacturers of caskets and to cemeteries. Now they manufacture oversize caskets to accommodate fatter and heavier corpses. Of course, their charges increase a lot more than warranted by the increase in size. Some companies are now specializing in producing oversize caskets with extra bracing and reinforced hinges.

Goliath Casket manufacturing Co. of Lynn, Indiana, has now specialized in manufacturing oversized caskets with triple width, extra bracing and reinforced hinges to accommodate corpses weighing up to 700 pounds. Woodlawn cemetery in Bronx, N.Y., has increased size of their standard burial plot from two feet to four feet[97]. In 1980s, they were selling triple wide caskets no more than once a year. Now they are shipping four or five a month. Our bulging waistline is a great economic opportunity to the funeral business. Isn't it a blood curdling thought? Isn't the problem serious enough? If it is not a sufficient reason to diet and take a few hikes, what else is?

Airline industry has been advised to revise their weight estimates because travelers are heavier. It is a matter of time before airlines will begin charging not only by distance but also by the weight of passengers. Perhaps then we will realize how serious is the overweight problem.

FOOD INDUSTRY'S RESPONSIBILITY

American food industry's role in the outbreak of obesity cannot be overemphasized. If the effort by the food industry to increase consumption is only by advertising, it cannot be questioned legally but the food industry makes many behind the scene-maneuvers in the Congress, Federal agencies, courts, universities and public organizations to promote consumption of products designed for profit rather than health. Worst of all, food industry indoctrinates school children not only to make them consume more but also to make them their spokesmen and spokeswomen in future. Food industry's main goal is to induce the public to eat more.

Even the role played by the government organizations like USDA is not praiseworthy. USDA plays a dual role providing dietary guidelines to Americans on one side (mainly through food pyramids) and promoting consumption of American agricultural products on the other by considering major producers of beef, dairy, sugar etc., as it's primary stakeholders. Therefore, health conscious individuals should not accept all recommendations of the USDA without a pinch of salt. Perhaps it would be better if the job of providing food guidelines to public were transferred to the department of Health and Human Services. That does not guarantee total elimination of the problem of conflict of interest but it can mitigate the problem to some extent.

Action Plan

It is understood that in deciding how much to eat, you must depend on overweight as a factor. Although overweight alone is not obesity, it is indeed the determining factor. Therefore, the first step in the action plan is to know the ideal weight you ought to maintain. Ideal weight is not the same for all persons of equal age; it depends on your height and size of your body frame. If your body frame is smaller than your friend Mary's, you should weigh lesser. In order to determine your ideal weight you should first decide what type of bone structure you have – small, medium or large. To determine your bone structure, follow the procedure outlined below.

MECHANICS OF TRANSLATING DIETARY RECOMMENDATIONS TO MENUS

Take the measurements of your height, and wrist. Measure your height in stocking feet correct to ¼ of an inch and note. Next, measure your wrist at the joint just forward of the bone. Hold the tape snugly round the wrist and note the measurement correct to 1/8 of an inch. Using these two measurements, you can determine your frame size from Table (A9-1) in appendix.

HOW MUCH SHOULD ONE WEIGH AND WHAT IS OBESITY

One of the best sources for finding the ideal weights is the data issued by life insurance companies. For over half a century Americans relied on the tables published by the Metropolitan Life

Insurance Company, the latest being the tables published by the company in 1983. In these tables, they combined the data from 25 other insurance companies covering 4 million policyholders. The ranges of weights given in these tables were associated with the lowest mortality among the policyholders.

To know your ideal weight, using the frame size you obtained in the previous step, enter height column in Table (A9-3), locate your height row, move to the right, and find your weight appropriate to your sex and frame. For example if you are a woman of medium frame with a height of 5'-4", your ideal weight range is 113 to 128 pounds. If you are a man of large frame with a height of 6'-0", your ideal weight range is 164 to 184. If your actual weight happens to coincide with your ideal weight found in the table, you are so much lucky. Guard it carefully from severe fluctuations. If your actual weight is a few pounds lesser and you feel healthy don't bother to gain. If your weight is more than required, you are in need of slimming.

It is beneficial to be slightly underweight than overweight. Obesity should be recognized not only from excess weight but also from the deposition of adipose tissue in some parts of the body. You can find it by taking measurements of those parts and comparing them with the ideal measurements given in the Table (A9-2) in appendix.

If your weight is five or more pounds above your ideal weight, it is a warning signal you have to heed. If it is more than 10 pounds above the limit, you have no time to lose. You have to start slimming down before it is too late. Having determined your ideal weight you should then go to Table (A9-5), which gives you your energy requirements.

Energy Balance - Weight Watching

Body fat is designed by Nature as a store of energy. All energy intake supplied to the body over the needs of metabolism is stored as fat to meet the emergencies like starvation. Suppose you want to maintain your present weight without starvation or fattening. To avoid either extreme you must maintain energy balance. Maintaining energy balance means taking in exactly as much energy as you spend for your body metabolism and other activities. All extra energy is deposited in your body as fat. The body converts every 3500 kilocalories (KC) of energy into 1 pound of fat. Let us call the result of this accumulation as positive energy balance or credit balance. Similarly, when there is a deficit in the energy needed by it, the body converts a pound of energy into 3500 kilocalories of energy to meet the deficit. Let us call the result of this exhaustion as negative energy balance or debit balance.

Therefore, the secret of losing weight safely is to maintain deficit balance in your energy account until your target weight is achieved. After reaching your target, maintain zero energy balance in the account so that your weight may remain stable afterwards.

For many people energy balance prevails automatically without making deliberate effort. I am one of those fortunate in this respect. You may wonder how it happens. My eating habits may yield the clue. I eat only if I am hungry. Otherwise, I do not. I never eat to kill my time. This explanation is simple and intuitive but it may not be understandable to people who are already overweight because they might have lost the distinction between hunger and habit. They can follow the mechanics of weight control provided here.

Calorie Counting

While discussing the RDA for proteins, we noted that the Food and Nutrition Board fixed them rather high on the assumption that protein deficiency is more critical to the individual than surplus. Contrary to that assumption, evidence suggests, that excess protein intake has been more harmful to the U. S. public. Like protein needs, energy needs of individuals vary. F.N.B./N.R.C. fixed their recommendation for energy intake at the mean of sample on the assumption that excess energy is harmful. It should be noted that this recommendation is for total energy intake and for energy coming from any particular source-protein, carbohydrates or fat, all of which contain energy. Therefore, if you satisfy your daily requirement of protein that will automatically satisfy part of the energy requirement. The remainder can come from carbohydrates and fat.

Based on these considerations, a scheme is provided below that will lead you step by step to determine your optimum energy requirements corresponding to your height, weight and sex.

CALORIE CRITERION:

Determining optimal mix of foods providing the required nutrients and minimizing the total calorie output is a mathematical programming problem that cannot be discussed here. Several studies made in this regard found that foods yielding anywhere between 1500 to 2000 calories per day would be required to supply the necessary protein and other nutrients. It is also realized that if we take care of the protein / calorie ratio, other nutrients would, in general, be taken care off automatically.

To illustrate this method let us take a female of median height 5'-5" and medium frame. From Table (A9-5) she needs between 1755 to 2010 calories per day. The average of these numbers is 1883.Deducting 30% from this, we get **1318** calories per day.

For a male of median height 5'-9"and medium frame, the calorie range is from 2130 to2340. The average of these numbers is 2235. Deducting 30% of this we get **1565** calories per day.

Table (A9-5) gives the caloric requirements based on heights and frame sizes of women and men separately. It may be seen that three values labeled 'Low', 'Mean' and 'High' are given for each height and frame size. If you are overweight, choose the low value to bring it down. If you are active and your weight is close to the ideal, you may choose the mean value and maintain your weight. Never approach the upper limit unless you are an athlete or do some strenuous work regularly. People with sedentary habits and occupations should choose the least of the three values.

After having determined your daily calorie requirement from these tables, you are ready to plan your meals. This consists of choosing the foods, which supply your body, the daily requirements of nutrients. There is a slight complexity in this. When you try to combine various foods with a view to make up your caloric requirement that may involve oversupplying or under supplying proteins. If you try to make up a specific amount of the total protein content, it may result in missing the caloric requirement. Therefore, you have to establish a balance between protein and energy contents of foods you ingest.

Now, in order to establish energy-protein balance you have to look for the calorie to protein ratio appropriate to you based on nitrogen conservation efficiency. Assuming you are a man of average nitrogen conservation efficiency with 5'-9" height, a medium frame, your mean ideal weight is 150 lb. If you are a vegetarian, you need 32gm./day of usable protein and 1566 calories/day. That means you should choose foods, which contain 32gm. of usable protein and 1566 calories per day. That combination yields 49 calories per gram (1566/32=49) of usable protein. An average woman of 5'-5" height (125lbs) and medium frame, needs 1318 calories/day. As a vegetarian woman needs about 27 gm of usable protein this ratio comes to 49 calories (1318/27=49) per gram of usable protein. Using similar calculation, the ratio comes to 49 for non-vegetarians. These values were recorded in Table (9-1).

To remember easily you may take this ratio as **50** for both vegetarians and non-vegetarians of both sexes.

Now everything is simplified. All that you need to do is to determine your calorie requirement. Using these calorie/protein ratios as your guides, choose the foods you like from the table in Appendix, to make up the total calories and compose your menu. Let us now summarize the required steps for composition of ideal diets.

SEVEN STEPS TO IDEAL DIETS OF LONGEVITY SEEKERS

1. Measure your height and determine your frame size from Table (A9-1). [The prefix 'A' refers to Appendix.]
2. Determine your ideal weight from Table (A9-3)
3. Determine the normal caloric requirement corresponding to your ideal weight from Table (A9-5)
4. Reduce the normal caloric requirement by 30% and arrive at your ideal caloric requirement. (Recommended by Caloric Restriction Theory).
5. Know your usable protein requirement based on your weight, from Table (9-1).
6. Determine your personal calorie /protein ratio from Table (9-1).
7. Choose your foods from table (A9-6) having less than or equal to the above ratio value to make up your total caloric requirement and make up your choice menu. Restrict the foods as

advised in the previous chapter. (Caution: Don't forget to consult your medical doctor).

What kind of Diet?

Our problem now is to discover what kind of diet is good for human consumption from the point of view of individual health and longevity. From the forgoing discussion, it is clear that one of the criteria to qualify as good diet is avoiding excess protein. Another criterion about which there is not much controversy is a plan to keep obesity at bay.

What kind of diet plan we have to follow? The contending major classifications are of course omnivorous diets and vegetarian diets.

It is in this context we have to study the relative merits and demerits of different kinds of diets without any preconceived notions. That is our position in further study of diet problem in this chapter, neither promoting one variety nor indicting of the others. If vegetarianism is proved injurious to health, we will not hesitate to abandon it. Similarly, if it is proved otherwise, let us curtail our meat consumption. If avoiding meat is necessary in the interests of health and longevity, we should say "Good bye" to it.

Vegetarianism

Vegetarianism has never been without its critics or without its advocates during any period of history. While critics simply call it in names like fad or faith, its advocates support it on rational grounds. From Aristotle and Pythagoras to Gandhi and Bernard Shaw, several thinkers pointed out the evils of meat eating. Arguments based on naturalness, cleanliness, respect for life, ethics and religion have been used by a galaxy of thinkers including Aristotle, Plutarch, Pythagoras, Diogenes, Benjamin Franklin, Tolstoy, Shelley and Bernard Shaw to expose the shortcomings of meat eating habit, each in his inimitable style. Rousseau who inspired the French revolution and Tolstoy the famous Russian writer went a step ahead of others and castigated meat consumption as an evil that spawns violent minds, which eventually lead to wars.

Vegetarianism in varying degrees has always been a way of

life for peoples in many eastern countries. In the west, it seems to be an outgrowth of industrialization prompting many young people to look for uncontaminated pure food and some spiritual values to sustain life.

HISTORY OF VEGETARIANISM

Vegetarianism in the East has been an ancient tradition. The origin of vegetarianism in the West can be traced to Rev. Cowherd, curate in Manchester, England. Rev. Cowherd grew skeptical about the correctness of eating meat and turned to the Bible for guidance. Bible told him clearly in some passages to eat grain and fruit. Some of the passages that inspired him are:

"Meat commendeth us not to God"

"Be not among the riotous eaters of flesh"

"He that killeth an ox is as if he slew a man"

Any loophole, ambiguity or opportunity to escape the intent of the holy writ was removed by the sixth commandment, which says:

"Thou shalt not kill."

Imbued by the true spirit of the bible, Rev. Cowherd established his own church in the year 1800. The cornerstone of this church, which was known as "Bible Christian", was non-violence and vegetarianism. The chapel of this church is still in King Street, Salford, Manchester. The torch was next passed on to Rev. Metcalfe.

Influenced by Cowherd, Rev. Metcalfe renounced bottle and roast in September 1809. It was he, who brought the mission of vegetarianism to America. As a pastor, he led a band of Bible Christians from Manchester to Philadelphia. In 1890, the 'Bible Christians erected a chapel in Park street, Philadelphia. Metcalfe founded the first vegetarian society in America and edited its magazine 'The American Vegetarian.' American Vegetarian society came into being in 1850, in Clinton Hall, New York City. After Metcalfe died in 1862, Henry Club took over the leadership of the society.

In England, A. F. Hill published "Vegetarian Essays" in 1897. Anna Kingsford, a medical practitioner in England devoted a substantial time propagating the benefits of vegetarian diets on anatomical grounds.

In America, the vegetarian movement attained visibility with the experiments of Dr. Kellogg and the establishment of the Battle Creek sanatorium in late 19th century. It became prominent with the founding of Kellogg's corn-flakes company. The emergence of Woodstock generation in recent times, with its appeal to "Back to the Land" slogan gave a further fillip to its popularity.

RATIONAL BASIS OF VEGETARIANISM

With the turn of the 20th century vegetarian movement had to face a new generation of skeptics who were wont to call it a fad or faith without going into the facts established by science. It is easy to give names to anything you don't like because it is a labor saving device. To counter calumny is not that much easy as it requires precise arguments based on some criteria. However, vegetarianism has not been lacking in such arguments. Five criteria can be used to test on rational basis which dietary practice is better — vegetarianism or non-vegetarianism:

1. Naturalness of food to humans
2. Adequacy of diets from the nutritional point of view
3. Alleviating world food shortages.
4. Promotion of inter- temporal bio-ethics between generations with regard to prudent agricultural practices, conservation of natural resources, protection of life and making them role models for the future
5. Promotion of health and longevity of the members of the current generation.

If plant-based foods are chosen at least three times in answer to these five criteria, the superiority of vegetarianism is established. Otherwise, let us admit that it loses.

Let us now compare both the diet systems in the light of these criteria.

1. NATURALNESS

Meat is unnatural food for humankind. If not, would you be willing to go into the butcher's shop, cut a slice from a carcass and put it into your mouth? You would not hesitate do so in the case of a fruit or vegetable. Conduct the following test and see for yourself what kind of food is natural to human beings. Put in a room; a baby, a chick, a kitten and an apple and watch what happens next. The kitten pounces on the chick and the baby reaches for the apple. The natural inclinations of the human baby and the animal nature are thus revealed.

Let us see how the physiological and structural differences in the bodies of carnivorous and herbivorous animals of human body confirm the argument that meat is not the natural food of humans.

- Like herbivores, humans have long intestines and not the short intestines of carnivores.
- The lower jaws of humans hang so that they can grind their food on the other like cows or horses; and not fixed vertically like those of dogs.
- Carnivorous animals have long teeth and claws to hold and kill their prey whereas herbivores have short teeth and no claws. Human beings have neither such teeth nor claws.
- Flesh eating animals secrete large quantities of hydrochloric acid to enable them to dissolve the pieces of bone they ingest along with flesh, whereas herbivores secrete very little of that acid.
- Carnivores must lap the liquids (like cats and dogs) while herbivorous animals drink by suction. We don't lap liquids like cats and dogs.
- Carnivores do not sweat through their skin like the herbivores. They control their body heat by rapid breathing and extrusion of the tongue.

Therefore, we are more akin to animals like, monkeys, cattle and elephants but not to animals like cats, dogs and tigers.

It is well known that one of the important causes of disease is toxicity caused by the waste matter remaining in the body for a long time. It is also well known that the extent of toxicity in the body can be understood by examining the excreta. The excreta of flesh eating animals are by far the most dangerous of all. Notice that carnivorous animals are compelled by natural instinct to scratch holes in the ground and bury their excreta soon after defecation, whereas cattle droppings are left scattered all over the surface of land and are utilized as manure by all vegetative growths.

Let us consider for a moment the wonderful machinery that

Nature installed in our bodies for getting rid of waste matter. All vertebrates are built on the principle of a long hollow tube around which the muscular, circulatory and nervous organs are built. Food is put at one end of this tube and it passes along slowly while undergoing transformations and leaves at the other end of the tube as waste matter. As it passes along, the body absorbs various elements of nutrition by the chemical action of digestive juices and the residue is pushed through the other end of the tube. Broken down pieces of body cells are also swept into this tube which now acts like a sewage pipe. Little by little, the contents of this tube are pushed right on to the end of the tube and then eliminated. So simple in theory and excellent in practice, this machinery is built to work for a hundred years and more; but fools get it out of gear long before it is normally due to wear out!

For the proper working of this machine, the owner must put into it suitable fuels in a suitable way. The food put into the body should be:

- Suitable to the human body structure,
- In a condition suitable to be dealt with by this bodily machinery,
- Of suitable quality,
- In quantity that is in harmony with the requirements of the body,
- At intervals sufficiently long to allow time for rest and repair of the various parts of the body.

Meat does not satisfy the desiderata specified above. It is not a natural food to humans. Diogenes, biographer of Pythagoras issued the challenge:

> If you declare that you are naturally designed for such a diet, first kill for yourself what you want to eat. Do it, however, only through your own resources, unaided by cleaver or cudgel or any kind of axe.

Can you do that?

The English lyrical poet P. B. Shelley, a committed vegetarian, wrote in his essay ' A vindication of Natural Diet'

> Let the advocate of animal food force himself to a decisive experiment on its fitness, and as Plutarch recommends, tear a living limb with his teeth and pumping his head into its vitals slake his thirst with the steaming blood...then and then only would he be consistent.

2. NUTRITIONAL ADEQUACY:

Balanced Vegetarian diet is nutritionally adequate. We have seen in our discussion of nutrition theory earlier, that amino acid chemistry makes no distinction between meats and vegetable products as far as source of protein is concerned. It is indifferent between different sources. We have also shown that there are many plant based foods that can be substituted for meat or many dietary combinations of complementary plant foods can be made to attain the full biological value on a par with meat. In addition, we presented a tabular feature dispelling many misgivings about vegetarian foods. Therefore, there is not a shadow of doubt that plant based foods are nutritionally adequate. Remember millions of people have been living healthy lives for centuries on vegetarian diet.

3. ALLEVIATION OF FOOD SHORTAGES:

HOW MEAT HABIT GOT ENTRENCHED?

Historically, carbohydrates in staple foods like wheat and rice were at the center of foods of many communities with meat as a supplementary dish. The old dietary practice of our ancestors and working classes was far more reasonable and beneficial. Their meal consisted of the basic food wheat or rice, vegetables and milk products and stews of all kinds in which meat, in small quantities, played the part of a condiment to satisfy the palate. This is still the pattern of meals in Oriental countries. All that has changed now.

The industrial revolution of the 18th century is generally regarded as the harbinger of prosperity but what is not generally realized is that it also sowed the seeds of self-destruction of industrial civilization. It was during this time that the British gentry acquired a taste for fat-marbled beef and spread it like plague in the western countries. With the increase in craving for beef, the new machinery and the assembly line methods that came into vogue with the dawn of industrial revolution were pressed into service for large-scale production of beef in killer factories. The machinery, which was invented to wipe the sweat of thy brow, was pressed into service to massacre millions of innocent lives.

For over a century, a love affair with meat has been flourishing in America. As Americans became more affluent and steak houses and hamburger chains multiplied the amount of protein in American diet coming from animal sources has increased to nearly 69% while the protein coming from plant sources has decreased to nearly 48%.

Americans eat about 260 pounds of animal protein per capita per year. (260/365=0.712lbs=11.4 oz /per day). This involves ingesting nearly 100 grams of animal protein per day; more than double the quantity required by the body. If we eat only meat, fish and poultry and never see the face of other foods, we can cut short protein consumption by half without being worried in the least about protein deficiency. However, we do eat other stuff. According to the estimates of USDA, 58% of our total protein intake is supplied by non-meat sources. Then viewing other way round, since 58 grams per day is more than enough, we may altogether eliminate the remaining 42 grams, without any alarm about protein deficiency, by giving up meat, poultry and fish.

Now we have to consider the proposition that switching over to vegetarianism will help alleviate the current shortages of food in several parts of the world. To determine the validity of this proposition we have to answer two related questions.

The first is: Are the plant-based foods acceptable to the vast masses of humanity? The second question is: Are the plant-based foods more economical relative to meats?

The answer to the first question is straightforward. To some vegetarians meat eating may be a taboo but there is none to whom eating plant-based foods is a taboo. Even the most voracious meat eaters do eat plant-based foods.

The second question deals with the relative economy of producing meat foods and plant-based foods. Whichever category is more economical and better, serves the purpose of alleviating the food shortages around the world.

Malnutrition is a prominent and widespread cause of ill health and shortened lives. Some people use the word mal- nutrition synonymously with under-nutrition. The prefix 'mal' could be used to mean a state that is either below the optimum or above it. From this point of view, most of the people in the world can be said to be suffering with malnutrition — some with under nutrition and some with over nutrition.

The geographical distribution of over nutrition related diseases as heart disease stroke and cancer on the one hand and the under nutrition related diseases as marasmus and kwashiorkar on the other reveal that the overall health of people could be greatly improved by simple redistribution of food. Community health can be greatly improved if those who fall sick due to overeating can be persuaded to eat less and allow others who fall sick due to lack of food to eat a little more. The consensus among nutrition experts is that over indulgence is as much detrimental to health as starvation. If a little is good, more is not always better. In the field of nutrition the true maxim is 'Just sufficient is the best' – neither more nor less.

Economics of Production

It is well known scientific fact that the efficiency of transfer of energy from one trophic level to another is less than 100 percent. Only a small percent of sun's energy that is transformed into chemical energy by the green plants (original producers of food) is actually passed on to the herbivores that eat plants as food. Various carnivores that prey on herbivores, receive in turn, decreasing shares of the energy budget. To use a familiar example, a pound of canned tuna fish must have used 10 pounds of herring, which in turn must have used 100 pounds of prey, which must have lived on, using the same proportion, 1000 pounds of algae. When the herbivores are eaten by the omnivores like humans, much less energy is passed on to them because the same energy distribution is applicable to the forage and ranch crops that are transformed into beef via cattle. It is thus an established fact that meat production in general and beef production in particular is very uneconomical relative to plant-based foods

Production of meat containing a given amount of nutrients per unit requires more inputs than production of plant based foods containing the same amount of nutrients. An acre of land devoted to production of cereals can produces 5 times more protein than the same land used for production of meat. An acre of land devoted to legumes can produce 10 times more and an acre devoted to leafy vegetables 15 times more. Even among meats, beef production is the least efficient for, it takes 16 pounds of soy feed to produce a pound of beef whereas it takes 3 pounds of soy feed to produce a pound of eggs or broilers and 6 pounds for conversion into pork. If production of beef is ecologically the least efficient, production of beef in feedlots is the worst of all, causing disaster to ecosystem, economy and the well being of future generations.

It was shown in the previous chapter that the grain fed to the livestock in USA could easily support five times the entire human population in the country.[98] It is estimated that by resorting to the indirect method of obtaining food by feeding grain to the cattle and taking their bloody flesh, we get back only 10% of calories and 6% of protein of what we would have got by consuming grains directly.[99] For every 16 pounds of grain and soybeans we feed to beef cattle, we get back barely one pound of meat; the rest is turned into manure. We lose 88% of protein by feeding grain to pigs and 83% by feeding to poultry.[100]

Translating these figures into land use patterns; we need 3¼ acres to feed one mouth with beef per year. To feed the same mouth with lacto-vegetarian food we need only 1/6 of an acre. In other words, one acre can support 20 times as many vegetarians as those eating modern American style meat centered diet.[101]

It is estimated that the world's cattle alone consume a quantity of food (grains) equal to the caloric needs of 8.7 billion people, nearly double that of the entire human population. The lion's share of this consumption is attributed to the cattle in USA alone.[102] According to the USDA statistics, one acre of land can grow 20,000 pounds of potatoes. If used for cattle feed the same land can yield less than 165 pounds of beef.[103] Marvel at the huge waste!

4. Promotion of Inter-temporal Bio-ethics

While comparing different food production schemes, it is necessary to consider the relative impact of their production systems on the environment and conservation of natural resources.

PROTECTION OF ECO-SYSTEM

There is a direct link between food production and environmental degradation. On the criterion of being friendly with Nature and preserving the checks and balances built into the ecosystem, vegetarianism comes out in flying colors as demonstrated in the previous chapter. Denuding the great timberlands of America, destruction of the great rain forests of the world along with their rare species of flora and fauna, topsoil erosion, depletion of water resources, and enormous ground water pollution can all be avoided and the deteriorating ecosystem balance can be restored if we moderate our meat habit. Thus, all the five criteria we set before us are satisfied by the plant-based foods. Vegetarianism therefore wins easily. In fact, there is no alternative to vegetarian food if we want to stop the rapid deterioration of our ecosystem. We owe it to our children and their children to protect our ecosystem.

Clinging on to non-vegetarian habit in spite of its weaknesses and its role in undermining health shows that it does not stand on any rational ground but only leans on the habit of people supported by the propaganda of vested interests.

5. PROMOTION OF HEALTH & LONGEVITY

On the question of which category of food promotes health and longevity there is abundant learned body opinion on the undesirability of meat-based foods.

The incidence of some fatal diseases and deaths typically indicate the geographical mal-distribution of food supplies. Marasmus and kwashiorkor occur mostly in areas populated by people who are deprived of adequate food whereas the three great killer diseases, heart disease, stroke and cancer occur in affluent areas where people overeat. More than half the deaths in USA are attributed to heart disease caused by arteriosclerosis caused by concentration of cholesterol and tryglycerides. The concentration of these villains is definitely known to be higher in meat-based foods.

In the famous comprehensive prospective study conducted by the Seventh Day Adventists of California during a period of thirty years, several important facts came to light. The essence of their findings was that vegetarian Adventists enjoyed better quality of life and lived longer than other Californians. Fig. (9-2) illustrates this conclusion.

CONSENSUS AMONG LEARNED BODIES

In spite of divergent opinions about the risk factors involved in heart disease, there is consensus among leading scientific groups like American Medical Association and Nutrition Board of the National Academy of Sciences and American Heart Foundation in favor of regulating the quantity as well as the quality of food intake

In a joint statement, AMA and Nutrition Board observed that lowering of serum cholesterol:

> 'Can be achieved most practically by partial replacement of the dietary source of saturated fat with sources of unsaturated fat especially those rich in polyunsaturated fatty acids and by reduction in the consumption of foods rich in cholesterol'[104]

The American Heart Association offered the following recommendations.[105]

- Caloric content of the diet should be controlled and physical activity should be increased so that obesity will be prevented.
- The amount and type of the fat eaten should be controlled to avoid elevated serum cholesterol and serum triglycerides.
- Carbohydrates eaten should be primarily of complex type.
- Salt content should be controlled to avoid predisposition to high blood pressure.
- The NRC also recommended decreasing total fat, saturated fat and dietary cholesterol to decrease the risks of coronary heart disease.[106]

These recommendations reveal clearly the connection between over nutrition caused by non-vegetarian foods and the life-threatening diseases and consequent shortening of even the naturally allotted life span. It is imperative therefore to identify foods containing high concentration of these prohibited stuffs and try to reduce their usage immediately.

CONTAMINATION OF FOOD AT HIGHER LEVEL IN FOOD-CHAIN

Planet earth's human population grew rapidly demanding more and more food. Naturally, the increased number of hungry mouths led to the quest for more food. Links between dietary fat levels and arteriosclerosis have been known for a long time but recent research implicated dietary fat also in forms of cancer like colon cancer and breast cancer. Think of the situation if these fats are also contaminated with pesticide residues!

Since the industrial revolution, not only growth of population but also per capita desires has tremendously increased. This has led to the intensive cultivation of land by use of chemical inputs that in turn has resulted in contamination of foods.

Man's first earth changing invention was agriculture. For a long time, agriculture was only a subsistence operation, but improvements in agricultural technology changed that situation. Agriculture became so efficient that it began producing surplus. Thus, humans were freed from the biblical imprecation "In the sweat of thy brow should thou eat thy bread". Ironically, this began to disturb the ecosystem by compounding the growth of human population. This abundance is attributed not only to extensive and intensive cultivation methods but also to methods of preventing losses of harvests before they reached granaries. Most of the losses during production time were inflicted by competing herbivores. Remember that the word 'herbivores' includes, not only large hoofed animals but also rodents, birds, and insects. The modern technique of protecting crops is to spray poisonous chemicals on the invading pests to kill them. Among the chemical pesticides, chlorinated hydrocarbons commonly known, as DDT, Endrine, and Dieldrine, found widespread use all over the world, during the 60's. In the beginning they were hailed as one of the great discoveries of the 20th century based on their usefulness to increase food production, However opposition to their use began to build up gradually as their pervasive destructive effect came to be known. The three most important arguments against the widespread use of the pesticides are that they are: (1) universal poisons, (2) fat-soluble, and (3) they degrade slowly.

We should be very much concerned about the latter two arguments as they have a direct bearing on our health. Chlorinated hydrocarbons are soluble in fat but not soluble in water or blood. Substances having chemical affinity are mutually soluble and it is difficult to separate them. (Isn't it easier to separate sand from water than it is to separate sugar or salt?) If small amounts of chemicals enter animal body, they lodge themselves in the fatty tissues of the body and they cannot be washed out efficiently by blood. When animals eat contaminated plant foods, concentration of pesticide residues builds up over time in their bodies. The pesticides increase in their concentration as they move up in the food chain.

The mechanism of transfer works like this. Suppose a field is sprayed with pesticide. Pest insects feed on the poisoned leaves. Because of fat solubility, the pests do not excrete the pesticide. When the predatory insects eat the herbivorous insects, the accumulated pesticide is transferred into their bodies with higher concentration. When birds prey on the predatory insects, their diet has even higher concentration of pesticide poison than the insects they eat. Thus, poison concentration builds up as you move along the food chain from primary consumers to secondary consumers, secondary to tertiary and so on until concentration reaches maximum in meat eating humans who are the final consumers.

A study examined the successive pesticide concentrations of pesticide toxaphene in various parts of a marsh. The study found that bodies of small invertebrates in the mud contained 0.2ppm (parts per million) of toxaphene. The fish that ate them had toxaphene in their muscles up to 8ppm. The pesticide residue in the fish eating birds reached a concentration of 650ppm. Imagine what happens if we eat these birds.

Another example is that of Trout. Trout build up a layer of fat during summer when plenty of food is available. In areas where DDT has been sprayed, their fat contains large concentration of DDT. During winter, the fat is used as a source of energy. When fat breaks down, the DDT released into blood stream is known to kill the animals. In another case, DDT is known to have poisoned seven hundred thousand hatching salmon in their eggs.

The effect of constant accumulation of pesticide residues may range from negligible to disastrous. This level of concentration may not have detectable effects on metabolism, but the long-range effects may be disastrous.

LOW DEGRADABILITY:

The complimentary problem associated with the chemical contaminants of food ingested by carnivores is their low degradability. The half-life of these chemicals is a critical length of time that falls within the human life span. Half-life of a substance is the length of time during which its original quantity is reduced to half.

Let us see how the half-life of contaminating chemicals affects human health. Suppose a grazing ground is contaminated with a chemical substance with a half-life of say, one day. If a steer grazes the grass on this ground, it is not of much consequence to human beings because the chemical ingested by it will have degraded quickly in a few days to negligible level and the beef obtained from it would be free from poisonous chemicals in a few days.

On the other hand, suppose the grass is contaminated with another chemical that has a half-life of one year. Traces of chemicals will remain in the body of a steer, even after a decade, if it grazes there. Most of the pesticides have a half-life of 10 to 15 years. That means during the lifetime of cattle, hardly 50% of the chemicals ingested will have degraded. Imagine eating them as beef. All the chemical residues in high concentration are transferred into the human bodies and lodged there for the remainder of their lives exerting their destructive influence all along. If the biodegradable life of a poison lodged in human body exceeds the span of human life the result is obvious. The effected human being becomes a helpless victim of the lingering destructive effects of the poisons that cannot be washed out of her/his body.

Well aware of the health hazards, the Environmental Protection Agency and the Food and Drug Administration took certain steps to keep pesticides out of the diets of animals and animal products we consume; but most of their efforts were directed against direct contamination while many sources of indirect contamination were ignored. However, their controls could not eliminate the hazards of pesticides because, either the government was not able to monitor all the violators or the pesticides that already entered the food chain had not degraded. It is estimated that chlorinated hydrocarbons take about 40 to 50 years to degrade to tolerable levels.

While there is no question about the injurious effects of pesticide residues in our foods, there is no unanimity of scientific opinion as to what constitutes health hazard. The tolerance levels of various chemicals differ. For instance, while DDT concentrations reported in the above study were said to be within tolerance levels, there are some cases where other poisons like heavy metals reached alarming levels. Some species of ocean fish were so much contaminated with heavy metals like mercury, that eating them constitutes a real health hazard. It was estimated that some predatory fish like trout and ocean fish like blue-fin-tuna and swordfish contain 0.5mg/kg of body weight of methyl mercury, the maximum permissible tolerance level. If these fish are consumed at the average rate Americans consume meat, you can easily accumulate in four to five years 60 to 70 mg of methyl mercury, a quantity sufficient to kill you.

RISK-BENEFIT ANALYSIS

Strictly speaking, none of the diet systems can be said to be completely beneficial or completely injurious to human health. Every dieting pattern has its own strengths and weaknesses. What we can do to resolve this problem is to ignore all the considerations for a while, measure the risk-to-benefit ratios of the competing diet systems, and choose the one with the lowest ratio. This ratio can be defined as the proportion of persons subjected to risk divided by the proportion of persons benefited by following a given diet pattern. This number integrates all the risk factors and benefits involved in a diet system. The smaller this number, the better it is as our guide for choice.

From our present state of knowledge, we can say that the health status of an individual does not follow a linear relationship with food i.e. we cannot say that health increases or decreases in proportion to the food ingested. If nutrition is deficient, health suffers from deficiency diseases, if it is excess, then it causes different kind of diseases. A well-known but misplaced criticism against vegetarian diets is that they are deficient in nutrients. This has led to the public health dilemma of choice between the two kinds of diets. It was mistakenly believed that some trade off between meat foods and malnutrition is inevitable.

In spite of the best efforts of oncologists and surgeons, Americans die of breast, colon and prostate cancers at the rate of five to thirty times the rate of people in other parts of the world. Many experts agree that diet and lifestyle have a lot to do with this alarming disparity. In a comprehensive analysis published last year, scientists of the American Institute for Cancer Research and those assembled at the World Cancer Fund concluded that poor eating habits account for a third of all cancer cases (roughly the same proportion attributable to smoking)[107].

For a long time, emphasis in biomedical research was placed on the health risks associated with the nutrient deficiency associated with consumption of vegetarian diets. This may be due to various reasons. First, the presently developed nations, when they were in the development stage, were concerned more with adequacy of nutrition rather than optimal intake. Secondly, nutritional research endeavors followed until recently clinical approach rather than public health approach. In clinical approach, reports are compiled from the medical complaints brought to the clinics by vegetarians and they dealt with short-term health effects of diet. Public health approach requires researchers going into the community and studying the long-term relation between vegetarian diet and chronic diseases and longevity. Thirdly, industrialized nations that were funding research, were mainly composed of non-vegetarians. Finally, a cultural bias against meatless diets resulted in exaggeration of their potential health risks. Until late 1970's, those following vegetarian diets were considered even anti-establishment and avoidance of meat by them was considered to be for reasons other than health.

Advances in nutrition research during the past two decades have changed our understanding of the contribution of vegetarian diets to human health. Dozens of epidemiological studies relating to human nutrition have documented the quantifiable benefits of vegetarian foods, namely reduction of risk of suffering chronic degenerative diseases and mortality. A paradigm shift seems to be on its way. Diets based on well balanced plant foods are viewed as more effective in improving health than as causing disease. Several studies documented the association between certain foods and their risks and protective effects.[108] It is argued that a replacement of a quantum of meat with equivalent amount of nutritious vegetarian food would result in health benefits that would overweigh any ill effects attributed to alleged loss of nutrient value.

Another significant fact that has emerged recently is that the increased risk of cancer and cardio vascular diseases experienced by meat eating populations may be due to not only consumption of excess energy, and saturated fat but also due to deficiency of phytochemicals and other compounds found abundantly in plant foods. It is considered that the risk of phytochemical deficiency among those who follow meat-based diet is greater than the classic acute nutrient deficiency diseases in some vegetarians.

In a number of studies conducted by the Seventh Day Adventists, it was found that some varieties of cancer were associated with meat eating. On the other hand, ingestion of whole grain, bread, legumes, nuts, soymilk, and fruits was found to exert protective influence.

MEAT IMPOSES EXCESS MEDICAL COSTS:

Meat eaters endure unnecessary financial burden. In a recent study by 'Physicians Committee for Responsible Medicine', it has been shown that substantial medical costs in America are attributable to meat consumption.[109]

The study was conducted on two groups, omnivores and vegetarians while isolating the effects of smoking, alcohol consumption and other relevant factors by suitably controlling the experiment. Medical costs attributable to meat consumption were considered the fraction of direct treatment cost that corresponds to the excess prevalence of disease among omnivores, compared with vegetarians. Based on the excess prevalence of various kinds of diseases among the omnivores relatives to vegetarians, Barnard et al estimated excess medical costs attributable to meat consumption. Their findings are displayed in Table (9-4).

Table 9-4: Estimated Medical Costs (in $billions) Attributable to Meat Consumption

Disease	Excess cost attributed To meat eating. ($ Billions)
Hypertension	$ 2.8----8.5
Heart disease	$ 9.5
Cancer	$0---16.5
Diabetes	$14---17.1
Gall bladder disease	$ 0.2----2.4
Obesity	$ 1.9
Food borne illness	$ 0.2---.5.5
Total	$28.6--61.4

Source: N. D. Barnard et al, "Medical costs attributable to meat consumption, Preventive Medicine, vol.24:6 (1995), pp 646-655 Reproduced with permission.

These extra costs can be eliminated if we emancipate ourselves from our slavery to meat habit.

Apart from promoting general health, vegetarian foods have the distinction of helping the practitioners to prolong their lives, as it is easier to implement the principle of caloric-under nutrition with the vegetarian foods than with the non-vegetarian foods. Even on the criterion of promoting health and longevity, vegetarianism wins.

At what price are these mountains of meat obtained and what is the result? The price we pay is degradation of the ecological system that shelters us. The result is ruined public health and shattered lives.

All the five criteria we considered above exhort us to get rid of our meat habit.

The issue is finally clinched by the extensive epidemiological study and findings of the Adventist Mortality Study of California. The prospective study conducted over a period of thirty years, found that the life expectation of vegetarian Adventists was consistently higher than other Californians at all ages and they enjoyed better quality of life than others. These findings are summarized in Fig. (9-2a and 9-2b) displayed here.

Fig. 9-2a: Survival of Adventist women compared to non-Adventists of California.

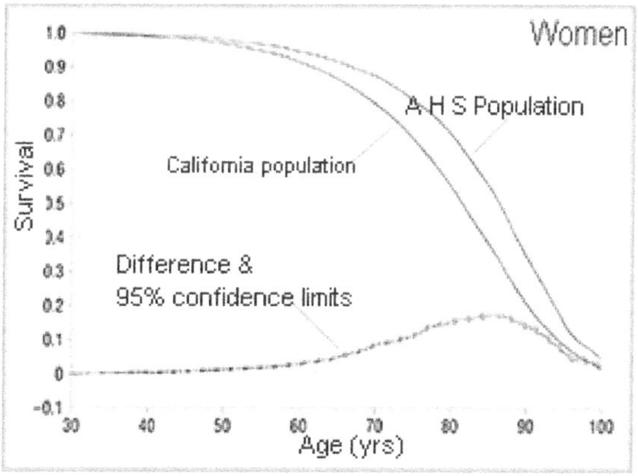

These graphs show that the proportion of surviving women in the Adventist population is higher than other population, at every age.

Source for Figs. 9-2a and 9-2b: Fraser GE, and Shavlik DJ, "Ten years of life: Is it a matter of choice?" Archives of Internal Med", 161:1645-1652, Copyright, 2001, American Medical Assn., adapted with permission.

Fig. 9-2b: Survival of Adventist Men compared to Non-Adventists of California

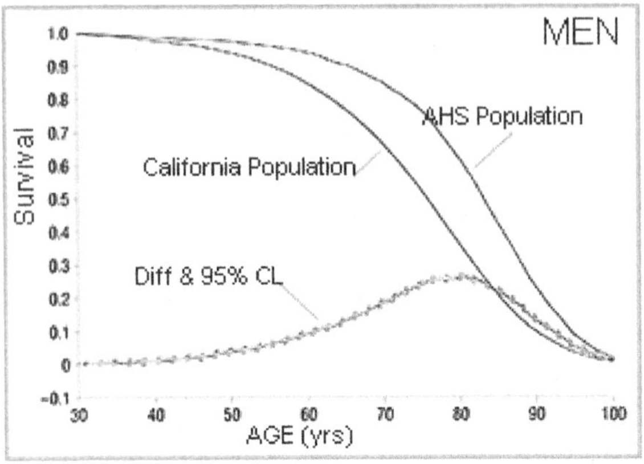

These graphs show that the proportion of surviving men in the Adventist population is higher than the proportion in other population, at every age.

REGAIN LONG-LOST PARADISE

We can make a hell of heaven or heaven of hell depending on our attitudes. We lost the paradise on earth by our misplaced priorities, unhealthy habits and the tenacity with which we cling to them. We can regain the lost paradise if we modify our habits instead of trying to rationalize our perverted habits and attitudes. Turn back to Nature and rediscover the paradise.

EAT LOW IN THE FOOD CHAIN.

Eating low in the food chain is better for health. Some questions asked in this connection are (1) why should we not eat these foods because there is no conclusive evidence that these foods have been causing heart disease and cancer (2) Whether we can eat those foods with toxic residues less than the levels fixed by government. Such questions only reveal ones needless attachment to the palate. When you know the water is deep, you don't have to jump and drown yourself to test. Secondly, trying to test the tolerance levels fixed by government and still trying to eat the same contaminated foods shows excessive attachment to the palate rather than a desire to live healthy life. The trouble with such an approach is that the standards fixed by government are not infallible due to several reasons. First, these tolerance levels were fixed by experimentation on guinea pigs other than human beings and the effect of these tolerance levels on human beings is not known exactly. Secondly, these tolerance levels of toxicity have not been stable as the scientists tended to adjust them from time to time.

Last and perhaps, the most contentious issue is the long-term effects of short-term exposures. Added to these, the whole issue of the impact of toxic chemicals on human wellness has been clouded by the subjective judgments of rival groups like environmentalists and industrialists. The difficulty of determining what constitutes health risk in the use of chemicals for various purposes "has become a real gremlin in the work of federal regulators, chemical companies, scientists and even judges". That is why the committee on Risk and Decision Making of the National Academy of Sciences has recently started work on preparing a report on methods of risk assessment of chemicals. Therefore, the questions posed above can be answered by parrying the questions by asking, "What is the harm in avoiding foods known to be toxic and hence harmful?" Loss of purity of most of our foods is the price we are paying for heavy industrialization. We have approached the stage where contamination is only a matter of degree rather than yes or no. What is to be done in circumstances so vexing as these?

Common sense requires that we should try to make the best out of a bad situation. In the complex social and economic systems now we have, it is hard to live close to Nature. Yet, we can minimize the harmful effects of contaminated foods by making rational choices. By making proper choices of foods we eat, we can minimize the harmful effects of commercial production. You stand to gain whoever may lose in the endless empirical debates. You can definitely win by choosing to eat low in the food chain. Eating low in the food chain not only minimizes the health hazards discussed above but also tends to increase the overall well being of society. You can avoid most of the aforementioned risks by eating low in the food chain. Giving up meat, which is the highest in the food chain, can help you to avoid the risks easily..

Prejudices are Self defeating

LIVING EXAMPLES OF FIT VEGETARIANS

Prejudices do not die easily. In spite of all that is said, we come across popular misconceptions and prejudices in the matter of food practices. One such prejudice is the prejudice against vegetarians. So, let us examine whether the dominance of the prevailing prejudices can be logically justified. Prejudices sound like truth when they carry enough people away. Prejudices spread faster and wider because they are labor saving devices. People who cannot or do not care to dig up facts are attracted to such laborsaving devices. One such prejudice is that vegetarians risk their health by not eating meat, but a number of scientific studies revealed a contrary story. These studies published in a number of scientific journals expose the presumption that meat guarantees more strength and stamina to those who eat it, as baseless.

Irving Fisher, well-known statistician and professor at Yale University designed a series of experiments to compare the strength and stamina of vegetarians with meat eaters and published the results of his study in the *Yale Medical Journal*[110] He divided his subjects into three groups: meat eating athletes, vegetarian athletes and sedentary vegetarians. The results of his study did not lend any credibility to the notion that meat eaters have superior physical strength and stamina. He says:

> Of the groups compared, the flesh eaters showed far less endurance than abstainers (vegetarians), even when the latter were leading a sedentary life.[111] The average score of vegetarians was found to be over double the score of meat eaters even though half the vegetarians were sedentary people, while all the meat eaters were athletes. After analyzing all factors he concluded:
> ... The difference in endurance between flesh eaters and the abstainers was due entirely to the difference in their diets.... There is strong evidence that a non-flesh diet is conducive to endurance.

In another study, Dr. J. Ioteyko of the Academie de Medicine de Paris[112] compared the endurance of vegetarians and meat eaters from all walks of life in a variety of tests, and found that vegetarians scored two to three times more in stamina over the meat eaters. Even more remarkable is the fact that vegetarians recovered from exhaustion in only one fifth of the time their counterparts took.

Physicians in Belgium compared the number of times meat-eaters and vegetarians could squeeze a grip meter. The vegetarians won scoring 68 while the meat eaters lagged with an average score of 38 only. Even in muscle recovery time vegetarians recovered much faster than the meat-eaters[113].

There are many more such studies but let us now switch over from the laboratory to the field and see some examples that proclaim vegetarianism is certainly not a bar to superb physical development and marvelous athletic performance.

As a sports enthusiast, you might have heard the names of Dave Scott, Sixto Linares, and Edwin Moses. Dave Scott of Davis, California won Hawaii's legendary Iron man triathlon four times in all and three times in a row, establishing a record of 2.4 miles of ocean swim, 112 miles cycling and then a 26.2 miles run, all in a day. He is a vegetarian. He calls the idea, that people, especially athletes, require animal protein a "ridiculous fallacy."

Linares, who became a vegetarian, against the wishes of his parents, while in high school broke world records in one-day triathlon by swimming 4.8 miles, cycling 185 miles and then running 52.4 miles. Linares experimented with lacto-ovo vegetarian diet for a while but finally gave up eggs and dairy products and remained a pure vegetarian.

No man has ever dominated track and filed events as Edwin Moses did in sports history. He was the Olympic champion of 400 meters hurdles event for eight years. *Sports Illustrated* gave him 'Sportsman of the year' award with the citation:

> *"No athlete in any sport is so respected by his peers as Moses is in Track and Field"*

Moses is a vegetarian.

Have you heard of Bill Pickering and Murray Rose, the great swimmers? Bill Pickering, the vegetarian, set a world record for swimming the English Channel. He bettered his own performance by setting a new world record at the age of 48 for swimming the Bristol Channel.

The latest addition to this galaxy of sportsmen is Prince Fielder, the baseball star who gave up meat. He says he has a lot more energy since he turned a veggie and that his body feels cleansed. The individual list goes on like that but let us see how teams fed on vegetarian diet fared in games.

You know the Japanese are as serious and crazy about the game of baseball as the Americans are. In 1981, a guy by name Tatsuro Hirooka was hired as the manager of professional baseball team called Siebu Lions, which finished, in the last place during previous season. After getting acquainted with the personal habits of the team members, he told the players that meat and other animal foods increase the susceptibility to injury and decrease their performance and sprang up surprise on them. He told them, like it or not, they were going to be on vegetarian diet. In the 1982 season, the Lions took quite a drubbing prompting a rival manager to sneer that they were only 'eating weeds'. Soon after that, the heckler had to eat his own words when the Lions beat his team in the Pacific League Championship matches. Then the Lions went on to defeat the 'Chunichi Dragons' in a series that is deemed equivalent to the 'World Series'. Lest anyone should think this was merely a fluke, the Lions came back next year and trounced the opposition winning once again the League and National championships.

This narration of these stupendous achievements of vegetarian athletes is not to say that all vegetarians can perform these feats or anyone can perform similarly by becoming a vegetarian. It is rather to emphasize that for individuals having innate biochemical composition and other physical abilities, vegetarian food is certainly not a drag to develop strength and stamina. To such individuals it is better than animal-based foods.

It is interesting to recall how Bernard Shaw retorted his detractors on his vegetarian habit. Shaw's doctors warned him that the vegetarian diet would kill him. When he grew old, he was asked why he didn't go back and show them what good it had done him. With his characteristic humor he replied, "I would, but they all passed away years ago."

There are many eminent vegetarians; the list is too long to enumerate all. It is enough to know that there are a lot of them in all occupations

We have lost the opportunity of making paradise out of this small planet when we started misusing the benefits of industrial revolution for mass production of cattle in feedlots and for their destruction in killer factories. This pernicious beef eating habit has not only promoted the mass destruction of life but also resulted in the decimation of great rain forests, drying up vast aquifers, distortion of resource-use and diseased populations. There is no wonder if the gains we made in life expectation after the industrial revolution will gradually be lost in course of time if the present trend of our food habits is not reversed. The beginning of the long decline of civilization started when the feedlots and the killer factories were established. The paradise lost can be regained only if we emancipate ourselves from our slavery to beef. That should be enough to convince anyone about the importance of food in longevity planning.

The question that now arises is: How to plan our diets? This has been one of the great public health debates going on for decades. The dieting schemes that have lured public in the past few decades are now furnished for the readers' ready reference.

How To Plan Your Diets?

A Banquet of Diets

Americans are always fond of ready-to–use-programs that tell them how to become slimmer and healthier. Among these programs, those that tell them how to stay slim and healthy while eating all they wish to eat are more popular. Using these predilections to their best advantage, numerous diet gurus emerged throughout the 20th century with their own prescriptions. Some of them became quite rich. Some are no more.

Many diet programs have been in vogue. The ten programs listed below have been well known to the public.

- Dr. Peter's diet
- Rubenstein's food for beauty
- Zen Macrobiotic Diet
- Calories don't count
- Dr. Atkins's Diet
- Stillman Diet
- Weight Watcher's Diet
- Scarsdale Diet
- High fibre Diet
- Pritikin Diet

U.S. Government Position

In 1968, a bipartisan senate select committee on Nutrition and Human needs was formed under the chairmanship of senator George McGovern. After a long study and deliberations, the committee at last released in 1977, its landmark report "Dietary goals for the United States." Its conclusion that cholesterol-rich foods may be dangerous to health and the recommendation to reduce consumption of high fat foods raised such fierce controversy that the committee had to issue an amended report accommodating the demands of food industry lobby.

Dietary Goals for the US (1977) was intended as a set of guidelines to nutrition. The committee concluded that as far as dieting is concerned, calories do count and that caloric intake must be reduced below maintenance level to lose weight. In general, McGovern committee recommended increasing consumption of complex carbohydrates and naturally occurring sugars from 28% of caloric intake in the average U.S. diet to 48%, reduction of fat from 40% to 30%, reduction of refined sugar intake from 20% to 10% and protein making up the 12% of total caloric intake. The committee's original recommendation to reduce consumption of meat was modified in the final report to decreasing consumption of animal and saturated fats. The recommendation to use 3gm of salt per day was raised later to 5gm.

Lower fat and protein consumption and greater reliance on complex carbohydrates on a national level would promote health; reduce medical expenditures and some of the energy involved in food processing industries. However, the editors of Consumer Guide suggested that high-protein legumes replace some of the carbohydrates.

Overview

A careful review of the banquet of diets reveals the following points. Of these, the FDA has discredited Zen Macrobiotics diet as grossly inadequate and dangerous. The promoter of 'calories don't count' diet was charged with fraud for selling his pills along with his book and was convicted. Dr. Atkins's diet Revolution program was condemned by many experts in strong terms like unethical, without scientific merit and as a plan for self-aggrandizement. Stillman diet is considered anachronistic as high-protein-diet should be regarded as high-calorie-diet also and it is impossible to lose weight on such a diet. Further, there is no room for other nutrients when protein is increased to 90%. Scarsdale diet was found to be only a temporary solution to a chronic problem because one has to follow that regime under close medical supervision. Once the monitoring is stopped the dieter tends to become a glutton, as the diet did not prescribe the sizes of various portions of diet. The other diets had a mixed reception while many people viewed Pritikin program scored high overall though some people called it vegetarianism in disguise. Judged from the success of Pritikin program we consider that remark, as a compliment to the age-old practice of vegetarianism.

Judged from the history of these dietary experiments, the "Dietary Goals for the U.S. (1977)" issued by McGovern Committee appear reasonable. Food in these days is not allowed to be the sole concern of people who eat it, but it has become money making business. This is typical case of producers creating the demand. Profit maximization goal does not brook interference by other considerations like ethics and welfare of people. So, by the very nature of it's reasonableness, the report fell foul with the food industry, which has a vested interest in creating demand for meat and high fat foods and kicked of a storm of protests from the food industry. Yielding to the pressure, the committee had to issue some modification to its original report making some concessions to the protestors.

Unfortunately, the food industry continues its battle against any sensible dietary guidelines with various kinds of publicity including occasionally sponsored articles in journals, which appear to be scientific but cleverly take on their adversaries with sarcasm and ridicule to protect their interests.

It is for the intelligent reader to decide now whether vegetarianism is a bane deserving condemnation or a boon given to humanity for surviving longer on this crowded planet.

Composition of right food of longevity seekers, involves following the action plan outlined earlier with the additional requirement of reducing the total caloric intake by 30%. Let us now summarize the required steps for composition of ideal diets.

Food Groups

This choice involves mixing foods providing protein, carbohydrates and fat. The question then arises what percent of the total energy should come from protein group, what percent from carbohydrate group, and how much from fat group. This is an unsettled question. Various people suggested various programs. A number of canned diet programs that took their rounds in U.S. were discussed earlier under the heading 'A banquet of diets.' There were wide disparities in their recommendations and the aim of some of them was self-aggrandizement of the sponsors. In such a situation, the public naturally expects the government to advise them on this issue. As we have seen in the preceding pages, the government constitutes expert committees now and then to study the issue and advise the people but the food lobby brings a lot of pressure to modify them in their favor. It is a regular tussle between two opposite interests. Therefore, regardless of all checks and balances in the governing system, we still face a caveat emptor (Let the buyer beware) situation on the food front. The recommendations of the senate select committee were by and large fair with the caveat that they were meant for improving the current state of sub optimal diets eaten by large groups of Americans and not sufficient for increasing life expectation. The major part of the diet plans presented in the banquet of diets involved shuffling the percentages among proteins, carbohydrates and fat followed by some subsidiary prescriptions like avoidance of animal products, exercising and fasting. It is imperative therefore that we should first delineate an ideal proportion of the basic food groups to be maintained in the total calorific intake.

IDEAL FOOD GROUP PROPORTION

From a review of the various points of view and earlier experiences, we consider the following combinations, as ideal mix to be helpful in longevity diets.

- Up to age 40: 50% Complex carbohydrates, 25% protein component and 25% fat content
- From 41 to 60: 60% Complex carbohydrates, 20% proteins and 20% fat.
- After age 60: 65% Complex Carbohydrates, 20% proteins and 15%fat.
- The important point is you must keep reducing fat content and replenish it with complex carbohydrates.

Food Pyramids

The Federal government has been involved in developing nutrition guidelines for the American public since the middle of nineteenth century, when United States Department of Agriculture was established. These guidelines had a dual purpose of preventing nutritional deficiencies and of promoting consumption of U.S agricultural products. In the past several decades, the emphasis has shifted from preventing diseases due to nutritional deficiency to prevention of diseases due to over-nutrition and nutritional imbalances. It has been quite difficult to develop these guidelines with total objectivity due to the lobbying of interest groups. Nevertheless, federal agencies have been issuing since 1970's, nutritional guidelines dealing with prevention of chronic diseases. Two of these better-known guidelines are the Recommended Dietary Allowances (RDA's) and Food Guide Pyramid. Food guides provide conceptual framework for diet planning. They translate recommendations on nutrient intakes to types and amounts of foods, which together are supposed to provide a nutritionally satisfactory diet. One such pyramid developed by the U S D A is illustrated below

Fig.9-3:U S D A Food Pyramid (Adapted from USDA's Web-Site)

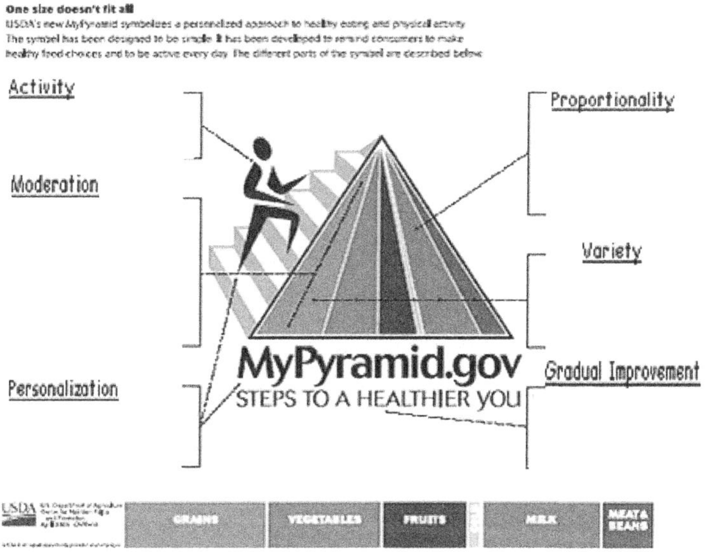

This pyramid is an improvement over its earlier counterparts. We feel gratified to see that it has recognized the point of view we have been advocating all along and repeatedly emphasized in this book. It has shifted the emphasis from meat and excessive protein to plant foods and moderate protein

Anatomy of 'My Pyramid'

One size doesn't fit all
USDA's new MyPyramid symbolizes a personalized approach to healthy eating and physical activity. The symbol has been designed to be simple. It has been developed to remind consumers to make healthy food choices and to be active every day. The different parts of the symbol are described below.

Activity
Activity is represented by the steps and the person climbing them, as a reminder of the importance of daily physical activity.

Moderation
Moderation is represented by the narrowing of each food group from bottom to top. The wider base stands for foods with little or no solid fats or added sugars. These should be selected more often. The narrower top area stands for foods containing more added sugars and solid fats. The more active you are, the more of these foods can fit into your diet.

Personalization
Personalization is shown by the person on the steps, the slogan, and the URL. Find the kinds of amounts of food to eat each day at MyPyramid.gov

Proportionality
Proportionality is shown by the different widths of the food group bands. The widths suggest how much food a person should choose from each group. The widths are just a general guide, not exact proportions. Check the Web site for how much is right for you.

Variety

Variety is symbolized by the six color bands representing the five food groups of the Pyramid and oils. This illustrates that foods from all groups are needed each day for good health.

Gradual Improvement
Gradual improvement is encouraged by the slogan. It suggests that individuals can benefit from taking small steps to improve their diet and lifestyle each day.

ONE SIZE DOESN'T FIT ALL

This shift can be seen by the relative bandwidths assigned to various food-groups in the pyramid. In addition, food requirements are not the same for everyone. Everyone has to continuously modulate one's requirements. One size doesn't fit all.

The number of servings recommended by the USDA is based on the RDA. We showed earlier that these recommendations were biased on the surplus side.

Therefore, the reader should take into consideration his/her own personal requirements based on weight, height and individual calorific needs explained earlier. Further, this pyramid is not suitable for vegetarians as their nutritional needs are served by a different set of food-groups. A different pyramid is required for the guidance of vegetarians and the would-be-vegetarians. In the absence of a specifically designed food pyramid, some aspiring vegetarians are tempted to remove the meat foods from the USDA pyramid and adopt the rest but that does not afford nutritionally adequate diet. Recognizing this need for separate guidelines for vegetarians, the International Congress on Vegetarian Nutrition, developed a vegetarian food guide comprising of a graphic and an associated explanatory document. The graphic is shown in Fig.(9-4) with some modifications..

A healthy vegetariandiet is not one, which merely excludes animal foods, but it is a diet comprising of a variety and abundance of plant foods. Plant foods include: grains, legumes, vegetables, fruits, seeds, nuts, vegetable oils and spices. Daily consumption in sufficient quantities of variety of foods from these groups can satisfy the nutrient needs of humans except vitamin B-12 and possibly, vitamin-D.

One guideline to follow by vegetarians is to eat, as far as possible, unrefined or minimally processed foods as refining deprives them of some nutrients and bioactive components. Though vegetables, fruits, nuts and legumes can be consumed with minimal refinement, it is not the case with grains. They are refined due to storage and commercial considerations. To compensate the nutrients lost in processing, vegetarians have to resort to food supplements. In addition, vegetarians should practice making up menus with complimentary combinations of foods. Some of the suggested combinations are provided below for the benefit of the reader. A vegetarian food Pyramid is provided here keeping in view the special needs of the vegetarians. The following combinations of foods are helpful in composition of menus suggested by the vegetarian food pyramid.

Some Suggested Complimentary Combinations

(1) Wheat + Rice + Soy beans	(2) Rice + Beans+ peanuts
(3) Rice + Peanuts	(4) Wheat + legumes
(5) Wheat+ Cheese	(6) Wheat + Milk
(7) Wheat + Rice + Soy beans	(8) Rice + Sesame seeds
(9) Wheat + Cheese	(10) Rice + Soybeans
(11) Soybeans+ peanuts +Sesame	

Fig.9-4: Vegetarian Food Pyramid

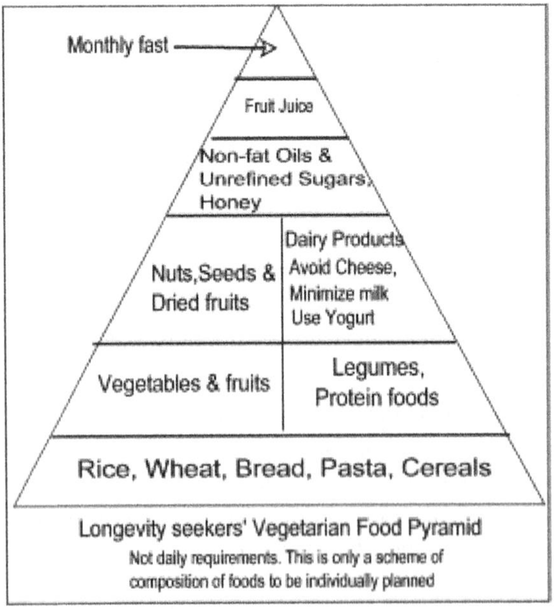

Observe the differences of this Vegetarian Pyramid with USDA Pyramid.Meat group (comprising of meats, poultry, fish and eggs) is replaced with protein supplements group comprising of soy products, nuts and seeds. Complimentary combinations fortify the basic grains group. Fats/Oils/Sweets group (rich in saturated fatty acids, trans-fatty acids and refined sugars) of USDA is eliminated and simple oils group is added. Monthly fast serves to cleanse the system and rest it for a day.

If you learn to manage fat, doing the rest of the chores suggested above is not difficult. There are abundant nutritious grains and cereal foods from which you can choose. There is the cornucopia of beautiful, delectable vegetable foods and luscious health giving fruits. It is your turn to choose and eat well. While choosing, you should first pay attention to the kind of fat that is mixed in your menu.

How to Manage Fat

As observed earlier a number of times it is a curious phenomenon of nature that many things which have acquired bad name for injuring life are also necessary for sustenance of life. Cholesterol, which is hated as the culprit, which causes heart disease, is actually necessary for the body in small quantities. Free radicals, which are now blamed for damaging the cells causing death, are also necessary for the body in limited quantities. Similarly, fat is an essential component of body, giving it many benefits but when it exceeds limits it can do havoc. It is the nutrient whose ups and downs are visible on the surface and it is recognized as an all-rounder, which causes many diseases. Therefore, it is identified as a nutrient that should be controlled.

None, even committed vegetarians, can avoid eating some fat but by making careful choice they can reduce the bad effects of fat largely. There are four or five varieties of fats.

Transfats:

Transfats are generally found in packaged foods. They are artificially processed poly-unsaturated. They are very bad to the heart. Try to discard them whenever possible.

Saturated fats

Saturated fats are notorious for their role in promoting heart disease. They are found mainly in animal products like meat, lard and cheese and butter. Avoid them.

Poly unsaturated fats

Linolenic fats (also called Omega-6): This is an essential fatty acid. The body requires it in small amounts. It is also bad because tumor cells thrive on it- found in corn, soy and some vegetable.

Alpha-linolenic fats (Omega -3): This is considered the best of all fats. It is found in flax seeds and cold-water fish. It may reduce cancer and heart disease.

Monounsaturated fats

These are considered good fats. They don't do harm to the heart like saturated fats. They don't promote cancer. They are found in avocados, olives and peanuts.

Now, new slogans like *"Eat all you can and live healthy"* are emerging. How is that? Is it an invitation to rapacious eating? Wait, it is quite the opposite. If you observe temperance in eating, choose your foods carefully, and follow other prescribed regimen, yes, the odds are in your favor to beat the Big Two: Cancer and heart disease.

Food Supplements

After regulating the quantity of food intake, you have to take care to see that the quality of food you ingest is not only nutritious in general but also it contains some special ingredients that protect and strengthen some vital functions in your body which fight out disruptive forces. Where such vital ingredients are not available (generally they are not, in modern foods) in foods we eat, we have to resort to food supplements. Food supplements are necessary because the food we eat is either unbalanced or deficient in some nutrients or we cannot absorb all the nutrients required from the food we eat. To compensate for those deficiencies we have to supplement our natural foods. Some of the **food supplements** used are:

1. Vitamins

Vitamins are our good old friends who have stayed with us through thick and thin. As food supplements they are well known to everyone. Deficiency of B-complex group of vitamins is a universal trait found in South Indian people whose staple food is rice. They eat polished rice, which lacks vitamin-B. Whatever may be the objective of polishing, the result is people are deprived of the life sustaining B-vitamins from their staple food. Therefore, health conscious people are left with no option but to use synthetic vitamins to make up for the losses they suffer in eating modern foods. Many people use B-complex vitamins to fight the deficiency disease beriberi or to fight general debility. C-vitamin is used to fight scurvy or as a general precaution. The well-known scientist Linus Pauling spent a considerable part of his career popularizing and advocating the use of vitamin-C to tone up general health. Vitamin-E is believed to have anti aging properties. Vitamin-E and mineral selenium are supposed to have the special property of reducing the oxidative stress and promoting longevity. There is no doubt therefore, that food supplements are necessary for protection of health in an imperfect world where not all people can ingest all the required ingredients through foods they eat.

However, overuse of vitamins and other food supplements can cause toxicity and defeat their very purpose. Vitamins and food supplements are the most misused articles on the shelves of drug stores.

Therefore, they should be used with discretion and with the help of your physician. There are a few more special products known to help preserve our health.

2. DHEA (De-Hydro-Epi-Androsterone)[114]

DHEA is a hormone precursor. Hormone precursor means this substance is naturally converted into necessary hormones in the body. It was found that DHEA is the most prominent hormone in the blood. Adrenal glands make DHEA from cholesterol in the body. (See, cholesterol is not the real culprit, it is our eating habits) Levels of this hormone increase through adolescence and peak at age 25. Thereafter it steadily declines. By the time we die, it comes down to 5% of what it was at 25. It produces two major classes of hormones namely androgens and estrogens. Androgens are male sex hormones while estrogens are female sex hormones.

DHEA is reported to inhibit fat formation, slow down free radical formation, slow proliferation of pre-cancerous tissue and reduce synthesis of mutagenic cells. The various benefits conferred by DHEA are said to be; increase of energy and libido, restoration of memory, rejuvenation of immune system, taming stress, reduction of body fat and prevention of heart disease, amelioration of the after effects of menopause and smoothening of wrinkles-an impressive list.

3.SOD (Super Oxide Dismutase)[115]

Super oxide is a free radical that attacks body tissue and can lead to disease and death. Researchers found that inadequate levels of SOD result in accelerated aging. Human body tries to protect itself from the super oxide molecule by producing SOD, an enzyme that attaches to super oxide and renders it harmless. If enough SOD is produced, it keeps the super oxide level low and eases the aging process. As people age, SOD production in their bodies, diminishes. Further, due to differences in lifestyles, genes, diets and age, the levels of this anti-aging substance vary from person to person. If the body is not producing, enough of it can be taken as an oral supplement to rid the body of free radicals. SOD cannot be well absorbed in oral form unless certain minerals like; zinc, copper and manganese are also present. That is why it is important to see, while purchasing SOD formula, that it contains chelated minerals including zinc, copper and manganese are present.

4.Melatonin[116]

Endocrine organs secrete hormones into the blood stream, which are then carried throughout the body. Their effects are varied and far ranging. Melatonin is a hormone secreted by Pineal gland, a tiny cone shaped body, which is situated deep in the middle of the brain, near the optic nerve.

In recent years, Melatonin received a lot of attention for its ability to prevent diseases, to rejuvenate the body and to promote sleep. Melatonin is produced in darkness following what is known as 'circadian rhythm.' Once light is reflected on the retina, pineal body ceases to release this hormone. Thus, blood levels of this hormone are low during day and high during night. As people grow older and older, their bodies produce less and less of Melatonin. By age 85, the body produces only 5 percent of what it made at 25. This leads to the presumption that melatonin bestows anti-aging effect on the body. The better cousin of sleep in producing repose, restfulness and rejuvenation of body is meditation. Meditation is now universally acknowledged an elixir that rejuvenates body and mind. Melatonin enthusiasts claim that this rejuvenation comes through the medium of this hormone, which is replenished by meditation. You may try to supplement it with food supplements on the advice of your physician. Better still if you cultivate meditation to aid the body to produce it.

5. RESVERATROL:

Resveratrol, a substance found in red wine has acquired some reputation of its ability to delay aging. In experiments conducted at Harvard Medical School and the National Institute on Aging, researchers found that mice fed on a daily dose of resveratrol, were perkier than the their counterparts without the substance. Preliminary findings indicated that they might have achieved a life extension of as much as 20%. Scientists who used the substance on fruit flies and yeast reported similar results.

However, the ratio of this substance available in red wine is so small that in order to get as much substance as the mice's daily dose, all from red wine alone, one has to drink 300 glasses of wine a day! One way of avoiding this difficulty is to manufacture resveratrol in concentrations required for therapeutic human use. However, one should not be hasty in using it and wait until further research confirms these tentative results and assures us of no adverse effects.

A suggestion worth considering in this respect is to consider whether the traces of resveratrol available in red wine and other food supplements may really be sufficient to give the desired result. If the Theory of Hormesis (the theory which says that toxic substances and conditions which normally injure human body would stimulate protective responses if taken in too tiny doses.) is valid, the traces of the substance available in red wine may be adequate and regulated use of red wine in **medicinal quantities** may be useful. Researchers should seriously consider this suggestion and investigate the role of hormesis, in this context.

We will now proceed to examine in the light of the current knowledge, the extent to which lifespan can be increased.

CHAPTER TEN

Prospects for Life Extension

> Hope springs eternal in human breast:
> Man never is, but always to be blest

---Alexander Pope (An Essay on Man, Epistle1)

Human beings are never satisfied with what they are already blest. They always want more. They are not satisfied with a hundred years of life. They want more of it. Is that possible? Scientists and demographers are now fretting and fuming about the possibility of extending the human life span.

The Controversy

We mentioned earlier (Ch.3) about the controversy raging between two groups of demographers about the prospects of life expectation. Analyzing the historical data of deaths in the previous century one optimistic group argues that life expectancies have been steadily rising as a consequence of the decline in mortality rates. Extrapolating this trend into the future, they argue that there is no upper limit to life expectancies. The other group counters this argument saying that the trend of rising life expectancies cannot be sustained forever. They admit that the life expectancies may rise further for some more time but argue that they will do so with deceleration. That means they will eventually level off.

Even as gerontology came out of the backwoods of magic and spells and appeared to be moving into the mainstream of biology, differences developed among the biologists dividing them into two opposing camps, those who stick to the age-old dictum that "Man is mortal", and those who believe that aging and death can be controlled by human intervention.

The latter group characterizes the former as people who miss opportunities due to lack of vision while the former accuses the latter of misleading the public by raising false hopes and by trying to promote expensive longevity treatments whose worth has not been proved. Controversy in science is not new and sometimes it is useful as far as it serves to bring out truth, but in this case, nothing of that sort seems to be happening. The debate has often taken bitter turns and no efforts have been spared to support one's own theory and to denigrate the other. Hyperboles, jugglery of numbers have been freely used to support their points of view. A small sampling of these conflicting opinions will give the reader the flavor of this controversy. A member of the American Medical Association says:

> "Every blade of grass, every bacterium, every person, we all want to live as long and as well as we can."... Now we are getting the capability of living as long as we desire, not just hundreds of years but thousands."

It is easy to make such wild claims because people who make them know that they will not be there after some time to stand scrutiny of their claims and that knowledge makes them comfortable to make any unreasonable claim.

Michael Rose an evolutionary biologist predicted years ago that,

> "... Eventually we will double human life span-and when we do, we will be ashamed that we did not do it much sooner."

Wilmoth, a demographer and his co-researchers claim:

> "Our analysis refutes the common assertion that human life span is fixed and unchanging over time. Although maximum has increased much more slowly than the average, the entire distribution of ages at death has been shifting upward for more than a century in Sweden and presumably in other countries as well. Reductions in death rates at older ages has accelerated in recent decades seem likely to continue and may gradually extend the limits of achieved longevity even further."

Jim Oeppen and James Vaupel, demographers say that their

> "...Mortality research has exposed the empirical misconceptions and specious theories that underlie the pernicious belief that expectation of life cannot rise much further."(Science, vol.296, May 10. 2002)

Another protagonist of limitless life theory is James R. Carey, an entomologist, who uses bio-demographic models in his research. The ideas around which his theory is woven are: that (1) there exist linkages between reproduction and longevity. (2) Mortality response of males and females is context specific, (3) there are no species specific lifespan limits and (4) that studying the causes of death provides little or no insight into the nature of aging.

Based on the recent deceleration in death rates in older people and perhaps on his own research, Carey believes that the traditional Gompertz model of mortality should be overturned. Gompertz model states that the rate of increases of mortality with age follows the exponential law. Carey suggests that longevity extension is "self reinforcing process" which will lead to the decommissioning natural selection so that individuals can make choices conducive to engineering a long life.

OTHER SIDE:

On the other side are ranged, for example, people like Leonard Hayflick, professor of anatomy and noted researcher in cell biology, Bruce A. Carnes and S. J. Olshansky, demographers and a few others. Hayflick who established the current cellular paradigm, in which cells can divide only a finite number of times, is cautious and refrains from making sweeping assertions but has a dismal view of extending longevity. Olshansky and Carnes hold the view that though gains in life expectation may continue for some time, those future gains will diminish due to entropy in life table.

Demographic Scenario in perspective

PREDICTION OF LIFE EXPECTATIONS

The answer to this enigma lies, perhaps, in the interpretation of the experimental findings with a proper perspective view. These findings are extrapolated from the past using a mathematical technique known as regression analysis on the assumption that the future will behave like the past. The gains made in the previous century were primarily due to steep decline in the infant mortality rates and eradication of many infectious diseases. The optimistic view is that the remaining diseases can also be eliminated. However, what guarantee is there that future will behave like the past? Then the gains made have to be viewed as short-term gains. The analysts are not able to show any other specific endogenous factors that will sustain this trend forever. If the range of data is changed, different results can be obtained and different inferences have to be drawn from them. Some argue that it is unrealistic to conclude that the trend of declining death rates will continue forever until they become so insignificant as to lead the life expectation at birth, to infinity.

Life Expectation of a cohort, at a point of time is the average number of years an individual in that cohort is expected to live hence. It is not the same as maximum life span. Life-Expectation, by definition, is less than the maximum possible life span and is generally the same for all in the group from which the individual comes. Maximum life span of a species is estimated by observing the age at which the oldest individual in the species dies. Life expectations are computed from life tables containing age-specific death rates. Most of the statistical predictions are made using the technique of extrapolation of the present trend into the future. Underlying this is the assumption that the state of affairs that gave rise to the data used to characterize the present trend will prevail. In the view of optimists, predictions based on such an assumption and the laws of probability have proved to be quite trustworthy in the past.

Most of the organizations connected with public policy decisions involving age, use this method. Insurance companies and huge organizations like Social Security Administration that are committed to make entitlement payments based on age, relied on this method and found it useful. Insurance companies who placed their bets on this method, maintained their financial integrity and have remained successful for a long time. Therefore, they ask, is it not reasonable to consider the prospects of increases in life expectancy of human populations with the same degree of confidence? But the trouble with this argument is, that life expectancy calculations depend entirely on the type of data we use and if the data contain a few bunches of more than average deaths, the life expectation will automatically decline.

Moreover, it is misleading to drag in the argument of success of life insurance companies in this context because what the insurance companies bet upon, is the certainty of death and not the certainty of continued life. They invoke the laws of probability to compute life expectation based on death rates in order to earn maximum number of premiums and pay minimum number of annuity payments to policyholders. If people do not perceive risk, they don't insure at all and insurance companies go out of business. It is the certainty of death that induces people to insure and the insurance companies to stay in business.

Life expectation increases or decreases depending on the

fluctuations of age-specific mortality rates, which in turn depend upon many factors, which include external factors affecting life, like sanitation, environment, personal hygiene, and the availability of medical services to the population under study. However, this change in life expectation may not be in proportion to the change in these factors. This phenomenon was empirically demonstrated by demographers S. J. Olshansky etc. al in 1990[117] They showed that as life expectancy at birth rises, it becomes less sensitive to death rates. That means the rise in life expectancies does not keep pace with declining death rates. In other words, it means that for every additional unit of bettered life conditions, the corresponding rise in life expectancy will be smaller than the previous rise. This phenomenon is known as the law of diminishing marginal returns in economics. It is analogous to entropy in physics, and demographers call it "entropy in life table."

To understand the scope of entropy in life table, let us consider these facts. When life expectancy at birth is 50, raising it by one more year requires a 4.1% reduction in mortality rate at all ages. Raising life expectancy from 80 to 81 requires a 9.1 % reduction in total mortality at every age. Thus, one more year at every higher age requires a lot more improvements in life promoting conditions than earlier.

Life expectancies at birth rose rapidly in developed countries, throughout 20th century due to dramatic decline in death rates in all age groups. Life expectancy of females in USA increased from 48.9 years in 1900 to 79.0 in 1995, a gain of 30 years. If life expectancy were calculated again applying the same declines to the 1995 mortality figures, the resulting gain would be 10.1 years and not 30 as before. A second repetition of this procedure would yield a gain of 6.1 years only.[118] In fact, we cannot expect such large-scale reductions in death rates in this century because the kind of diseases we are facing now are of different nature. This means that much larger reductions in mortality rates at higher ages are required for life expectancy to rise significantly. For example, the 1995 death rates would have to decline by more than 50%, in order to push life expectancy in USA to 85 years.

Table (10-1) illustrates the pattern of mortality reductions in selected countries and the time required to achieve life expectancies of 85 and 100 based on those mortality patterns.

Table 10- 1: Changes in Death Rates & Projected times by which given life expectancies are attained

Country	Annual average percentage change in death rate at ages						Projected year for reaching life expectation of	
	0-19	**20-39**	**40-59**	**60-79**	**80-99**	**0-99**	**85**	**100**
France (F)	2.7	-0.8	-1.3	-2.2	-1.6	-1.7	2014	2106
France (M)	-3.2	0.0	-1.6	-2.0	-1.3	-1.6	2052	2138
Japan (F)	-0.5	-1.4	-1.5	-2.2	-1.9	-1.5	2010	2118
Japan (M)	-1.5	-1.2	-1.8	-0.9	-0.7	-1.2	2060	2182
USA (F)	-1.1	+0.7	-0.7	-0.6	-0.3	-0.4	2125	2485
USA (M)	-1.1	+1.0	-0.5	-1.4	-0.1	-0.4	2239	2577

Source: S.J. Olshansky, "Prospects for human longevity", Science Compass vol.291, p1492, Feb 2001, Copyright (2001) by AAA, Reproduced with permission.

This table shows that Japanese females will be the first to achieve a life expectancy of 85 as early as next decade. American males will not be able to do it before twenty-three decades. Americans will not be able to achieve life expectancy of 100 years before fifty-seven decades, while the French and the Japanese are expected to achieve it in about 100 to 200 years from now. Look at the reduction in death rates of US males and females. They are the least among the countries given in the table. Death rates are increasing in the age group 20-39 because of our sinister policy of constantly waging unnecessary wars and sacrificing our boys.

Monumental changes in health and hygiene took place in the 20th century giving a quantum leap to life expectancy. No such changes are in the offing to expect a repeat performance. Neither telomere therapy nor gene therapy, nor any other genetic engineering procedure is in a position to promise another quantum leap. The U.S. Social security administration estimated that the life expectancy at birth would rise to 79.3 for males and 83.9 for females by the year 2070. Considering this prediction too pessimistic, an advisory group recommended that these figures be raised by another 3.7 years. Such a rise requires doubling of the decline in mortality rates already projected by the Social Security administration. This shows how inelastic are life expectancies to changes in mortality rates.

As repeatedly observed in this book, it is a strange irony of life that biochemical mechanisms themselves, required to sustain and operate the machinery of life, sow the seeds of its destruction. Biologists should recognize this phenomenon, before making tall claims of achieving immortality or adding hundreds of years to human life every time they get some impressive result in their experiments with a worm or a small creature.

What is the significance of this to gerontologists? It is bad news to them because the gains in life expectancy will bc made at snail's pace, no matter what the improvements are, in living conditions. Reputed demographers computed that life expectancy at birth will not exceed 85 years in the near future, unless some extraordinary inventions take place and drastically reduce the death rates of large groups of people. The question is whether they will ever take place.

Those that contend that there are no limits to human life span base their assertions on the recently observed trend of deceleration of death rates and consequent increase in life expectancies in some advanced countries. They use this fact to extrapolate the trend into future using mathematical models and claim that there is no upper limit to human lifespan. However, even lay people know that a mathematical trend depends on actual biological conditions and does not determine them. We can never say a trend is not reversible. A trend can flatten or reverse. There are strong grounds to argue that there is no guarantee of the up-trend in life expectancies rising further, far into the future to enable us to predict that human longevity will be extended to thousands or even hundreds of years.

An extrapolation model is not an abstract model. It is based on ground realities and tries to predict the future in the belief that the ground realities will continue their movement in the same direction but that belief may be belied if our model does not dynamically keep pace with the changing realities underlying the model. The dramatic reductions in infant mortality and control of infectious and parasitic diseases in the 20th century contributed a lot to the recent gains in life expectancies. That is a very brief special period in the history of humanity. Unless science succeeds in repeating such momentous events like large-scale reductions in infant mortality and decimation of infectious diseases, we cannot expect the recent trend to continue. When we can find at least twenty such trends in the secular graph, we may revise our opinion, not until then. We should not lose sight of the fact that mortality is an independent biological phenomenon and not a property of any mathematical model.

One may also notice that all the factors that led to the recent extension of life expectancies, viz. improvement in public health measures, decimation of infectious diseases and controlling infant mortality were exogenous. The momentum of these beneficial exogenous factors is waning. Not much could be done to control the endogenous factors like diseases and degradation that contribute to deaths.

NEGATIVE FACTORS

Gompertz model postulates that human mortality increases exponentially with age i.e., the trend builds on itself. Some scientists have started repudiating this model based on recently observed deceleration of mortality in older cohorts in some regions. Those scientists who have gone to the extent of interpreting the observed deceleration in death rates as failure of the Gompertz model ignore two factors. One is that Gompertz himself said that his model does not apply to isolated old cohorts. Second, Gompertz model, when applied to heterogeneous populations with sub groups, they may exhibit deceleration of mortality. When it is applied to homogeneous sub groups separately, the distributions retain the exponential nature. This means that when the tempo of living longer continues for some time the shorter living sub groups gradually vanish by merging into the few longer living groups, the exponential law comes back as populations gain some homogeneity. Hence, it is not easy to discredit Gompertz model unless monumental life-saving technologies are developed.

Apart from the aforementioned public health improvements that led to the recent increase in life expectancies, there is a metaphysical proposition worth considering as it is supported by some historical findings. Nature operates an equilibrating mechanism that governs the sizes of populations in different times and places. Well known German philosopher Schopenhauer, in his book —'*Die Welt als Wille und Vorstellung*' speaking about palingenesis says:

> —It must not be neglected that even empirical grounds support palingenesis of this kind. As a matter of fact, here does exist a connection between the birth of the newly appearing beings and the death of those that are worn out. It shows itself in the great fruitfulness of human race, which appears as a consequence of devastating diseases. When in the fourteenth century the Black Death had for the most part depopulated the old world, a quite abnormal fruitfulness appeared among the human race, and twin births were very frequent.

Casper in his '*Ueber die wahrcheinliche Lebensdauer des Menschen' (18*35), confirms the principle that the number of births in a given population has the most decided influence upon the length of life and mortality in it. This always keeps pace with mortality; so that the deaths and the births increase and decrease in like proportion always and everywhere. He places this idea beyond doubt by an accumulation of evidence collected from many lands and their various provinces. Recent empirical evidence gathered by the UNO reveals that it is not easy to refute that proposition.

Fig.10-1a: Life expectancy and Fertility rate

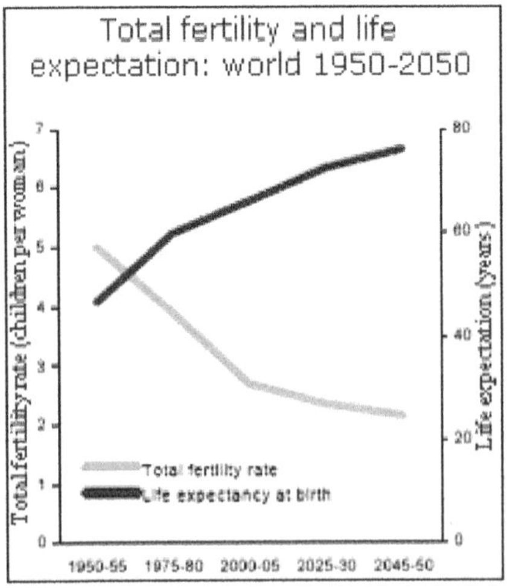

The UN report on "Population ageing, 1950-2050", reveals that fertility rate is the most important determinant of life expectancy. The adjoining graphs show that life expectancy bears an inverse relation with fertility rate. (Fig.10-1a). Fertility rates in several countries have indeed been declining as shown in (Fig.10-1b). Hence, it is premature and naïve to allow us to be lulled into the belief that the recent rise in life expectancy is an indication of the success of science in vanquishing death, as that phenomenon is not a secular trend.

A perusal of the data used by Jim Oeppen and J.W. Vaupel in their regression analysis reconfirms the foregoing observations as the death rates used by them come from the countries in which there had been significant declines in birth rates. This decline in birth rates is in conformity with the worldwide projections (Fig. 10-1b) given in the report mentioned above.

Fig. (10-1b): Projections of Fertility rates:

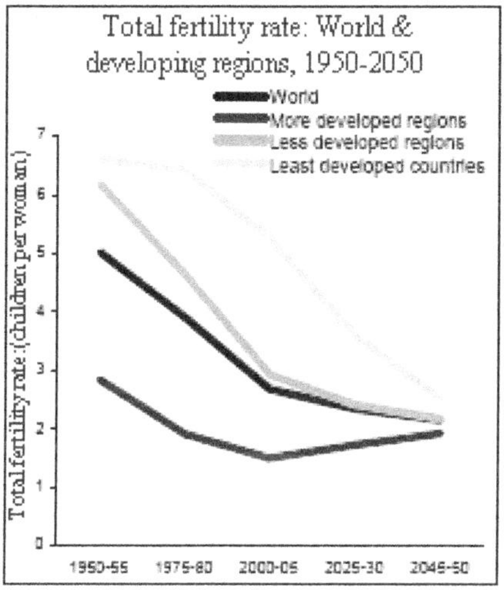

Although we may notice some further creeping up of life expectancies due to anticipated benefits of some biomedical technologies in the pipeline, unless and until we have the technologies in place for **slowing down** the **mortality rate,** we cannot expect life expectancies to make quantum leaps from present levels. No such technologies are in sight at present. All that we can expect therefore, is that life expectancies may settle down around 100 with marginal adjustments.

Even investigators who use holistic ideas like "Vitality" or "Reserve capacity" cannot avoid the dread of that vitality or reserve capacity declining. Vitality is understood as overall ability of body to function efficiently. This efficiency increases gradually with the bodily growth and then declines with further aging. This efficiency is the resultant of a number of measurable characteristics like basal metabolism, maximal breathing capacity, nerve conduction velocity, standard renal plasma flow and cardiac index etc. These attain maximum value at some point of time in young age and then they begin to decline. Studies by Falzone, Shock and

Norris found that the reserve capacity declines to about 80% of maximum by age twenty and to 20% by age eighty.[119] This is reinforced by the recent researches of Bruce Carnes and S. Jay Olshansky who found pathology burdens increasing progressively with age, providing incontrovertible evidence that age takes its toll on organisms.[120] They argue, justifiably, that if there were no limits to maximum life span there should not be pathological degradation at any age but the available evidence confirms the existence of such degradation.

Although many biological processes are handed down to us as a genetic legacy, those necessary for unfading youth and extreme longevity were not among them. Death and aging continue to be inseparable partners of birth and growth like the sides of a coin.

When we probe into the biological processes that underlie the extrapolation models of demographers, we have to recognize the dichotomy of the causes of death. Death is inevitable because, not only it is an evolutionary necessity but also because nobody has absolute control over external causes of death like natural disasters, starvation, homicides, infectious diseases, accidents and suicides. Even if all the endogenous causes like genetic defects, usual diseases and degenerative diseases were eliminated; can we say that death can be avoided from exogenous causes?

We are not anywhere near eliminating these diseases, as our bodies are not designed to last forever. Unless substantial strides are made in medical technology to enable us to save the lives of people on a scale that happened in the context of infant mortality, maternal mortality and infectious diseases in the 20th century, we cannot expect the demographic projection models to hold good. Hence, the probability of extending significant gains into the future is very low. The claim of longevity extension being a self-reinforcing process is based on the belief in the prognostic efficacy of extrapolation models. The trend in extrapolation models can be extended indefinitely if projected little by little, but that is unrealistic when not justified by corresponding biological improvements. A trend lasts as long as there is momentum in it. When it is exhausted, it falls either flat or reverses unless it is given further momentum by external forces.

Apart from these theoretical considerations, we have epidemiological evidence accumulated by the Adventist Health Study of

California that reinforces the idea of limit to human life. This study analyzed several risk factors involved in routine living and their effect on longevity of the people under study. It revealed that if we modify our life style in such a way as to avoid all known risk factors, life span could be extended utmost by ten years. The length of life that can be gained by avoiding differing risk factors is displayed in the bar diagrams of Figures (10-2a) and (10-2b).

Fig. (10-2a) Risk factors and corresponding ages at death of California Adventist Women

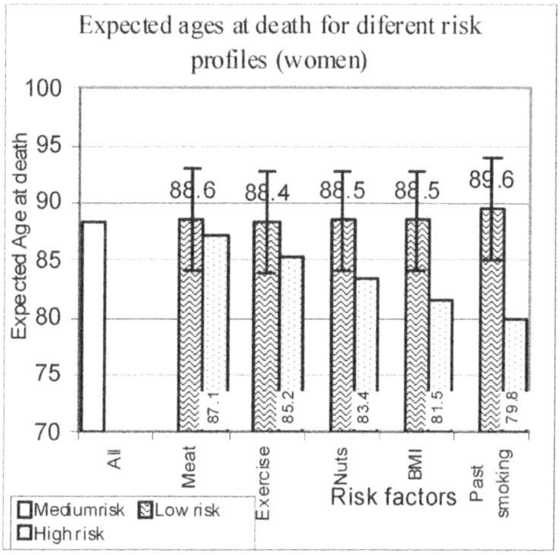

Each couple of bars denotes the ages at death of the subjects corresponding to low and high values of a risk factor. Successive bars toward the right indicate cumulative effects of all the factors to the left. Thus, the rightmost couple of bars indicate the cumulative effect of risks due to meat eating, lack of exercise, lack of nuts in diet and unfavorable BMI.

Fig. (10-2b): Risk factors and corresponding ages at death of California Adventist Men

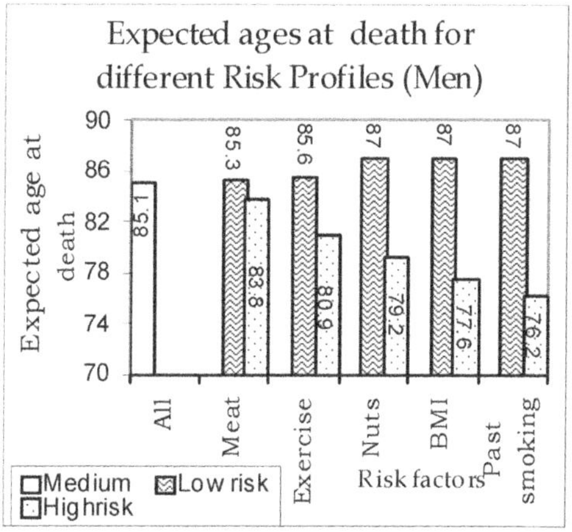

Source: Fraser GE and Shavlik, "Ten years of life: Is it a matter of choice," *Archives of Internal Medicine*. Copyright (2001), American Medical Association, Reproduced with permission.

In effect, the study indicates that if all the indicated risk factors are avoided, males can gain 10.8 years and females 9.8 years of extra life at the most.

Potential decline in Life Expectancy

Encouraged by the trends of recent past, predictions are made that life expectancy would continue to rise and in course of time scientists would be able at least to engineer what they call "negligible senescence". By that, they mean that they can at least stop aging by keeping the age specific death rates constant for all ages defying the exponential nature of Gompertz curve. That is a fine hope but we don't have the necessary technology anywhere in sight to reduce the death rates on a mass scale. Extrapolation models that ignore the present state and depend only on past data turn out to be myopic as the volatile aspects of future happenings are unknown. If the present state of our national health is any guide, the future does not look so rosy as some people try to paint.

The Challenges

Obesity has turned out to be a challenger of the protagonists of the expanding life expectation theory because of its association with the risks of diabetes, heart disease, cancer and other complications. The effect of body weight on mortality has been studied extensively. In a study of more than a million American adults, the lowest death rates were found among men with BMI's of 23.5 to 24.9 and among women with BMI's of 22 to 23.4. Obesity in young age was found to be associated with greater number of years lost. Using data from six cohort studies, Allison et al estimated that obesity causes approximately three hundred thousand deaths per year in United States. The Centers for Disease Control and prevention put it at one hundred and twelve thousand. If this trend continues unabated, its adverse effect on life expectation could be substantial.

Olshansky et al estimated that, life expectancy in USA would be reduced by one-third to three fourths of a year (4 to 9 months), if the current death rates associated with obesity remain constant in this century. There are several indications to suggest that this will continue into future if left unchecked. They predict that because of the potential of obesity to produce morbid conditions, life expectation in USA would level off or even decline in the first half of this century.

Apart from obesity, there are other threats like AIDS, infectious diseases, and the ubiquitous influenza. If wars and epidemics strike, they can wipe out the health gains of a century within less than a generation. Unless we develop policies to check the galloping obesity threat at the population level, all the gains in recent decades may be lost or they may even be reversed. So, let us talk of stopping obesity and other morbid conditions to save our gains instead of boasting of stopping death and reversing aging.

EXTENSION OF LIFE IS AN ART

Does this all mean dashing our hopes of extending longevity? There is not much hope, if we view life extension as an industrial product that can be manufactured on the assembly line. After all, extending one's life is one's own choice. If one is willing to sacrifice other things to achieve life extension and work hard towards that goal, it is possible to extend it within probable limits. That does not mean expectation of life can keep pace with the maximum life span that may be achieved by some select individuals who use the available tools diligently and skillfully. Life extension is not an industrial process. It is an art.

All the same, we should realize that extremely long life is not such a rosy option as some people think. Jonathan Swift's graphic descriptions of the lives of Struldbrugs forewarn us about the travails that attend upon never ending lives. Perpetual youth is a misnomer. Oscar Wilde showed us how obsession with this misnomer led two characters in his novel, "Picture of the Dorian Grey," to their disastrous end. Shadows can be lengthened only by the setting sun. Do we want the sun to set on this world forever?

IN DEFENSE OF OLD AGE:

Biology considers aging as a process leading to utter uselessness. Aging ends in death but aging is not death just as destination is not the same as the route leading to it. As a destination may have many routes, death has many routes leading to it. Normal aging is one of them and perhaps better than others. Disease or no disease, death comes anyway. We cannot look upon aging as a disease, for effect cannot be avoided when its cause cannot be avoided. We can stop aging only when we can stop time.

If you decouple aging from disease you begin to realize that old is not a concept to be condemned. Oldness is present in many phenomena we admire and use, like old seasoned wood, old wine, and vintage articles to name a few. The energy we use is extracted from very old, fossilized organic matter and we use the wisdom of old people. Oldness is useful too.

CHAPTER ELEVEN

Intimations of Immortality

The Art of Living Long

Our birth is but a sleep and a forgetting
The soul that rises with us, our life's star
Hath had elsewhere its setting
And cometh from afar;

----William Wordsworth.

Wordsworth's idea of immortality implies timelessness. If a person believes in his immortality, he feels that not only he will never die but also that he has always been alive, that he had an existence on this earth before his present life. Wordsworth did not realize this existence through any scholarly reasoning or logical analysis but through instincts, hints and suggestions – "intimations.". The problem faced by the modern biologists who are pursuing the idea of extending life span beyond the life expectation predicted by demographic trends is similar. They do not enjoy the support of any plausible theory to buttress their endeavors. They have hints and instincts based on empiricism. Moreover, they carry the drawback of championing the cause of a movement that is generally considered against nature. Wordsworth's intimations were in the nature of his realization of the immortality of soul. He did not restrict the hopes of mankind to one life and exclude the possibility of transmigration of soul from the meaning of immortality.

Break with tradition – Beginnings of Integral Approach

Scientists following the reductionism approach have been under the illusion that piecemeal research at various levels of biological organization – genes, cells, molecules and tissues– can solve the mystery of aging and help determine longevity. This will perhaps remain an unsolved problem in biology unless the organism is viewed as a whole and we begin questioning how the assembled parts work together. Unless we solve this mystery first, we cannot have a clue to determine how long they will work together. Though integral approach has not taken roots in biological research, there is evidence that some investigations were carried in that direction without explicitly mentioning the idea.

The first break with the entrenched tradition came with the discovery of transposable genes by Barbara McClintock. If some genes can excise themselves from a chromosome and move to other locations in the genome, this movement can have major effects on the timing and development. Other genes can make their own copies and allow these daughters to tread novel paths by making new combinations, while they themselves adhere to the duty of development and maintenance of the organism. That means the idea of static and rigid genome has given place to a fluid genome. With the discovery of a substance called "reverse transcriptase," which can go back, read RNA into DNA, and insert new materials into genetic programs, the central idea of Watson-Crick representation of genome as a helix of beads with its one-way street, where information was restricted to flow only from code to product, collapsed. A set of new ideas such as fluidity of genome, free flow of information, interaction and feedback control have emerged lending credibility to the holistic approach.

Some of our difficulties in longevity research have arisen because of our attempts to accommodate certain mutually incompatible ideas in our theories. We want to believe that Cartesian reductionism is the only method of acquiring knowledge; though it cannot explain how assembled structures like human bodies work with consummate integration. We want to believe in the immutability of Darwinian interpretation of the role of evolution in explaining length of human life though it fails to explain why humans have been living long after the time allotted to them by that theory, expired.

How can we use science which envisages short run directed changes to change the fate of our lives, supposed to have been under the hold of immutable Darwinian changes which are supposed to be infinitesimal, unobservable, directionless and so slow that they do not attain perceptible dimensions even in a thousand generations? Hence, it is time we have to reconsider our positions in life sciences.

Gerontologists claim to be making impressive progress in their endeavors to extend maximum lifespan of humans but are frustrated by the inability of biologists to reconcile their conflicting ideas. Should they repudiate the Darwinian theory? If they abandon it, they face the obligation of providing theoretical underpinnings to their experiments, without which they cannot stand. However, it is worth reconsidering whether we can transcend the Darwinian legacy to gain fresh insights into the problems of life and death. That will only enable us to look at our research from different angle.

Interpreting Darwinian theory in a different way is not going spell any disaster to our world but adding hundreds of years to our lives will have disastrous consequences to the humankind. Therefore, it behooves the research community to develop a solid theory on which to base their experimental research and to draw long-range plans to cope up with the consequences of possible success in their research.

As of now, scientists have not succeeded in discovering a method either to reverse or to stop aging. They have not been able even to decelerate aging. The much-touted technique of telomere manipulation stopped short of practical application. Some trials were made with gene therapy but they did not make much headway. Some gains were made in genetic engineering, but not in the context of aging. Coaxing stem cells into desired parts of body is a promising research but they are no ready-to-use results in the pipeline now. The only viable method that has stood many scientific tests in conferring substantial health and longevity benefits is caloric restriction.

We will endeavor to show in this chapter how to manipulate and combine caloric restriction with the three other elements of the longevity formula to produce the elixir of life and use it to live longer than the expected life predicted by demographers. It is an art to be performed by the interested individuals; it is not an industrial manufacturing process that can mass-produce elixirs. If you really prefer to live a healthy and longer life, you should learn the art and practice it systematically.

You should make proper choices, as your actual choices and not your desires, will reveal your real preferences. If you prefer healthier and longer life, you should choose a lifestyle that accords with such a goal and stick to it. The art of living long is for those who prefer it, choose it, and live by it conscientiously.

The Art of Living Long

SOME CLUES FOR LONGEVITY OUTLASTING DEMOGRAPHIC PREDICTIONS

It is in this background we have to assess the prospects of extending life span. Having been sobered with the realization, that, in spite of extensive search, science has not been able to find a viable and reliable method of extending life it is time we come out of our utopias. Various methods were tried at the cellular level in model organisms and were either inconclusive or not capable of being extended to human level. Barring caloric restriction there is no other known method that has yielded satisfactory results at the organism level.

It may be possible to gain some rejuvenation and add a reasonable number of years if we look for a combination of some pragmatic procedures and adopt them with diligence. One such method we propose consists of a four-variable model in which **caloric restriction** and **modified exercise** constitute two variables. The other two variables in this model are **mind** and **choice**. The first two variables have gained a fair degree of acceptability in scientific community, as determinants of health and longevity, while the other two have not yet been fully explored. This model is oriented towards individuals and not a mass technology model, which, anyway, is not desirable.

A Basic Formula for Human Longevity

(Right Food + Right Physical Culture + Right Mental Discipline + Right choices) ⇨ Longevity

No Quick Fixes:

The world of immortality is not a modern city where you can

revel in wine and women, dance and disco, and hot dogs and hamburgers. Disciplining the mind and restraining the senses are prerequisites to enter the frontier of Immortality. It is a place where some virtues are practiced in pursuit of everlasting peace. The route to it is not paved with cement concrete on which you can speed away at 80 mph. You have to cross many hurdles all along the path. The so-called "fountain of youth" is not an artesian well from which the Soma juice wells up without effort so that every passer by can drink it carefree. Like liquid gold, it has to be extracted with great effort from the depths of strata in which it is embedded. Nobody can set up a rig to extract this gold for you; you have to do it yourself. If you strike it rich that is all yours, no one can steal it.

From a review of the state of the science, a careful reader can surmise that there are no easy fixes for the aging problem because it is difficult to isolate any single cause that leads to aging. It is more reasonable to assume that it is the result of a bunch of interrelated causes. Many thinkers believe that every individual in a species begins with a number of life preserving assets that may be grouped under the collective name "Life Maintenance endowment." The observed phenomena of varying degrees of aging and deaths at varying ages suggest that the rate of exhaustion of this endowment is a random phenomenon. We can assume that the probability of exhaustion of this endowment is low until the individual attains reproductive age and then increases rapidly with chronological age. The exhaustion of this endowment depends on factors like genetic & epigenetic factors encountered by the fetus during prenatal development and diseases in the post natal development, some of which in turn may be caused by improper life style and yet some may be symptoms of unavoidable temporal loss of integrity of organic systems.

This theory however should be limited to decay and death due to endogenous causes. It excludes the exogenous causes of death like predation, accidents, deprivation, suicides and environmental assaults.

At individual level, not much that can be done to counteract environmental assaults, but every individual can do something to prevent or reverse at least partially, the internal damage occurring to the body.

Internal damage increases in proportion to the increase in metabolic activity because increased metabolic activity augments generation of destructive oxidants. Inadequate stress response compounds that problem. Oxidative stress has especially profound influence on the telomere metabolism and cellular senescence. Thus, we can identify the rate of metabolism, efficiency of stress response and efficiency of internal regulatory mechanisms as some of the limiting factors of longevity. From this point of view, we can speculate that any process or a program that reduces metabolic rate increases the efficiency of stress response. Optimization of the working of internal regulative systems can help us in achieving the goal of conquest of old age and extension of life span.

We have seen that, so far no single method, be it genetic manipulation, reduction of core temperature to influence metabolism or restoring the dwindling telomeres to their pristine state, has been perfected to be of practical use. Even if such devices become successful, they are all invasive methods and it is not known exactly what would be their overall consequences to the individual and to the society. The only practical way seems to be to sustain and promote all the factors that maintain the so-called "Life Maintenance Reserve." One of these factors is undoubtedly right food-intake about which we discussed at length. Right food-intake implies not only right quality but also right quantity of food we ingest.

An Art that uses science

A non-invasive method known as "Caloric Under-Nutrition" that is helpful to the individual in promoting health and longevity and helpful to promote the general welfare of society, has recently attracted the attention of scientists. To put it in the language of mathematics, **Right Food** is identified as one of the four independent variables, the others being, **Right Exercise**, **Right Mental attitude** and **Right choices.** They act together and produce various grades of health and longevity with innumerable hues in the quality of life. So, is this a science? We cannot say that, at least for now, because this model has no repeatable properties and leans heavily on the individual performance. It is an art that makes use of science.

That much sought after anti-aging recipe or fountain of youth is within us waiting to be recognized. There is no use of searching for it outside. Votaries of this recipe can prolong their health and life spans without the vexed biological, medical and social problems discussed above. You may wonder: "These are our everyday chores, we do them everyday what is special about them?" That is the paradox. You miss them thinking they are worthless because you think they are commonplace. Yes, these four actions, viz. eating, moving our limbs, thinking and making decisions are our routine chores. Therein lies the advantage of this method, namely we have been familiar with all these chores. Because they are routinely done, we assume that they are done properly. Therein lies the deception, for they are not done properly and efficiently. May be, some of us are right in some but perhaps not in all. By and large, they are done improperly and inefficiently. We do not eat properly; we do not even sit properly let alone properly exercising. We don't think properly and quite often, we make wrong choices. Let us examine one by one.

1.Right Food

The concept of Right Food – what is right in quality and quantity – was discussed extensively in the previous chapter. The figures in the various tables given there are aimed at normal maintenance of health. If you aim at maximum life span extension, they need to be modified using the concept of caloric under-nutrition.

You might have already realized that the importance of food couldn't be overemphasized. As is your food, so is your mind. As is your mind, so is your conduct and as is your conduct, so is your health.

Nature has seen it fit to make eating and drinking enjoyable. They are necessary actions, which recur very often, as often as every few hours. The pleasures of the palate are such that few of us can resist always the temptation to pander to our acquired tastes and live for their satiation. In doing so, we forget that we must eat and drink to live and not live to eat and drink.

Food and Aging

We are supposed to eat in order to live but most of us eat as if we live to eat. Such indiscriminate and omnivorous eating habits bring in their wake all sorts of health problems as discussed earlier.

The four activities mentioned above are like four ingredients of an apothecary's secret formula. If these ingredients are not of good or not mixed in proper proportions, the recipe may not help. Otherwise, it can yield the desired result. Unfortunately, we don't Do we have such an apothecary who can dispense it for us? Each of us is an apothecary who has to do the mixing and each of us is the patient who has to follow the regimen of treatment.

Some guidelines are provided here. Yet, we all may not achieve extra long lives, because every one of us may not fully succeed in cooking up this recipe. but there is every hope that in proportion to our effort we may achieve lasting youth and longer life spans. This recipe will also avoid the pitfall of some mass-produced longevity pills placed on the shelves of drug stores. By the very difficulty of procuring this product, it ensures the success of only those people who try for it with discipline of body and mind, industry and perseverance. In other words, those who pay the right price will get it. Now let us see what is meant by the word 'RIGHT' used here.

Caloric Under-Nutrition

My grand mother used to fast on every 11th day of the lunar phases, and occasionally on some important religious days. She was strict in not eating miscellaneous stuff in between the two meals she used to eat. When asked why she was so strict about eating habits she simply used to tell me that such austerity was good for life both in this world and in the other. On occasions, I used to imitate her but used to break the fast after a while due to childish temptations. She lived a healthy life for ninety years in poor Indian conditions and she was alert until the last. My father, who himself conducted many experiments with dieting, used to tell me that if I wanted to live a long and healthy life, I should emulate the example of my grandmother. When asked why it was so, he used to tell me that most of the diseases are consequences of incomplete digestion and assimilation resulting from overeating. Periodical fasting gives an opportunity to the overworked digestive system to direct its efforts at destroying the toxins accumulated in the body by overeating. Though I never could follow my granny's example completely, I used to respect and admire her austerity of life in general and her food habits in particular. That is why I was amazed when I first came to know that scientists began experimenting with caloric under-nutrition as a possible means to retain youth for a long time and to prolong life span.

A careful reader might have noticed that all available evidence to date points in the direction of dietary control as one of the most important components of any natural life extension program. Implementing the scheme of caloric under-nutrition is not a big task if you make a determination. It becomes easier by renouncing meat. Dietary control in this context means limiting total intake of calories without loss of critical nutrients. As discussed in chapter six, scientists have started realizing the value of under nutrition as a viable alternative to invasive surgical methods and unpredictable gene manipulation methods for life extension.

Experiments conducted on small worms and then rodents have yielded conclusive results in favor of caloric under-nutrition. As the results obtained on small model organisms cannot be straightaway extrapolated to higher organisms like human beings, these experiments are being repeated on primates. Intermediate results obtained on primates throw favorable light on some of the biomarkers and it is hoped that after completion of the experiments, caloric under-nutrition will be fully vindicated. Experiments conducted at the Massachusetts Institute of Technology have also lent scientific legitimacy to this method by discovering the biochemical link between caloric-under-nutrition and longevity.

Any procedure that results in inadequate development and function of brain is not recommendable. Fortunately, in all these experiments brain weight increased at the same rate in the controlled group as in the uncontrolled group. Diet restriction not only increased the average life of these animals but also retarded the aging of brain cells. Even in intelligence tests, restricted mice were as good as the fully fed mice were or even better than them. Restricted mice, which are 50% older displayed more alertness and intelligence than the fully fed mice did. Adult initiated diet restriction regime was found to keep the brain clearer and to retard it's aging.

The key element of dietary control is caloric restriction. High protein diet is best during early growing period. Progressive reduction in middle age and low protein regime in later part of life should follow this. Following a regime of caloric restriction reduces the oxidative stress dreaded by our cells, reduces damage to cells, slows down the metabolism and restores the equilibrium of internal regulatory processes. A practical rule of thumb to implement this program is to reduce the daily intake of calories by 30% of what is normally indicated for your age, height and weight. Periodic fasting for a day (absolute fasting – nothing but a few ounces of fruit juice) once in a fortnight helps to give rest to the overworked digestive and eliminative organs and promotes their efficiency while adding alertness to mind. In addition, such discipline tones up your head and heart. Therefore, right food constitutes the first ingredient of this recipe. Composition of right food involves following the action plan outlined in previous chapter with the additional requirement of reducing the caloric intake by thirty percent.

2.Right Physical Culture

Right physical discipline consists of keeping the body clean and well exercised to keep the muscles toned and the ligaments elastic. No amount of dietary discipline can keep you healthy if it is not coupled with physical discipline also.

People in the West are by now familiar with what is known as "Yoga system" of exercise which is qualitatively different from the body building exercises, This system aims at maintaining and promoting elasticity of limbs and spine and integrating the body and mind.

Proper Exercise Prolongs Life

We have enough empirical evidence to assert that proper exercise prolongs life through its effect on primary aging process. Just as lowering body temperature in cold-blooded animals and some insects, hibernation in warm-blooded animals retards aging. Caloric restriction and proper exercise retard the primary aging in humans. Fig (11-1) displays the results of experiments made with rodents to find the effects of exercise and caloric restriction.

It is interesting to note that there is significant difference in survival rates of sedentary and exercising rats when food intake is not restricted (observe curves A and B). On the other hand, when food intake is restricted (Curves C and D), the difference is not significant.

Fig.11- 1: Effect of Caloric restriction and Exercise on survival of rats

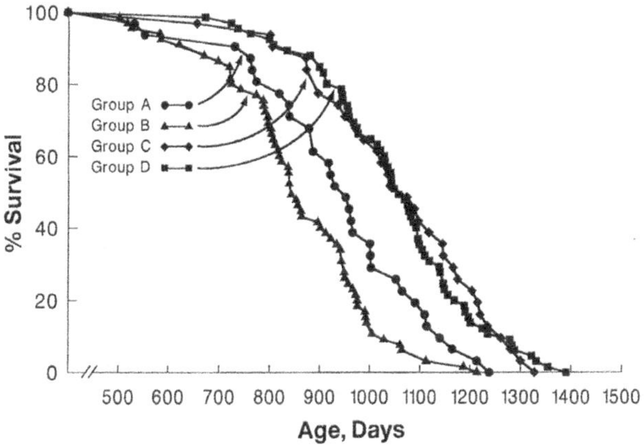

Curve marked 'Group B' (lowest) – Sedentary and eating (Control group)

Curve marked 'Group A' (second from left)—Eating but running

Curves marked 'Group C' (third from left)—Food restricted runners Curve marked' Group D' (rightmost)—Sedentary but food restricted.

Source: Holloszy, John O. "Mortality rate and longevity of food restricted exercising male rats: a reevaluation" J app. Physiol.82: 399-403, Copyright (1997), J App. Physiology. Figure reproduced with permission.

. Based on this, we are inclined to think that in the case of humans, food restriction coupled with aerobic exercise will further enhance the survival rate. *Prāṇāyāma* is the supreme aerobic exercise.

EXERCISES TO MAINTAIN ELASTICITY OF LIMBS

Right Exercise promotes longevity through its effects on secondary aging. After changing your food habits, controlling your food intake and carefully watching that all required nutrients are ingested, the next thing to do is to learn the proper exercise and follow it with a well-regulated program. Remember that no exercise helps if you do not control your voracious eating habits. Physical exercise can augment the benefits already set in motion by diet control. Human body is an incredible machine. Ordinary machines wear out by work. Human machine wears out by keeping it idle. This ancient secret is rediscovered recently and led to fitness revolution.

Different types of exercises produce different results. They fall into two categories broad categories and it is very important to understand the difference between the two. One method known as calisthenics aims at building and strengthening muscles. The other method known as aerobic exercise aims at building up endurance, nervous strength and general wellness. Some may involve both. Gymnastics and various types of ball games, for example, improve agility, balance and coordination. Weight lifting and resistance training increases muscle mass and strength. Endurance training like cycling, long distance running and swimming induces adaptive changes in both the skeletal muscles and the cardiovascular system. Yoga practice is unique in the sense that it not only promotes flexibility and agility of body but also integrates body and mind. It has the potential to prevent many diseases if regularly practiced.

The individual has to decide for himself (herself) what type of exercise he (she) needs. Although everyone needs some muscular strength, a highly muscular body is not necessarily a healthy body. Just as you would want to follow a prescribed regime, of eating and exercising if you want to become a wrestler or a weight lifter, so also you should follow a suitable program of eating and exercise if your avowed objective is to live a long and healthy life. Gymnastic is aimed at making the body graceful and agile, weight lifting is aimed at building particular body muscles. These belong to the first category. Swimming, jogging and dancing are aerobic. These activities and a combination of aerobic and muscle building activities belong to the second category.

The ancient Indian yoga system is an integral system aimed at promoting health and longevity. The yoga system promotes endurance, nervous strength and overall wellness. It involves static stretching of limbs, and regulated breathing exercises. In the parlance of yoga theory, the former is called *Yoga-Aasanas* (3rd limb in the eight-limbed yoga system) and the latter is called *Prāṇāyāma.*

Prāṇāyāma. (4th limb of the system)

Before you embark on such a program, pause for a while and consider the physiological processes involved in various kinds of exercises and their results. Almost all the modern exercises are designed to develop muscles by violent mechanical movements of limbs. As opposed to the fast and violent movements of limbs, there is the ancient yoga system, which emphasizes slow and systematic stretching of limbs with controlled respiration. Muscular development of body does not necessarily mean healthy body as commonly assumed by many people. Health is a state of perfect functioning of all bodily organs under the intelligent control of a placid mind.

TACKLING OXYGEN DEBT:

The principal aim of any good exercise should be to increase the oxygen intake and maintain of healthy blood circulation. As exercises are meant for increasing circulation through the motion of skeletal muscles and oxygen intake, let us compare the function of muscles during moderate and heavy exercises. When muscles contract during exercise, glycogen breaks down into lactose and additional energy is released. One-fifth of the lactic acid so produced is oxidized into carbon dioxide and water releasing energy again. This energy is used for recycling the remaining four-fifths of the lactic acid back to glycogen. If muscles are unable to receive adequate oxygen to complete the process of converting lactic acid into oxygen, fatigue results. When too much lactic acid accumulates, the muscles experience a temporary disability to contract. During strenuous exercise, we are unable, even though respiration is faster; to breathe in sufficient oxygen to meet muscular demand. Consequently, oxygen debt is created.

Oxygen debt is the difference between the amount of oxygen actually required by the active muscles and what is actually received by them. That is why, after completion of violent exercise we continue to breathe deeply and faster in order to repay the oxygen debt. There is a limit to the accumulation of lactic acid in the muscles and blood. Exercises prescribed in the yoga system overcome this defect by regulating inhalations and exhalations. Another important function of exercise is maintaining the elasticity of ligaments.

ELASTICITY:

One of the most important reasons for drifting into old age is the gradual stiffening of the spinal column and the ligaments.

Yoga exercises protect the flexibility of spinal column and maintain the elasticity of ligaments thus keeping the whole body supple. Elasticity (elasticity of spinal column, elasticity of ligaments and elasticity of arteries) occupies a preeminent position in the yoga system of physical culture. This keeps you youthful but unfortunately, it is almost completely ignored by the modern practitioners of physical culture.

Strenuous exercises lead to abnormal accumulation of fat in muscular system, which results in hardening of muscle tissues. Hardened muscle tissues make the body less agile. Elasticity of arteries plays an important role in preserving good health because it maintains even blood pressure between heartbeats. With inelastic arteries, the flow of blood will be intermittent. If elasticity is maintained, blood flows steadily along the arteries and veins. This quality is enormously important. Loss of elasticity sounds death knell.

Vertebral column is the pillar of support of the trunk and the cranium. Four different curves are noticeable in the vertebral column viz., cervical, thoracic, lumbar and pelvic. These curves lend resilience and spring to the vertebral column enabling the body to walk, bend and jump. The skeleton in addition to supporting the framework provides space for attachment of muscles, tendons, and ligaments. Because of these the body frame can not only stand but also bend, swing to the sides and can twist to some extent. These movements are restricted to many people due to biological shortening of ligaments.

If bodily posture and balance are good, ligaments remain elastic for a long time. Otherwise, they cause discomfort and pain while stretching and bending. The average individual, even at the age of 30, cannot bend and touch the floor with fingers while the knees are kept straight. As one grows older, the backbone stiffens because the ligaments become tighter. Excessive stiffness can be due to several causes but especially due to faulty body alignment and poor balance that cause shortening of ligaments. It should be remembered that the restriction on mobility of any part of body would lead to general immobility because the structure of ligaments is continuous. For long lasting and elastic ligaments, correct upper spine posture and balance are necessary. Shortening of ligaments in the vertebral column can be noticed in people with sedentary habits. This is because a person in sitting posture thrusts his neck and head forward and causes the spine to bend backward to compensate for the forward thrust for maintaining balance. Therefore, exercise should be designed to keep the proper curvature of the spine and to increase its flexibility by stretching its ligaments. With proper knowledge of the spine and its movements, the body can be trained to maintain its flexibility late into old age. In correct posture, the long axis of the head and neckline should continue to be the long axis of the trunk.

To determine the correct posture of body, try this test. Drop a plumb line from the front of the head to the forward part of the feet. In a good posture, the bodyline should approximately be parallel to the plumb line. Another method of finding the extent of deformity is by standing the individual against an upright pole. By performing this test, one can detect three different types of deformities.

- Gorilla type, in which the abdomen protrudes, chest is sunken and head is thrust forward.
- Round back posture in which the back is bent in such a way that the hollow in the small of the back is eliminated.
- Chest is pushed forward and lower spine overextended making exaggeration of lumbar curve.

Extended (Yoga) Discipline

Right Breathing exercises

As long as there is breath in the body there is life. When breath departs, so too does life. Therefore, regulate breath. --

Hathayoga pradeepika

The importance of breathing cannot be overemphasized in the study of aging. It is perhaps by far the most important biomarker of aging. Though it has not received much attention in the modern world, the few scientific studies available on this topic are revealing.

Normally, the volume of air in the lungs depends upon body size and build but it also varies with the body posture, decreasing in the sitting posture and increasing in the erect standing posture. To test the respiratory function, measurement of respiratory volumes is used as a clinical test. The amount of air that can be forcibly expelled in one second (Forced Expiratory Volume) is often used as a predictive variable of fitness. FEV is found to decline invariably with age. The higher this volume, the better it is for fitness. This phenomenon lends strong support to the technique of *prāṇāyāma* as a protector of health and promoter of longevity. These volumes are illustrated in Fig. (11-2). As the total lung capacity of an individual is known to be constant, any increase in the residual volume occurs at the expense of FEV and the vital capacity.

Fig.11- 2:Diagram showing pulmonary capacities and volumes

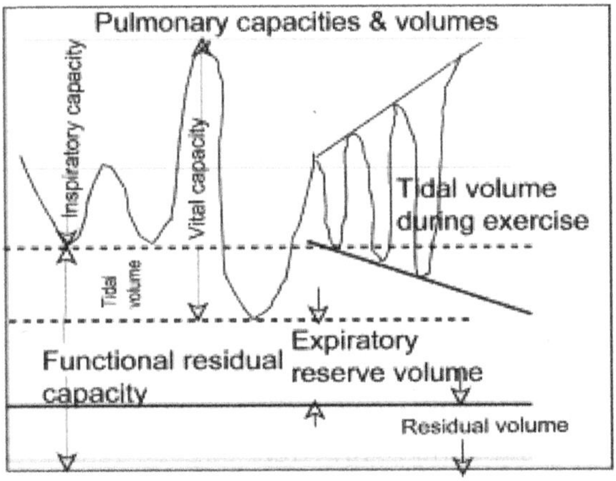

Decline in maximal oxygen intake capacity of individuals is found to be an inevitable concomitant of primary aging. The following figure (11-3) displays the results of oxygen uptake test performed by both men and women of different ages. The rate of decline displayed is almost same though men have higher uptake capacity. That seems to be the reason for women becoming frail sooner than men.

According to conventional thinking, dwindling breathing capacity of individuals is an inevitable consequence of aging. Therefore, we are programmed to think that breathing is totally an involuntary action and its efficiency is at the disposal of aging process. Opposed to this mode of thinking, the view prevailing in the hoary tradition of India claims that breathing can be voluntarily controlled. If breathing can be controlled, the drive of causation can be controlled. In other words, if we can control breathing we can control aging.

Figure (11-3) shows the decline in maximal oxygen uptake capacity (VO2-max) with advancing age independent of any disease. This fact is verified by the increases in maximal oxygen uptake capacity (VO_2) capacity in response to training as shown in Table 11-1.

Fig.11- 3: Relation between Maximal Oxygen Uptake Capacity and Aging.

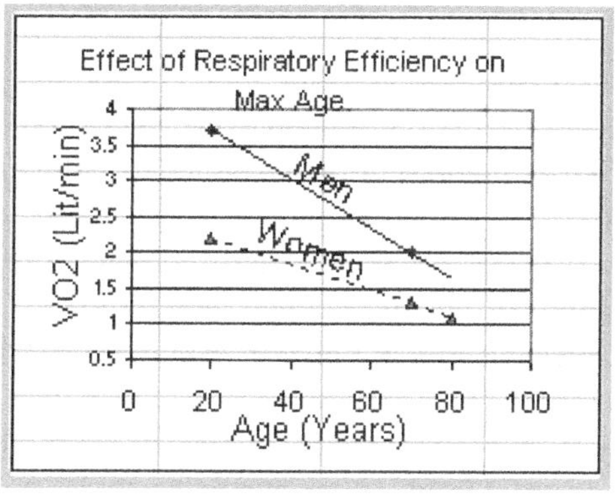

(Source: Holloszy J.O. and Kohrt W.M)[121]

Table 11- 1:Increase in Maximum Oxygen Uptake Capacity by training

VO-2-max (ml/kg/minute) in 65+/-3 year old men and women		
Gender	Initial	Final
Men	27.5+/-4.2	34.6+/-5.7
Women	21.6+/-2.9	26.5+/-3.5

Source: Kohrt WM et al,"Effects of gender, age and fitness level on response level of VO2 max, to training in 60-71 yr olds", j Appl. physiology, 1991:71, pp2004-2011. Adapted with permission

The figures in the table illustrate how maximal oxygen intake capacity can be increased by effort. If ordinary aerobic training can show significant increases, we can be confident that the benefits accruing from a specially designed and time honored system, as *prāṇāyāma* will be substantial.

Prāṇāyāma: This is a compound word consisting of the words "Prana" and "Ayama". Controlling "Prāna" is called. *"prāṇāyāma."* Aāyāma means to stretch, to extend, to prolong or to control. Therefore, right Breathing exercise literally means controlling this "Prana.".

PRANA

Prāṇa may be thought of as the vital energy that pervades the universe, the potential energy hidden everywhere and in all beings. It may be viewed as the hub of the wheel of life. *Prāṇa* is in the air but it is neither oxygen we breathe nor the carbon dioxide that plants breathe. It is what makes us and the plants breathe. It is in sunlight but cannot be equated to heat or light energy. *Prāṇa* is in all forms of matter but it is not matter. It is the force that animates matter. *Prāṇa* is not even consciousness as we know it, but it is the form of energy used by the soul in its material and astral manifestations. *Prāṇa* in an individual is a part of the universal *prāṇa*. It is manifest in the living beings as "breath." Living beings are enabled to breathe by the action of *prāṇa*.

Prāṇa is manifest in human beings as involuntary movements. Wherever there is life and movement there is *prāṇa,* its grossest manifestation being involuntary movements in the body like the pumping of heart, expansion and contraction of lungs, its finest manifestation being the thoughts and thinking process. The various life processes in the body are performed at the behest of *prāṇa*. Life is maintained by absorbing this *prāṇa* from the food we eat, the water we drink and the air we breathe. Yogis and individuals, who are more spiritually evolved, do this absorption more efficiently by their control of this *prāṇa* than the common folk.

Controlling and regulating breathing can gain control over this subtle energy, *prāṇa. Prāṇa* is an auto energizing bioelectrical force harnessed by the act of *Prāṇāyāma.* It is a psychosomatic exercise that controls and regulates respiration in human beings.

Prāṇāyāma has a hoary tradition in India. The central theme of yoga theory and practice is *prāṇāyāma.* It is through being established in the practice of *prāṇāyāma* that one attains the status of a yogi. It is widely believed in India that the key to the conquest of immortality is in *prāṇāyāma* because it is said that every individual has a finite number of breaths to exhale and that as the last breath is exhaled one's mortal coil falls down. If I release my breaths faster and exhaust my quota, I finish my innings faster. Conversely, if you are able to conserve your breaths and release them slowly, you live longer. While your given endowment is fixed, you are not under any obligation to exhaust it quickly. There is a method of stretching the endowment as long as you wish, if you are willing and capable of implementing the technique. The art of *prāṇāyāma* depends upon this principle but fortunately; it is not a mass production technology. It is based on individual effort.

Before we describe the **art** of *prāṇāyāma*, it is necessary to recapitulate the mechanics of respiration. The trunk is divided into two parts, thoracic cavity and the abdominal cavity. Respiration takes place through the elastic movements of the lungs and the activities of the sides and bottom of the thoracic cavity. The thoracic cavity is occupied mainly by the heart and lungs and is bounded by the breastbone at the top, spinal column at the back and by a muscular partition at the bottom. There are twelve pairs of ribs emerging from the spinal column to each side. The upper seven pairs are called true ribs and are fastened to the breastbone while the lower five pairs are known as floating ribs. The muscular partition that separates the thoracic cavity and the abdominal cavity is called diaphragm, which plays an important role in respiration. This muscular partition is convex upwards towards the thoracic cavity.

In the process of inhalation, the ribs are moved outwards by the inter coastal muscles and the diaphragm is pushed downwards towards the abdominal cavity. The outward movement of the ribs and intercostals muscles and the downward movement of the diaphragm enable the elastic lungs to expand. As soon as the lungs expand, a partial vacuum is created and the outside air rushes filling the vacuum in the lungs.

The art of *prāṇāyāma* starts with proper control of diaphragm and the respiratory muscles, to enable absorption of maximum amount of life-giving energy from air. An average person takes in about 500 cubic cm. of air with normal passive inhalation. With conscious deep inhalation, this increases by as great as six times, to 3000 cubic cm. The mechanics described above suggest that diaphragm has an important part to play in deep breathing. The descent of the diaphragm increases the volume available in the lungs as it pushes down the abdomen.

Normally, respiration is involuntary, that is you don't send any mental signals to inhale and exhale but normal breathing is not very efficient because it does not make full use of the structure of respiratory system. That is why ancient seers invented the technique of *prāṇāyāma*. Without going into the details of *prāṇāyāma*, let us first, see how the structure of respiratory system can be utilized to make breathing efficient. Try the following tests yourself.

Test No.1: Sit erect with your head, neck, and spine straight. Contract the abdomen and draw it towards the thoracic cavity. Now take a deep breath by raising the shoulders and the collarbones while keeping the abdomen contracted. Repeat it a few times mindful of how much air you are breathing in.

Test No.2: Sit erectly as before. Keep your diaphragm still and don't allow your abdomen to expand. Now expand your chest and take a long breath. As the diaphragm is in neutral position here, the breathing is done through the action of intercostals muscles connected with the ribs. Repeat it a few times observing the difference between the previous test and this as to the duration of inhalation and the quantity of inhaled air. By careful observation, (you may count the number of seconds it takes to fill up the lungs) you will realize that duration of inhalation is longer and the quantity of air inhaled is more in this case.

Test No. 3: Sit as before, keeping your head, neck and body vertically straight. Relax your abdominal muscles. Do not raise your chest or shoulders. Now take a long and deep breath while allowing the diaphragm to descend. (Count the number of seconds with a stop-watch if possible) When the diaphragm descends towards abdomen, its dome shaped center becomes flat increasing the cubical capacity of lungs. Here the inter-costal muscles and ribs are at rest. Practice this test several times comparing the duration and volume of air breathed. You will definitely find that this method of breathing is even better than previous method and hence the best of the three. Because of the action of diaphragm in expanding the capacity of lungs, this method is called diaphragmatic method of breathing or simply deep breathing. This is the basis of *prāṇāyāma* but we will not go into the details of the *prāṇāyāma* technique.

The respiration of people suffering with respiratory disorders is generally high, as described in the first method. The breathing in ordinary people is generally restricted to shallow breathing described the second method, either due to lack of knowledge or due to incorrect postures of sitting or standing or both.

The first step in *prāṇāyāma* is to restore breathing to its natural state as described in the last method by allowing full play of the diaphragm. Several physical and mental benefits accrue from constant practice of *prāṇāyāma*.

What happens in *Prāṇāyāma?*

The physiological basis and the biochemical pathways through which *prāṇāyāma* exerts its influence on longevity were explained in an earlier chapter. We will now see the variety of benefits that accrue to the practitioner.

Benefits of *Prāṇāyāma*

The foremost benefit is that it neutralizes the fluctuations of mind and acts as a springboard to eventual emancipation.

It is interesting to know the physiological changes produced by *prāṇāyāma* and how they benefit the practitioner. In *prāṇāyāma*, the quantum of oxygen taken in is greater than the output in a given amount of time. Respiratory rate which is normally 14 to 18 per minute can be reduced to as low as 2 to 3 per minute. With increased retention of oxygen per cycle of respiration, the efficiency of energy production in the body increases and the longevity promoting genes are activated. With reduction in respiratory rate, the metabolism is reduced and the body is reduced to a state of temporary hibernation with all its benefits.

RESPIRATORY SYSTEM

Prāṇāyāma is not mere deep breathing as is commonly misunderstood by some people. In the latter, adequate time is not given to the cells of the lungs to get soaked in the inhaled oxygen. In the former, deliberate slowing down of the respiratory movements gives adequate time for every alveolus to soak in oxygen.

DIGESTIVE SYSTEM

The rhythmic use of the diaphragm and the abdominal muscles stimulate the peristaltic and segmenting movements of the intestines as well as promoting intestinal circulation. It helps intestines in absorbing food and disposing of solid wastes, more efficiently. Gall bladder and liver are massaged. Constipation is relieved.

As spleen is seated just below diaphragm, *prāṇāyāma* stimulates splenetic blood circulation. It helps to maintain flow of pure blood which tones the cardiac muscles, spinal chord, brain and nerves.

CARDIOVASCULAR SYSTEM

Due to the constant change in the chamber size of heart during *prāṇāyāma*, cardiac muscle is properly exercised without strain. Autonomic heart rate is reduced and rested. This helps in more efficient functioning of heart, even as metabolic demands of the body are curtailed. *Prāṇāyāma* maintains the elasticity of the arch of aorta thus preserving and promoting coronary flow. As more capillaries are opened, microcirculation is enhanced. Enhanced supply of nutrients to cells increases their longevity.

All this happens without increase in pulse rate and blood pressure.

NERVOUS SYSTEM

Prāṇāyāma gathers the runaway mind and makes it tranquil. It reduces excitability of sympathetic nervous system thus stabilizing the neuro-physiological activity. *Prāṇāyāma* balances sympathetic and parasympathetic systems thus strengthening the neuro regulation of circulatory and endocrine systems. As a result, great energy flows and the vital age of the body is maintained.

Brain

Brain is the miracle part of human body. About 100 billion nerve cells are packed into the brain. Each one of these receives inputs from 10,000 other neurons in the brain and sends messages to 1000 others. What potential capacity does this give the brain? Astronomer Carl Sagan said that the human brain could hold information that *"would fill some twenty million volumes, as many as the world's largest library."* James Watson of DNA fame said, "The brain is the most complex thing we have yet discovered in the universe." Neurologist Richard Restak, who resents the comparison of computer to brain, says, *"The brain's uniqueness stems from the fact that nowhere in the known universe is there anything even remotely resembling it."*

Neuroscientists estimate that we are capable of using only a small fraction of our potential brainpower, only about 1/10,000 or 1/100th of one percent. Is there any way to release at least a little more of this awesome power of brain? As of now, nothing is known to conventional science as to how to do it. The yoga system tells us that *prāṇāyāma* can perform that feat. *Prāṇāyāma* can extend the reach of senses and it confers many extraordinary powers on the sincere practitioner.

General:

By simulating a state of hibernation, sympathetic overdrive is reduced. As a result, energy is conserved.

Prāṇāyāma protects the internal organs and cells by supplying optimum amount of oxygen. It neutralizes lactic acid, which causes fatigue thus enabling one to endure longer hours of work or to recover fast from fatigue caused by work.

As the mind is kept attentive to regulation of breathing, the frontal brain, the seat of intellectual activity is made quiet and complete physiological relaxation ensues. *Prāṇāyāma* increases vigor, vitality, and acuteness of perception, memory and sharpness of intellect.

Prāṇāyāma is said to have conferred on the yogis of the lore, the power to retain their bodies as long as they wished and to relinquish them at will. Best of all, it awakens the dormant energy called *Kundalini* and opens up the vista of endless possibilities including the beatitude of *samādhi*, which is the culmination of all human efforts.

According to Ayurveda, the body is made up of seven constituents[122] and three permeating humors (doshas).[123] They keep the body immune from infection and disease. They are churned together in *prāṇāyāma* for the production of nectar of life.

Proper exercise of inhalation and exhalation creates *Prana*, the vital energy of body. By the contact of the vital breath with the seven constituents of the body and the ten types of vital energies, the elixir of life is generated in body.

The name given to the individual soul, ***ātma***, in Sanskrit, is closely related to the biological aspect of man's nature. ***Atma*** literally means that which breathes. ***Atma*** is derived from the root 'An' meaning 'to breathe'.[124] This process of breathing goes on throughout life, in waking, dreaming and dreamless sleep. Along with material body, and mind (manas) that uses the five sensory organs and the five organs of action, the principle of breath *(Prāṇa)* which regulates unconscious activities, forms an integral part of the individual soul or Jeeva.

Yogis realized long ago this link between the breathing principle and longevity. According to their theory, life is consumed not by the number of calendar years but by the number of breaths. We drain our life force constantly by acting, talking and even thinking. It has to be replenished by recharging the body with *prāṇa*. This recharging is done partly by ingesting food and partly by rhythmic breathing.

Most of us think that this replenishment is done by food intake as we breathe involuntarily. That is where the error lies. Just as any type of food is not adequate to absorb *prāṇa* efficiently, so also casual respiration is not enough to absorb enough *prāṇa*. The automatic respiratory mechanism of the body is not fully efficient. Ancient yogis argued that the endowment of respiration is nearly equal for everyone and life span is the duration of its usage. If you use it faster, you exhaust it faster. If you run faster, you reach the destination quicker. If you are slow and steady, you can extend that duration of life's travel. We can see the force of this argument by comparisons of rates of respiration of various species and observed maximum lifespan from Table (11-2).

Table 11- 2: Association of Respiratory rates with observed Maximum life spans of some species.

Animal	Ave. respiratory rate.	Maximum Life span
Mouse	50 times/ minute	3 to 4 years
Monkey	30	24
Dog	28	20
Horse	16	40
Human	15	100
Tortoise	03	250

According to ancient Indian thinking, the lifetime of a living being is supposed to be determined by the rate of respiration. If the rate of respiration is faster, the span of life is shorter and vice versa. A yogi who is adept in *prāṇāyāma* is said to be capable of stopping time, by *Kumbhaka prāṇāyāma* by which he conserves breaths while others are wasting it. *Kumbhaka* is inflating the lungs like a pot and carefully retaining the inhaled air inside as per the rules prescribed.

The testimony of ancient saints who spent selfless lives developing extra sensory perception in their yoga laboratories, never craving for riches or fame is as valuable as the results we obtain in our laboratories with the limited abilities of our perceptions. Indian literature is replete with references to those people that crossed the barriers encountered by ordinary mortals and lived as long as they wished or at least longer than the ordinary.

Now we'll pause for a while to consider the importance of mind in successfully implementing our model.

3. Right Mental Discipline

We examine now the role of mind in our model. In any scheme of promoting health and longevity, neither body nor mind can be ignored. Unless we realize that body and mind form an integral system, we really cannot make any headway in our quest.

THE INFLUENCE OF MIND ON BODY

The reader might have observed that in all the theories mentioned above, mind was conspicuously absent. Body and mind are treated as if they are alien to each other or at least independent of each other. Modern medicine has been beset with that great drawback. Plato said "the greatest mistake physicians make is that they attempt to cure the body without attempting to cure the mind; yet the mind and body are one and should not be treated separately!"

One of the well entrenched but absurd notions of modern medicine is that mind and body are unrelated entities and that changes in one do not affect the other. This notion, though unconvincing, reigned supreme until recently as it formed part of a paradigm, which tried to explain away every phenomenon by the Cartesian reductionism.

However, the conceptual walls erected by the modern medical establishment between mind and body began to crumble in the 1970's with the empirical studies of psychologist Robert Ader and immunologist Nicholas Cohen. Before going into the scientific investigations of Ader and Cohen, let us recall the idea of severing heads and attaching them to different trunks mentioned earlier. As said earlier, the idea itself is not novel as it was explored in ancient Indian literature.

It is not exactly known whether the aim of the story was to impress the reader that attaching a severed and live head to another torso is within the realm of possibility, or to tell us that mind influences the body, or both. I am inclined to think that the idea is to illustrate the role of mind in influencing the body and to present the enigma of identity of recombined limbs. The exchange of heads in the story produced change of personalities. Interested readers may read the story in 'Transposed Heads" by Thomas Mann.

The ghastly scenes and the tragic pathos in this story may sway the reader, but that is not the aim of the story. It is rather to highlight the idea of transformation of body following cues from the mind. The aim of introducing here, the theme of this story is to point out the fact that the idea of transposing heads was conceived in ancient times and it has been told and retold for many centuries.

The discovery of Ader was accidental. Ader was conducting the classic Pavlovian experiment on rats to produce conditioned reflexes. The aim of the experiment was quite simple. It was to teach the rats to respond with aversion to saccharin-flavored water as a conditioned response. The design of the experiment was also very simple. The rats were given saccharin-laced drinking water and were immediately injected with the drug cyclophosphamide, to produce nausea. The idea was that the rats would thereafter associate nausea with saccharin water itself, and would show aversion to it.

However, the young and healthy rats began getting sick and dying unexpectedly. Ader realized soon that the drug he was using to produce nausea was also suppressing their immune systems by lowering the number of T-cells. After this realization, Ader conceived a brilliant idea. He stopped using cyclophosphamide and fed the rats only with saccharin water. Then came the great breakthrough. When the blood of these rats that drank only saccharin water was examined, it was found that the T-cells were lowered. The important point here is that the T-cells were lowered even without using the immunosuppressive drug. Ader was surprised but realized that the classical conditioned response had occurred but in a quite unexpected way. The taste of saccharin water triggered a response in the immune system by lowering the T-cells. Rats learned to associate the nausea of the immunosuppressive drug with the saccharin water itself and their consequent aversion triggered the suppression of their immune systems. Ader realized that conditioned response was, after all, the reason for the premature deaths of otherwise healthy rats.

The experiment of Ader was similar to that of Pavlov though the triggering causes and the responding organs are different. In Pavlov's experiment the sense of sound triggered salivation in the mouths of dogs. In Ader's experiment the senses of smell and taste triggered changes in the immune system of rats. Yet, Ader himself was hesitant to declare his results because they go against the existing wisdom of the biologists and the medical establishment. Therefore, he teamed up with the immunologist, Nicholas Cohen of the Rochester University School of Medicine and repeated his experiments, which produced similar results.

As recently as 1985, the prestigious 'New England Journal of

Medicine' criticized, in its editorial, the notion of mind having any effect on health. The editorial stated, "Most reports of such a connection are anecdotal" and concluded, with an air of finality, "It is time to acknowledge that our belief in disease as a direct reflection of mental state is largely folklore."[125] However, the same prestigious journal reversed itself, six years later, on this notion, which it declared disdainfully as mere 'folklore.' In the fall of 1991, the Journal published a study showing direct link between mental stress and susceptibility to common cold.[126] Though this study did not identify, by itself, the physiological mechanisms through which brain affects the health; other scientists began accumulating evidence that immune system might be the link between the brain and health. They have shown that nerve cells respond to chemical messengers secreted by the immune system and that immune system too responds to chemicals.

Hippocrates believed that his patients who had strong enough life force would be well in due course, if the healing mechanisms of their bodies were not adversely interfered. Doctors of the Hippocratic school of thought recognized that, in most case, bodily defenses (what we now call immune system) would, restore the patient to health, if the patient had a positive mental attitude. The existence of this phenomenon can be scientifically proved with what is known as 'placebo effect' in clinical studies.

Placebo Effect

For centuries, physicians have known what is called 'Placebo Effect'. This is a psychological phenomenon, which emphasizes the power of belief. Placebo is dummy medicine. Giving the patients some pills without any medicinal value but telling them it is medicine creates this effect. Believing that it is some medicine, some patients report later that their condition has improved. That is the power of belief. Mind can thus create self-fulfilling prophecies.

Although this published study did not identify the physiological mechanism that links brain with physical health, other scientists found convincing evidence about the link between the state of mind and the level of physical health. That link is the Immune System.

With this discovery, a new branch of science called Psycho-Neuro-Immunology began evolving and researchers got interested in using human subjects in their investigations rather than animals. Experiments were conducted on people whose spouses were lost recently, on medical students preparing for their examinations and on volunteers subjected to noisy and sleepless conditions, the degree of stress being in decreasing order. In all cases, the results provided convincing proof of the effect of mental stress on the immune response level, though the degree of stress varied.

The scientists who conducted the experiments lumped together all unpleasant situations as 'Stress' but we can analyze further and identify these mental states. The people who lost their spouses were suffering from grief and possibly bewilderment. The students who were preparing for the examinations were in a state of anxiety and the volunteers were kept in a mood of dejection and vexation.

Several other moods like anger, fear, jealousy, revenge and frustration produce physical stress. Many of these have been in our routine experiences. Who does not know that fear produces increased heartbeat? Nervousness makes an otherwise competent person inefficient. Anxiety produces dry mouth. Anger produces flushed faces and romantic feelings produce blushed faces! Jealousy and revenge are known to consume a person. It is a well-known fact that when the mind is in a worried state, even the normal physiological processes like digestion and assimilation of food are impeded.

Even the conditioned mental responses are known to affect the physical behavior. Everyone knows know about snakes and the consequences of snakebites. The fear of snakes is written on everybody's psyche. The following incident illustrates the point Once it so happened that in a crowded movie theater in India, somebody yelled suddenly "Snake, Snake." The moment the words "Snake, Snake" were heard, there was a sudden rush of people towards the gates. In the ensuing stampede to escape the "snake", several people were injured and a few were crushed to death, it was reported. When the lights were turned on and a search for the snake' was conducted, alas, there were only injured people and no snake! The conditioned responses created by the sound of the word "snake" wrought the havoc creating fear, trembling, mad rush in which several people were crushed.

The illusory snake is used as an analogy in Advaita (non-dual) philosophy to illustrate a point. It is a well known experience that when a man, steps on a curly rope lying in his path in dim light, jumps back in shudder thinking that he stepped on a snake. If he is weak in heart, he may even succumb to the shock as his bodily defenses yield. This frightful reaction is produced as a conditioned response to the soft touch and the curly image of the rope resembling the snake. This type of conditioned responses in people, even when occurring by accident, are known to have produced the same type of fearful consequences associated with snake bites. This analogy is known by the name *"rajju-sarpa-bhranti,"* (Sanskrit for 'rope-as-serpent illusion') in Advaita philosophy and brought to illustrate the defective understanding of the human about the nature of phenomenal world, and the effect of mind on body.

Can Mind Influence Matter?

This is a more general question than the previous question whether mind can influence it's own body? This question refers to matter outside, not necessarily mind's own body. Does mind which is generally considered spirit influence inert matter? To be more explicit, can some people influence the outcome of a physical occurrence?

Empirical Verification of Mind's Influence over Matter

A subject that is generally considered to be in the realm of metaphysics or parapsychology was taken over by scientists in the recent past. A group of mathematicians, scientists and engineers at Princeton University, working under a program called "Princeton Engineering Anomalies Research" (PEAR) has been trying to answer this question. To answer this question, one has to observe whether human volition can influence the outcome of a random event experiment. To be rendered scientific, the design of the experiment should be such that it is repeatable and the events of the experiment should be measurable. After initial experiments with various physical devices like simple pendulum and laser interferometer, they have eventually settled upon the use of Random Event Generator for their experiment. This simple device was designed to simulate the coin tossing experiment (Fig.11-4).

Coin tossing experiment consists of repeatedly flipping a coin and noting the outcome in a large number of trials. Normally, what do you expect the outcome would be when you flip a coin? The outcome will be either a head or a tail. Suppose you flip a coin 200 times. You can safely bet that, on the average, you will get 100 heads. Why? When you conduct the coin tossing experiment which has only two possible outcomes, and the experiment is repeated a number of times, we can predict, on the basis of the laws of probability that either head or tail will occur, on the average, half the number of times. Based on this idea, the simulator was designed to generate a series of ones and zeros to represent heads and tails.

In a typical experiment, an operator sits before the equipment and records the number of ones produced in a 200-millisecond interval. Normally, you expect that in a 200 millisecond experiment there would be 100 ones. If the REG equipment functions properly, the grand mean of a large number of experiments will settle down close to 100, but due to the random nature of the experiment the outcome may vary from about 80 to 120. If the numbers of occurrences for various values around the mean are plotted, we may get something like a bell shaped curve. The question now is how to make use of this equipment to find out whether human volition can influence the outcome of the experiment. It is simple.

The human operator sits before the equipment and desires strongly that the outcome should be one way or the other, say for example, for higher count of ones. A run of the experiment typically consists of a few thousand trials. The computer calculates the cumulative deviation of counts from the expected value 100, as the series of trials progresses. If the operator does not exercise any influence on the experiment, the cumulative output of deviations will be close to zero. Otherwise, there will be some positive and some negative deviations. Fancy the wonder – independent observers found that they could influence the outcome! Look at the Figure (11-4).

Fig. 11-4: Influence of Mind over Matter – Empirical verification

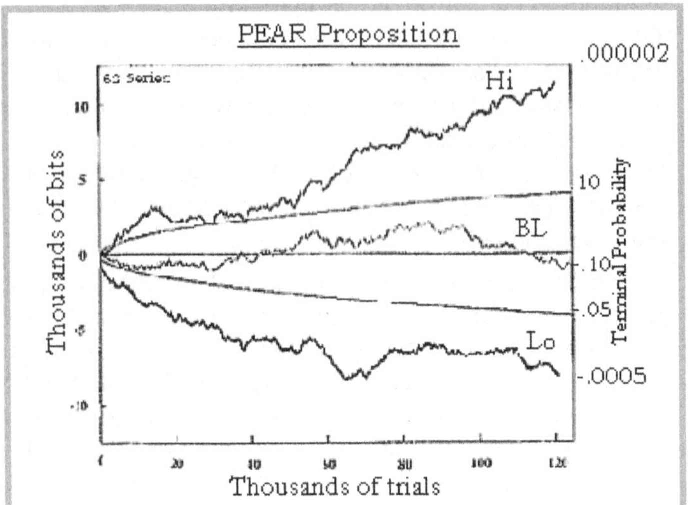

Cumulative deviations of all mean shifts achieved by an operator over entire data base of some 125,000 trials per intention.

The figure shows that the deviations went up when the operator strongly desired them to go up, and came down when he/ she wanted them to come down.

Source: Robert G.Jahn and Brenda J.Dunne,"The PEAR proposition", Jl of Science Exploration, vol.19, No.2, pp195-245, (2005). Reproduced with permission of the authors.

On the other hand, if the operator succeeds in influencing the out comes of the REG unit, the output of cumulative deviations from mean will steadily increase as the experiment progresses. If the operator wishes for less number of ones, the cumulative output of deviations should steadily diminish on the premise that the operator is influencing the outcome. Both the possibilities are shown in Fig. 11-4.

Over a 15-year period, PEAR program has clearly established that there is a statistically significant correlation between the pre-stated operator objective and the properties of the random binary sequences. That is to say, that mind does and can influence even inert matter.

The Role of Mind in Alternative Systems of Medicine

Alternative systems of medicine like Ayurveda and Homeopathy give mind its due importance in their therapeutics.

According to Ayurveda, every mind comes into being with three gunaas (Qualities) namely, *Sattva guna* (Serenity), *Rajo guna* (Activity) and *Tamo guna* (Inertia), in different combinations. An individual, in whom *Sattva guna* preponderates, is called *Saatvika;* one in whom *Rajoguna* preponderates is called *Raajasika* and the one in whom the *Tamoguna* preponderates is called *Taamasik*a.

A *Saatvika* is one who is mentally active and remains serene and calm in all circumstances, has a benevolent attitude towards others and bears a cheerful disposition. A person who is physically more active and mentally restless and irritable and has an aggressive) disposition is called *Raajasika.* A *Tamasika* is one who is slothful both mentally and physically and is brutish in attitude.

The equipoise of a *Saattvika* maintains an even rhythm of heart and blood pressure and his benevolent and cheerful attitude keeps his immune system well tuned. The restless and aggressive mind of a *Raajasika* exerts its turbulence on the heart activity and the blood pressure and his irritable disposition tends to throw his immune system off balance. The slothfulness of a *Taamasika* and his morbid mind slow down his entire physical apparatus, atrophy his vital organs and throw out of gear his immune system.

It is imperative that one who wants to master the art of life and live a healthy and long life must subdue the base impulses and try to be a *Saatvika.* A *Saatvika* should possess certain benign mental qualities as advocated by the great seers and is characterized by the absence of malevolent qualities.

In Homeopathic system of medicine, mind is given a place of pride. Homeopathic treatment is based on symptoms rather than on the name of disease. To elaborate a little more, patients diagnosed to be suffering from same named disease are not administered same medicine but different medicines based on symptoms. If one of two patients suffering from fever exhibits restlessness while the other appears drowsy, then a different medicine is indicated to each. When the physician is unable to decide upon a proper remedy based on physical symptoms, he clinches the issue based on mental symptoms. Such is the importance given to mind in homeopathic therapy.

A well-known fact is not publicized much due to apathy of medical profession. That is, treating the mind or changing mental attitudes can cure many ailments. The malevolent mental attitudes discussed earlier can exert their evil influence on the body.

Paradigm Shift in Allopathic Medicine

This paradigm shift in allopathic medicine was the subject matter of a conference held in May 1974 at the Brunel University in London. The subject of the interdisciplinary conference popularly known as 'May Lectures' was "New Approaches to Health— Individual and Social" and the participants came from various disciplines.

The conceptual framework that emerged out of the conference was anticipation of a profound change in science and western philosophy that in turn was bound to lead to a new medicine based on different perceptions of human nature in health and illness. The participants agreed that the paradigm shift in science would be from a mechanistic and reductionism point of view of human nature to a holistic and ecological point of view. They were extremely critical of the present system of allopathic medicine, which they believed reached its limits and was no longer able to improve or even maintain public health. Their discussions made clear that the future health care would have to go far beyond conventional medicine to deal with various phenomena that influence health. What is needed is relating the biological aspects of illness to the general physical and psychological conditions of human beings in their social and natural environments.

One of the new concepts recognized in the conference was the interdependence between mind and body both in health and illness suggesting a psychosomatic approach to all forms of therapy. The other was the realization of the inevitable interconnection of all human beings and the environment in which they live. Both kinds of interconnectedness were discussed in terms of energy patterns. The Indian concept of "*Prana*" and the Chinese concept of 'Chi' were mentioned as examples of traditional terms referring to these subtle 'energies.' In these traditional disciplines, illness is seen, because of changes in these energy patterns and techniques were developed to influence the body's energy system to help it restore the health of the individual.

Carl Simonton, an oncologist from Texas, expressed the opinion that there was abundant evidence in literature about the role of emotional stress in the onset and development of cancer. He says,

> "The question is not whether there is a relationship between emotional stress and cancer, but rather what is the precise link between the two." Simonton suggested the notion of cancer personality, i.e., of a certain pattern of behavior in response to stress that contributes substantially to cancer, just as another type of behavior is known to contribute to heart disease. He is reported to have said "I have verified the existence of these personality factors in my own research"

In Simonton's cancer treatment, the main effort was directed toward changing the patient's belief system about cancer. The widespread image of the disease is, that it is an external agent invading the body and setting in motion a process over which the patient has little or no control. In contrast to this image, Simonton's experience convinced him that belief systems of the patient and the physician are vital to the success of the treatment and can be used effectively to support the patient's potential for self-healing. Cancer is not an attack from outside but a breakdown within.

Simonton's psychosomatic model is based on the so-called surveillance theory of cancer according to which every organism produces occasionally cancerous cells. In healthy body, the immune system recognizes abnormal cells and destroys them. If for some reason the immune system is not strong enough, the cancerous cells reproduce resulting in a tumor. Emotional stress has two effects. On the one hand, it suppresses the body's immune system and on the other, it creates hormonal imbalances leading to production of malignant cells precisely at a time when the body is incapable of defending itself.

Since development of cancer involves a number of interdependent physical and psychological processes, they can be recognized and the sequence of events that leads to cancer can be reversed to lead the body back to a healthy state. To do so, the patient has to be helped to become aware of the wider context of his / her illness and identify the major stresses in his/her life. Then the patient has to develop a positive attitude toward the treatment and the body's defenses. Once the feelings of hope and anticipated recovery were generated, the body's system translates them into biological processes that begin to restore balance and revitalize the immune system using the same pathways that were used in the development of illness.

It is time that we give up the strange notion promoted by the modern medical establishment that mind does not have any effect on the body. In Sanskrit literature, an adage asserts "*Yadbhaavam tadbhavati*" which means, "As one thinks so one becomes". This idea is again illustrated by what is known as "*Bhramara-keetaka nyaya*" or the law of butterfly and the caterpillar. The caterpillar at one stage looking at its ugly body, longs to become beautiful and goes into hibernation with that strong desire in mind. As we know the dirty old caterpillar comes out of hibernation as one of the beautiful creatures of nature, namely the butterfly.

The sooner we accept the proposition that mind has a clinical role, the better for us. Just as it has a commanding role in the normal maintenance of the body, as we admit, so also we should admit that it has a role to play in causing distress to healthy body and a clinical role in restoring the body to normalcy in times of illness. Realizing the importance of mind, the psychosomatic rejuvenation system of ancient India advocated by almost all philosophical systems and religious groups, has given enormous importance to mind. Anyone desiring to understand the importance of molding the mind in the quest of immortality should understand and try to follow the code of conduct formulated by ancient Indian seers. This great psychosomatic code promises not merely healthy life but also liberation from finite existence to anyone who faithfully follows it. This code is based on a deep analysis of the human body-mind apparatus and its methods of interaction with its environment.

Psycho-Somatic Analysis of Human Personality

In the psychosomatic analysis of human personality, Indian philosophical systems are closely allied. Though their nomenclatures differs a little, their objectives are identical and methods of analysis are similar.

In all our activities, we are clothed in the garb of a 'Doer', 'Enjoyer', or 'Knower.' Every deed involves five elements; a doer, instruments for doing, an object of deed, a vehicle for doer and the field of action.

Take for example, the act of a painting. The doer is painter. The instruments are paints and brushes. Vehicle is the place where he positions himself and his painting materials. Object is the picture in his mind. The field is the canvas or the wall.

Next, let us take the case of enjoying a dinner. Here the enjoyer is the diner. Instruments are plates and spoons. Vehicle is the dining hall. Object is food. Field of enjoyment is palate.

Similarly take the case of acquiring knowledge by a biological scientist: Here, Knower is experimenter. Instrument of knowledge is, let us say a microscope. Vehicle is laboratory and other equipment. Object of knowing is, let us say DNA. The field is culture.

Thus, every deed involves the following five elements in general — A person, that person's instruments, vehicle or stratum for the action of person, an object, and a field. The first three of these factors, viz. experiencer, the instruments and the vehicle, constitute the subjective element in all our experience.

The first and foremost of these five elements is "Person". A person who acts, enjoys and knows, has an exact name in Sanskrit. That is "Jeeva". As it cannot be translated into English with suggestive effect, we use the word 'Jeeva' henceforth. Those who are not comfortable with anything that smells spiritual may continue to view Jeeva as the experiencing person, This Jeeva, in turns, is called the doer, the enjoyer and the knower as occasion demands.

This *Jeeva* has two kinds of instruments, viz. An Inner instrument and two sets of outer instruments.

The inner instrument is called *Antahkarana* in Sanskrit. Again, this has no English equivalent, the closest being the compound word mind-stuff. This mind-stuff has different modes. When it is in the mode of enjoyment, it is called Mind (*Manas)*. The same mind-stuff, when it is in the mode of knowing (discursive or intuitive) it is called Intellect (*Buddhi)*. The organs of cognition receive sense impressions and transmit them to mind, which experiences them or simply knows them according as mind (*Manas)* or Intellect (*Buddhi)* is active at that time.

The outer organs (*Bahya karanas*) are the five instruments of cognition and the five instruments of action. These are associated with the life principle called *Prāṇā,* consisting of five vital airs. Thus, Sukshma sarira comprises of seventeen elements including mind and Buddhi.

Though the faculties of perception and the faculties of action come under the category of outer instruments, there is an important difference. The faculties of perception *passively* receive the impressions from the external world and pass them on to the mind, whereas the organs of action take their orders from the mind and impress themselves on the external world. Thus, they act in opposite directions.

The gross physical body wherein the *Jeeva* takes his seat and exercises the above-mentioned faculties, is the vehicle. This becomes the vehicle of action when the Jeeva, as a doer, exercises his *Prana* and the organs of action. When this Jeeva, as enjoyer, exercises its mind (*manas)* and the organs of cognition, this physical body becomes the vehicle of experience. When this *Jeeva* as a knower, exercises its intellect *(buddhi or* the discursive faculty) and the organs of cognition, this body becomes the vehicle of knowledge. It is through the physical body the Jeeva achieves all its objectives.

The object is a limited portion of the field. For example, the canvas on which the picture is painted is the field and the picture itself is the object. The mural on the wall is the object and the wall itself is field. Thus, the object is not independent of the field.

The field, which is conceived of as the field of action, the field of enjoyment or the field of knowledge encompasses the entire universe. The field is not directly related with the person except as the material from which the *Jeeva* carves out objects with help of its instruments.

Because of this analysis, we have seen that the subjective element in all experiences can be divided into three factors, the experiencing person, his/her instruments and his/her vehicle. We have also seen that the *Jeeva* per se is undifferentiated but appears to have become differentiated into Doer, Enjoyer and Knower when it exercises the corresponding sets of instruments. We have also seen that the physical body is neutral but it becomes the vehicle of action, enjoyment or knowledge when the particular sets of instruments are active in the vehicle. Therefore, we can deduce that instruments cause differentiation in the Jeeva and the vehicle. These three factors put together make up what is called the body.

In addition, our efforts are based on the tacit assumptions (which may not be true) that we do not have Perfect Existence, Knowledge and Happiness and that we must obtain them.

Right or wrong, these assumptions are the basis on which the entire phenomenon of life rests. However, do we strive to obtain everything that we do not have? Obviously, we do not. Mere absence of a thing is not sufficient to impel us to action. We must have a desire to obtain it. Therefore, there are three stages for every activity.

- We realize that we do not have something. (Nescience)
- We want to have that thing. (Desire)
- We work to obtain it. (Activity)

By finer analysis, the concept of body can be sub divided into concepts of causal body (*Karana śarira*), Subtle body (*Sukshma śa rira*), and gross body (*Sthula* śa *rira*).

- The undifferentiated Jeeva is called Causal Body (Karana śarira)
- The instruments of Jeeva (17 in all) put together, are collectively called the subtle body. (Sukshma śarira). That is the seat of desire.
- The vehicle, the seat of action, is called gross body (Sthula Sarira).

The instruments of Jeeva, making up the subtle body, lend themselves for further division into three imaginary partitions, depending upon the state of the inner organ.

In an alternate nomenclature, these bodies and their finer divisions are given the name of vestments suggesting that one vestment covers the other effectively screening the innermost undifferentiated *Jeeva* from our experience. The composition and names of these vestments are given below with the corresponding Sanskrit names in the parentheses. In the alternate nomenclature:

1. The undifferentiated causal body is called the Blissful vestment (*Anandamaya kośa).*
2. The partition in which the inner organ, (in its 'Buddhi mode') is associated with the organs of perception, is called the Intellectual Vestment (*vijñanamaya kośa*).
3. The partition in which the inner organ, (in its 'Manas mode') is associated with the organs of perception, is called the mental vestment. (*Manomaya kośa*)
4. The partition, in which the five vital airs of Prana are associated with the five organs of action, is called Vital Vestment (*Pranamaya kośa*).
5. The gross body is given the name, 'Food vestment' (*Annamaya kośa*).

Thus, the three bodies correspond to five vestments as depicted schematically in Figure (11-5)

Live Gross body	= *Annamaya kośa* + *prāṇamaya koṣa*
Subtle Body	= *Manomaya kośa* +*vijñānamaya kośa*
Causal Body	= *Anandamaya kośa*

The Illusion: As the physical body is merely an inert vehicle, it is dependent on the Jeeva, its occupant. Jeeva is inactive by itself even though it is not inert matter. Activity is possible when Jeeva takes on the characteristics of Knower, enjoyer or doer and wields the instruments known as Intellect (Buddhi), mind (manas) and senses. Similarly, activity is possible in the physical body only when the characteristics of Knower, enjoyer and doer are superimposed upon it. Thus, in all activities there is, as it were, an apparent descent of the Jeeva to the levels of its partitions (*kośas),* while there is an apparent ascent of the gross body to the levels of partitions, resulting in an apparent fusion of the undifferentiated jīva with its instruments and producing the illusion of the body acting.

As the functioning of the physical body is subservient to the purpose of *Jeeva,* it has to vary with that purpose and that purpose in turn varies with the nature of the fusion. Therefore, all activity may be classified based on the purposes of the Jeeva.

A schematic view of these imaginary partitions is displayed

in Fig. (11-5).

Fig.11-5:Schematic display of five kośas (sheaths or partitions)

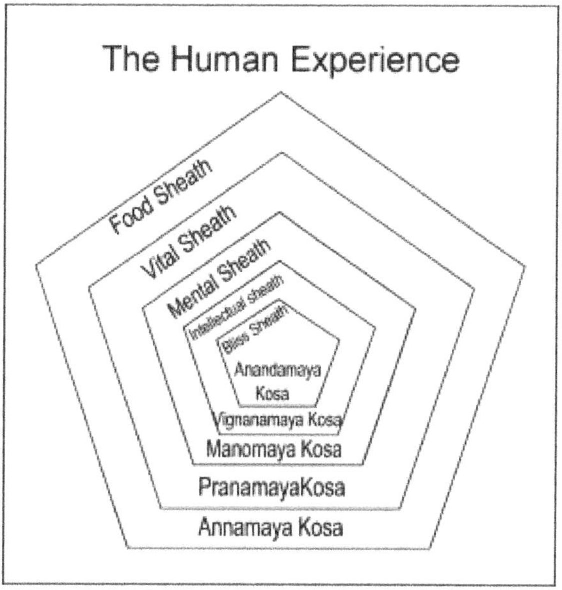

- When the *jīva* is reflected in the Food vestment (*Annamaya kośa*) its purpose is to ensure the well being of the physical body. (Related to State of health)
- When it is reflected in the vital vestment (*Prāṇamaya kośa*), its purpose is to exercise physical energy, recoup it or to enhance it. (Related to physical capacity).
- When the *Jeeva* is in conjunction with the mental vestment (*Manomaya kośa),* its activity is in quest of sensual experiences.
- When the *Jeeva* is in fusion with intellectual vestment (*vijñānamaya kośa*) it is in pursuit of knowledge. (Related to intellectual pursuits) Finally,
- When the Jeeva is fused in blissful vestment, (Anandamaya kośa) it is in peace. (Related to mental equilibrium and peace).

These activities may be related to our mundane objectives as follows.

The objective of the first activity is to maintain good health. The objective of the second activity is to promote the level of physical capacity. The third is in quest of comfort and sensual pleasures; the fourth is to acquire knowledge. The ultimate objective is to achieve the state of equilibrium and peace.

From this analysis, it is clear that these activities do not happen automatically and the three bodies viz., gross body, subtle body and the causal body have to be trained properly and tuned to achieve these objectives. Let us elaborate this a little more.

We can predicate of anything in this world, at a given moment of time, in one of six ways viz.,

1.It is (means it exits already)	2.It is born
3.It grows	4.It changes
5.It decays and	6.It dies.

Of these, the state of perfection can be attributed only to the first predicate. Nothing that has any one of the remaining attributes can be said to be perfect. It is obvious that our physical bodies have all the attributes listed from the second to the sixth. It is our common experience that all other objects also in this universe have these five attributes.

Therefore, we can say that this universe contains two categories of objects with opposing influences, those that promote life and growth and those that promote decay and death. However, these two categories are not mutually exclusive because what is good for one may not be good for another. Therefore, we need a code, which delineates which substances and actions are good, which are bad for promotion of physical fitness, which are good, and which are bad for promotion of the mental and intellectual faculties. You may say that this is not necessary as your physicians can advise you on that; but it is a deceptive assurance as the physician concerns himself only with preservation of the physical body and not with refining it or with maintaining and refining the instruments with which the body works. Here religious code comes to our rescue. Hindu religious code is so comprehensive that it encompasses all the aspects of maintaining and refining the body and the instruments with which it works. It helps to make the body a fit vehicle to attain the state of perfection.

In the citadel of *Jeeva* or Self, the weak points are the vestments, the self itself being immune to all influences. Naturally, the weak points have to be protected from decay and strengthened and refined to serve the purposes of Self.

The food vestment or the gross body is prone to be weakened or damaged due to improper upkeep or due to injurious external influences. Therefore it needs to be guided and regulated with regard its own upkeep and with regard to protecting it from undesirable external influences. The two kinds of training oriented towards these two aims are called *Yama* and *Niyama* respectively in Yoga System. They include almost all kinds of activities discussed earlier.

The vital vestment (*prāṇamaya kośa*) may be weakened or damaged due to misuse or excessive use. Therefore, the organs of action included in this vestment have to be regulated so that the gross body may, attain suitable and steady postures that are useful for maintenance of elasticity of limbs and are conducive to mental equilibrium. This process of attaining a comfortable and steady posture is known as *āsana* (3rd limb) in this system.

The five *prāṇas*, which form the other part of the vital vestment when left unregulated, waste away their great potential. When they are controlled by the process known as *prāṇāyāma* (4th limb), they can raise the individual to lofty heights of achievement. *Prāṇāyāma* that vivifies the body-mind apparatus occupies very important place in Yoga system.

The mental vestment may remain weak by its own doubts and vacillation or it may be weakened by outside distractions in the form of attractions and revulsion through the channels of the organs of cognition. That makes the individual restless and agitated. Training the organs of cognition to withdraw from external objects can control these distractions. This process is known as *Pratyahara* (5th limb). The steadying of the mind after such withdrawal is known as *Dharana* (6th limb) in yoga parlance.

The power of the Intellectual vestment is drained by waywardness of mind and lack of specific aim. This can be remedied by training the mind to concentrate on a specific idea. This process of uninterrupted concentration is known as dhyāna (7th limb).

When this process of uninterrupted concentration continues for a long enough time, so that the self-awareness of the subject dissipates and the consciousness of the object of meditation alone endures, that state is known as the samādhi (8th limb), the state of perfection and absolute bliss. In this state, the Jeeva is fused into blissful vestment (*Anandamaya kośa*), and that is the *summum bonum* of all human endeavors. This supreme state cannot be achieved without a proper training. That training of the body and mind is in strict observation of the psychosomatic code.

THE SEXTUPLET ENEMIES OF MAN

As a precursor to this training, the aspirant is advised to give up certain undesirable qualities. The whole gamut of undesirable human behavior is summarized into six specific qualities, which can wreak havoc with our bodies and minds. They are known as ariṣadvargas or the sextuplet enemies of mankind. They are identified as (1) *kāma* (relentless desire), (2) *Krodha* (Anger), (3) *Lobha* (Miserliness), (4) *Moha* (delusion), (5) *Mada* (Pride) and (6) *mātssarya* (jealousy).

How do the mental villains cause illness to healthy bodies? These qualities can run you down through the resulting fear, mental worry and nervous tension. Let us see how this happens

KĀMA (DESIRE AND COVETOUSNESS)

The fault is not in enjoyment but in craving for it and in that process to covet others' possessions. Coveting is bad because it creates avariciousness. When you get what you desired you desire again for more of it or something else. It creates strife for endless possessions.

Avariciousness or greed is a peculiar quality. It feeds on itself. If you allow it to take control of you, it begins to grow exponentially. As every acquisition feeds the greed, an avaricious person has no respite in life. He becomes angry with everybody and himself, tends to be untruthful, and ruthless if he does not succeed in his campaigns of acquisitions. A greedy man can never have peace of mind as it breeds other dirty qualities like hatred and jealousy which in turn are sure to destroy his peace of mind that is essential for all worthwhile avocations, including reaching the goal of longevity. Restlessness is associated with greed. You need a quiet mind to achieve anything great including success in yoga. An acquisitive mind cannot be peaceful. That is why the Upaniṣads advised "You can enjoy the whole world without coveting it." [127]

(2) *KRODHA* (ANGER AND PERTURBATION)

The mind, when such feelings as anger and hatred obstruct its view, cannot easily see what is right and no human being can hope to arrive at right decisions when his interests and passions are consulted at the same time. When the mind is freely exerted without passion, it's reasoning is sound; but if passion gains possession of mind, it becomes a tyrant, and renders reason powerless. It is a well-documented fact that anger, in addition to distorting the reasoning faculty of mind, produces visible malfunctions of the body.

Anger produces flushed face, slurred speech and twitched muscles and faster rate of breathing. It destroys circumspection and coordination of senses and eventually invites disaster. It increases blood pressure and we know of cases where fits of anger produced fatal heart attacks. Many angry men perished due to their irrational actions produced by anger. We know scores of instances where two people got into a simple altercation that developed into a duel and led to fatal consequences. Getting into an angry mood for a few minutes is said to destroy the *prāṇa* accumulated for three months. Control your anger. *śāntam* or lack of anger is a very precious quality that not only saves you from catastrophes at times but also promotes your healthy life at all times.

A perturbed mind cannot concentrate on any task and fulfill it. Equanimity in all circumstances is the hallmark of yogi. Patanjali, the author of Yoga aphorisms considered this quality so important that he made it the definition of Yoga[128]. You may be perturbed if perturbation solves your problems but it does not, on the other hand, it compounds your problems,

(3) *LOBHA (*MISERLINESS*)*

A miser is a person who himself neither uses, nor allows others to use the goods under his control. He is greedy and nothing satisfies him. The greed of a person does not depend on how much he already possesses. In spite of what one has, the greedy person desires more and is never satisfied. Such persons cause distortion of available resources and cause injury to the society.

(4) *MOHA (*DELUSION*)*

Moha is a state of confusion in which an individual attaches himself to false values. A deluded man attempts to do things to which he is not called upon by his station in life. Conversely, he forgets his station in life and fails to do things to which he is called upon. He also tries to justify himself with false reasoning. One form of delusion results from the feelings of 'I' and 'mine'. Unless one is cleared of this delusion, one cannot make any spiritual progress.

(5) *MADA (*PRIDE*)*

Mada is pride-- pride of learning, or pride of wealth and pride of power. One can have a feeling of contentment of having achieved something worthy but that should not make him arrogant. Such Pride goes before a fall. John Bunyan says: "Seest thou a man wise in his own eyes. There is more hope of fool than of him."

(6) *MĀTSARYA* (JEALOUSY)

Jealousy is perhaps the meanest quality of all. It indicates the impotence and inability of the possessor of that quality to achieve anything worthwhile. Unable to achieve anything useful; he begins to feel jealous of others' qualities, their wealth and status and develops needless animosity towards them. The central idea of the epic Mahabharata illustrates the consequences of unchecked jealousy–destruction of self and the unfortunate society that is controlled by jealous fools–. Shakespeare depicts in his great play Othello, the passion of jealousy in its worst form, he shows how it degrades the great character of Othello. Jealousy consumes the person who harbors it and never leaves him in peace. If you do not give up jealousy, it does not give you up until you are consumed.

HATRED

Hatred is a common quality prevalent in almost all people. This quality involves a reciprocal relationship. Good people hate bad qualities but not people in general, whereas bad people hate good people and their good qualities. This is one of the most difficult qualities to get rid of. Even the finest people hate certain things, bad thoughts, bad qualities of others and perhaps the bad people. It is difficult but possible to get rid of this quality by making a distinction between bad qualities and people who possess them. Even as you condemn bad actions, a bad person need not be hated if he reforms. You may condemn bad actions as such but not harbor hatred towards the actors because hatred breeds ugly thoughts and lands you in unpleasant situations that hamper higher mental pursuits.

It is generally admitted that these qualities disrupt the smooth functioning of social life but it is not realized that these qualities destroy one's peace of mind and thwart the higher pursuits of mind. Anyone who ponders over nature of these morbid qualities and makes introspection understands how difficult it is to get rid of them. Difficult as it is, it is not impossible to get rid of them.

FEAR:

How does fear arise? Fear, according to Vedanta philosophy arises due to the concept of a second entity. As long as a second entity is allowed, the causes of fear are innumerable. For any one who keeps at bay the sextuplet enemies and cultivates the sextuplet treasures, there is no fear, as he has no enemies. A seasoned *Vedanti* (a person who follows the philosophical system called Vedanta) does not fear even from non-personal sources like calamities. Even death cannot terrify him as he identifies himself not with the mortal body but with the immortal soul. For one who feels the entire world is permeated by one's own Self, where is the source of fear? That is why it is said that the hallmark of Vedanta is fearlessness. Fear is a constant source of worry.

WORRY

All the mental disturbances described above can be collectively called mental worry. Worry makes you tense and nervous and affects your stomach. Worry changes the gastric juices of your stomach from normal to abnormal and often leads to stomach ulcers. Dr. Joseph F. Montague, author of the book 'Nervous Stomach Trouble', says, "*You do not get stomach ulcers from what you eat but what is eating you.*" Look at the statement of Dr. W. C. Alvarez of Mayo clinic said to have been made after studying 15,000 patients treated for stomach illnesses. "*Ulcers frequently flare up or subside according to the hills and valleys of emotional stress.*" Worry can bring about all kinds of undesirable changes in the body. Few things can prematurely age a woman and destroy her good looks as worry. Worry and high-tension living can bring about that dreaded pain angina and heart disease. If you love life, and want to enjoy good health and long life, you have to keep peace of your inner self even in the midst of the tumult of the modern world.

How to conquer worry

Before we try to conquer worry, we should understand what exactly is worry. Is thinking or some other mental exercise worry? Mental exercise is obviously not worry. Worry is a sort of fear. Going around in circles over an issue without trying to solve the problem, or fretting and fuming with anxiety, or withdrawing oneself into a mood of depression brooding about life's problems, are all symptoms of worry. There are ways to avoid worry.

Worry is generally caused in two ways. We worry about things we have and we worry about things we do not have. We worry about things that are happening to us and we worry about things that are not happening to us. Bodily ailments give us worry. Spouse and children may also give us worry. If we have a good job, we are constantly worried about losing a well-paying job. If you have a business, you are worried about the failure in business. If you do not have a job, business, or money to lead a comfortable life, you are worried.

Let us see how to get over these worries. If we have some bodily ailments, they may be alleviated by taking some medications and not by worrying. There is plenty of evidence that worry aggravates a diseased condition of the body. If it cannot be cured, it should be endured without worry, because worry does not help us in any way. Softening your attitude and compromising with others can mitigate domestic worries. Do not think that you are always right. Do your best in your job or business and leave the results in the lap of god. If you do not have money, try to earn it by any honest means. Worry doesn't put one cent in your pocket.

After all, most of the things you are afraid will happen to you may not happen at all. Insurance companies make money based on our fears that some disaster will occur to us. They know that by the law of large numbers that these disasters do not occur so often as people imagine. Sometimes things happen the way we don't like. If you cannot prevent an adverse circumstance, or if you cannot reverse it after it happens, then it is better to adjust rather than resist it with your mind and wreck yourself with worry.

Philosopher William James said,

> *"Be willing to have it so. Acceptance of what has happened is the first step to overcoming the consequences of any misfortune."*

We are not advocating that you bow down to all the adversities that come your way. As Mahatma Gandhi said, compromise with evil cannot be defended in the name of non-violence. What we are saying is, let's fight as long as there is a chance of saving a situation. When we are up against something that is so–and cannot be otherwise– then, let us give up unnecessary mental resistance and follow the advise of Mother Goose.

> For every ailment under the sun,
> There is a remedy, or there is none;
> If there be one, try to find it;
> If there is none, never mind it.

Circumstances alone do not make us happy or unhappy. It is the way we react to circumstance that make us happy or unhappy. As the German philosopher, Schopenhauer put it

> *"A good supply of resignation is of the first importance in providing for the journey of life."*

It is also the essence of the code of conduct prescribed by the Hindu philosophers for achieving tranquility of mind and attaining immortality. A summary of the code is furnished here for the benefit of the reader.

Positive Disciplines

After cultivating the preparatory disciplines of ridding yourself of the negative qualities described above, you are ready for the positive steps with a clean and receptive mind. You are now asked to begin practicing some disciplines that make the body healthier and the mind quieter and adaptable to higher pursuits. Just as six enemies of mankind were identified, so also six treasures also were also identified and placed among these disciplines.

THE PSYCHO-SOMATIC CODE OF VEDANTA SYSTEM

Sri Sankaraachaarya of non-dual philosophy fame prescribed a group of four qualifications to one who is desirous of knowing the Self. This is a group of mental qualities, which should be cultivated by the advanced students. These are the required disciplines for one who aims at liberation.

1. *Viveka* (Discrimination)

The student is advised to start with practice of discriminating the permanent from the transient, real from the unreal. The objective of such discrimination is detachment.

2. V*airāgya.* (Non-attachment)

The aspirant should gradually cultivate to overcome all unnecessary desires and avoid attachment to unnecessary worldly objects, as desire and attachment are the root causes of sorrow and our finite existence.

3. *satka sampatti* (The Sextuplet Treasures)

The six treasures to be possessed are called Sama, Dama, Uparati, Titiksha, Sraddha, and Samadhanam. Their meanings arc explained now:

Name of treasure	Meaning
1.Sama:	Control of mind
2.Dama:	Restraining the organs of cognition form running after worldly objects,
3.Uparati:	Focusing the mind on the Ultimate Reality,
4.Titiksha:	Forbearance or enduring the pairs of opposites: like heat and cold, gain and loss, weal and woe, respect and contumely, without dejection and lamentation
5.Sraddha:	Devotion to cause, and
6.Samādhānam:	Compromising the mind for giving up trifles and concentrating on the Ultimate Reality.

4. *Mumukshatvam* (Strong Desire to attain divinity)

Attaining immortality is well nigh impossible without a strong desire and commitment to attain it. Persistent desire and commitment to attain that goal evolve the seeker from the animal existence to that exalted state of immortality from which there is no return.

The Yoga System

Yoga system is a life science technology. It does not make statements about any specific religious practice. As such, it can be used to support any religious faith. It is a comprehensive approach that aims at integrating body, mind and spirit. The author of Yoga Sutras has prescribed another group of preparatory disciplines, which collectively go by the names "Yama" and "Niyama," which constitute the psychosomatic code for the aspirant. These disciplines have special significance as they are regarded as universal values. Let me caution those people who shun anything that has the slightest flavor of 'spiritual', that these disciplines were formulated in ancient India when people would not give much credence to anything that was not spiritual. Knowing this disposition of contemporary people, Patanjali, like other great Indian seers, gave these psychosomatic disciplines, the color of spiritual. In fact temporal cannot be separated from the spiritual in his system. The eight limbs of his system are displayed with their meanings and explanations in Table 11-3.

Table 11- 3: Patanjali's Eight-fold Yoga system

1.Yama	Abstentions	Non injury, Truthfulness, continence, lack of greed and non stealing
2.*Niyama*	Observances	Purity, Contentment, Austerities, Study of Scriptures
3.*Asana*	Postures	Physical postures that enable you to remain steady for long periods.
4 *Prāṇāyāma*	Breath Control	Acquiring control over *prāṇa* and direct life energy at will.
5. *Pratyāhāra*	Sense withdrawal	Decreased reactivity to sensations
6 *Dhāraṇa*	Concentration	Bringing the mind to one point of target.
7.*D hyāna*	Meditation	Unbroken stream of thoughts on the target for extended period.
8. *Samādhi*	Identification	Ecstatic union with the Ultimate Reality

THE TEN COMMANDMENTS:

The Ten Commandments of Hindu religious practice are the first two rungs of Yoga, known by the collective nouns Yama and Niyama. Yama is a collection of five 'Dont 's, whereas Niyama is a collection five Do's.

1. Yama (Abstensions or Don'ts)

The group of five abstentions viz., non-injury, continence, lack of greed, non-stealing and truthfulness (not breaking the unity of thought, speech and action) constitute Yama. The five modes of abstinences constitute the first rung of the ladder of Yoga.

2. Niyama (Observances or Do' s]

This group of five disciplines, viz., Purity (śaucam), Contentment (*santosham*), Austerities (*tapas*), Study of scriptures (*svādhyāyam*) and Devotion to Divinity (*īśvara praṇidhānam*) constitute Niyama.

Yama and Niyama form an ethical code of a high standard to which the aspirant should adhere to the best of his/her ability. It is impossible to achieve any lasting success in Yoga if one knowingly lives contrary to the defined modes of abstinences and observances. If they cannot be followed to the letter, one should at least earnestly attempt to live up to the ideas embodied in them. A perusal of the details furnished below enlightens the readers as to why they are important not only for spiritual advancement but also for maintenance of a healthy and long life.

Non Injury (Ahimsa)

> Those noble minded persons who desire beauty, faultlessness of limbs, long life, understanding, mental and physical strength and memory should abstain from acts of injury....
>
> ------(Mahābhārata XVIII: 115.8.)

Non-Injury implies neither injuring others nor being injured. Avoiding being injured is very important as you may succumb to the injuries if they are fatal or they may cripple you if they are less serious injuries. Injuring others has to be scrupulously avoided because the injured party may retaliate you with injuries, and you may have to suffer the same consequences. Then there are cases when the injured party cannot retaliate. Even then, retaliation will come back to you in some form, sometime later.

The connotation of the word "ahimsa" (Non violence) is not injuring others, Non violence has never been more relevant to mankind than it is today. Mankind's survival is at risk now because the world is laden with destructive weapons of immense potential. If we want to live, we need new conceptual tools. One of them is non-violence. Strictly speaking, avoidance of violence completely is not possible. Activity is not possible without certain amount of violence. What we have to do therefore is to minimize violence to the greatest extent possible. Violence invites retaliation, that creates counter violence, and there will be no end for the sequence of violent actions and their reactions. The resulting orgy of violence threatens lives. Realization of this fact prompted the 20th century apostle of peace Mahatma Gandhi to advocate non-violence.

What is its relevance to the individual in this context? Being non violent to others is the primary requirement to protect your own body. In other words, do not become a victim of retaliatory violence by initiating violence. Even if your victim is unable to retaliate, your mind predisposed to violence can never be peaceful. Without being peaceful, you cannot succeed in concentrating and therefore you cannot pursue any avocation, successfully, much less life extension activity.

1.Practice of Yama

(a) Satyam (Truthfulness)

The correct meaning of "Satyam" is maintaining unity of thought, speech and action.

You may wonder why, on earth, should one be truthful in order to live long. The answer is twofold. In the first place, one's untruthfulness is meant either to gain undue advantage for oneself at the expense of others or to cover up one's faults and evade punishment. In either case, you injure life in the society in which you live and want to live an extended life. You cannot continue this deception for long, the society will get even with you, and will not let you live forever. Secondly, your untruthfulness blemishes your conscience and puts you ill at ease. The mind of a person with a guilty conscience becomes turbulent. A worried mind automatically leads to a diseased body. He can never concentrate his mind because the discriminating faculty of the mind-stuff has the peculiar habit of floating up to surface, the blemishes buried deep in his conscience and agitating his mind. Minds welled up with feelings of guilt and agitated minds are unfit for concentration. If meditation is accepted as a method of prolonging life span, one should be able to control the mind and that control of mind becomes impossible to one who is not truthful. This should not be interpreted as a purely spiritual discipline. Is there any law court in the world that allows testimony based on untruth? Truthfulness is a required quality in either secular life or spiritual life.

(B) ASTEYAM (NON STEALING)

Half the woes of the world are due to the tendency of people to steal the property of others. Whether it is stealing material objects or stealing time at work, it is all the same. This quality is prompted by greed.

(C) *BRAHMACARYAM* (CONTINENCE)

The strict meaning of continence is control of sex by self-restraint. This word is sometimes confused with the word Celibacy. Celibacy has a different connotation. Celibacy is strict and total abstinence from sex at any time, whereas continence is self-control in sex. An equivalent word in Sanskrit for continence is '*Brahmacharya*.' Etymologically it means the act of moving in the realm of Brahman.

Continence has a very important place in religious thought. Many ancient Hindu seers were not celibate, but they were strictly continent. It appears that celibacy as religious practice acquired significance subsequently due to the realization that married life creates many encumbrances and blocks spiritual progress.

The physiological importance of continence is obvious as sex life involves discharges of potent germ cells in the form of semen. Frequent discharges enfeeble a man. Retention of it is said to retain the *Ojas* (potency) in the body. This idea accords with the claims of some molecular biologists that their experiments with fruit flies proved that delaying reproduction could extend longevity.

Celibacy is possible only for a few people and only those called upon by their special station in life may remain celibate lifelong. Others should exercise restraint in their sexual practices.

Continence has acquired special importance in the present world, with a different context. World population has reached new high peak intensifying the competition for the limited resources. The ecosystem will soon breakdown unless population growth is stopped. The dreaded disease called AIDS has forced the modern man to change his methods of sexual behavior, which he would not do voluntarily. It has been established that sexual promiscuity is the only cause of AIDS. Scared by this dreadful disease, he is now willing to adopt mechanical methods developed in the context of population control. These mechanical methods may serve him to avoid contracting the disease but they do him no good otherwise as his potency is discharged as usual, because these mechanisms can do nothing to check the emotions of the partners. They are no way useful for personality development or for maintaining health. Therefore, it is necessary to repeat that longevity seekers should observe continence.

(D) APARIGRAHAM (NON-ACQUISITION)

The principle involved in this injunction is that we may have the necessities of life but not more. Most people want to accumulate a lot more than they actually need. This craze for accumulation creates a lot of unhappiness to the individual as well as his society. Non-acceptance of gifts is enjoined upon the aspirant as acceptance of gifts from anyone subordinates his will and freedom. The student should rise above misery-causing characteristic of greed. He can never remain calm so long as he craves for accumulation of unnecessary things.

2.Practice of Niyama

(a) Śaucam (Purity)

Cleanliness of body and purity of mind are the sin qua non of healthy long lives. It is because of improved sanitation that the life expectation in 20th century took a quantum leap. Perfect personal cleanliness and cleanliness of precincts is a fundamental requirement in the quest for healthful life extension. Unfortunately, modern civilization, which claims to have improved the lives of people, is itself turning against them. Individual efforts to live clean lives are no longer useful because of overwhelming environmental pollution.

In many developing countries, solid waste accumulates in mounds in the streets and neighborhoods due to lack of proper disposal. It is often thrown into open gutters blocking the flow of sludge creating terrible unsanitary conditions with unbearable stench. Industrialization has brought in its wake tremendous environmental pollution. Our rivers are fearfully polluted. In the olden days, you could take the water flowing in a river into your palms and drink it carefree. Now you cannot do that, as you can be sure that some factory upstream has discharged its effluent water into the river. With factory chimneys continuously gushing out smoke into the atmosphere and automobiles spewing exhaust fumes directly on the faces of people in the streets, we cannot even get clean air to breathe. If you ever visit Calcutta observe this: You go out into the city with clean white clothes and by the time you return home you can see a coat of soot on your clothes.

Individuals are helpless in such circumstance, to keep clean. It is imperative therefore that aspirants to longevity have to find some method of removing themselves from unclean surroundings to fulfill their dreams of living healthy long lives.

Internal cleanliness or keeping the purity of mind is equally important. Any morbid thoughts that enter the mind should be driven away promptly as morbid thoughts lead to disease of body.

(B) TRUPTI (CONTENTMENT)

A feeling of contentment will develop if the student focuses his attention on the favors life has bestowed upon him rather than harping on what he does not have. Happiness may be defined as the ratio of the number of desires fulfilled to the number of total desires entertained. To make the quotient larger, the numerator has to go up or the denominator has to go down. The numerator in this fraction can never be made equal to or greater than the denominator as it is impossible to any one to fulfill all his desires. As the denominator in this fraction increases, the possibility of happiness recedes farther and farther. The only available solution is therefore to reduce the denominator, (viz. total desires) to make the quotient larger. Therefore, one should learn to limit one's desires and cultivate contentment. You should count your blessings.

(C) TAPAS (AUSTERITY)

Tapas means willful curtailment of the excesses of life and practicing to live on bare necessities. It is an effort to bring the organs of one's body under one's control by deliberately restraining their activities. *Tapas* is not just an outward show; if it were so, it would be waste of effort. Living physically austere life, while dreaming of worldly riches is not austerity. Physical austerity should be combined with mental control. Austerity is an essential way of life to those who are intent not on the external riches of outer world but rather on the immense wealth of the inner life.

(D) *SVĀDHYĀYAM* (STUDY OF SCRIPTURES)

Svādhyāyam is devotional study. The aspirant should study constantly the texts that deal with great questions of life in this world and the nature of Self.

(E) *IŚVARA PRANIDHANAM* (DEVOTION TO DIVINITY)

Iswara pranidhanam means devotion to the omnipotent Divinity, which is untouched by the afflictions of life.

3. *Asana* (Physical Posture)

The third group of disciplines in the 8-limbed Yoga system is known as āsana, which formulates a comprehensive physical toning system. This third group of disciplines consists of various physical postures and exercises popularly but erroneously known as yoga in the west. This was discussed earlier under the heading physical culture. There are various physical postures designed to maintain elasticity of ligaments and keep the body nimble and agile. They develop endurance rather than build muscles. The reader is referred to various texts available[129] on this topic for details.

4. *Prānāyāma* (Breath control)

The fourth and perhaps the most important discipline in the Yoga system is *prāṇāyāma.* Although it is listed as one of the eight limbs of Patanjali's Yoga system, it is the lifeblood of every other Indian system of religion.

Prāṇāyāma as remarked earlier is a psychosomatic process. *Prāṇāyāma* and meditation are closely related processes. Unless you are able to restrain your mind from its wanderings and focus it on regulation of breathing, you cannot do *prāṇāyāma. Prāṇa* and *Chitta* are in constant association. Where there is *Chitta*, there *prāṇa* is focused and vice versa. The *Chitta* is like a vehicle propelled by two powerful forces *prāṇa* and vāsanas (tendencies). It moves in the direction of the more powerful force. Like a bouncing ball, the practitioner is tossed between *prāṇa* and *Chitta.* If *prāṇa* (breath) prevails, then the desires are controlled, senses are held in check and the mind is stilled. If the force of desires prevails, then breathing becomes uneven and the mind gets agitated. Success in *prāṇāyāma* implies success in training the mind to flow in one channel.[130] One, who is able to meditate and reach the state of *samādhi* transcends time. A casual reader may wonder how anyone can escape time, which is a universal phenomenon. A careful analysis of the concepts of space and time and its supposed universality can set our doubts at rest.

[5] *Pratyāhāra* (Restraining the Mind)

Pratyāhāra is withdrawing the mind from the objects of outer world and restraining the mind. When *pratyahara* is practiced the concepts of time and space will gradually vanish.

We begin this inquiry by trying to understand what we mean by the words 'space 'and 'time.'

Space: We become aware of the objects around us by the senses of sight and touch and to a lesser degree by the senses of sound and smell. We perceive these objects by different groups of our sense impressions. We recognize an object as a lump of clay and another as a piece of wood by the different sense-impressions received by us. When we receive the sense-impressions of an irregularly shaped and malleable object, we combine them with our stored impressions of softness when wet and hardness when dry and recognize it as clay. When we receive the impressions of a rugged opaque, grainy surface we combine them with our stored impressions of hardness and combustibility and recognize it as wood. In doing so, we associate a number of sense-impressions into one group and a different set of sense-impressions into another group and separate these groups. When we separate one group from another by perception, we consider they exist in space. When we shape a lump of clay into a pot, we consider there is space inside the pot. Space is thus a mental expression for the fact that the perceptive faculty has separated co-existing sense-impressions into groups of associated impressions. If the pot is broken, there is no pot-space. If the roof and the walls of the room in which I am writing are removed, there is no room-space. Pot space is mentally constructed by perceiving the shaped clay and the room space is mentally constructed by the perception of walls and the roof.

To say therefore, that a thing exists in space, is to assert that the perceptive faculty has distinguished it as a group of sense-impressions from other groups of sense-impressions, which actually or possibly exist. Space is simply a mode of perception. It has no existence if these objects are withdrawn, no more than a pot-space if pot is broken or no more than a room space if the walls and the roof are removed. Space is a mode in which we perceive things apart and it is limited by individual perception.

Time: Much that has been said about space applies to time also. Space and time are so familiar in character that if we call space the breadth of the field of perception, time may be called its length. Just as space is one mode in which perceptive faculty distinguishes objects, so time is the other mode. If perceiving the coexistence of perceptions at a time is like measuring breadth, perceiving marks of the progression of perceptions at a position in space is like measuring length. If there is no progression of perceptions, there is no time. While the individual does not experience progression of perceptions, time stops for him. Thus, space and time are creatures of mind.

(6) *Dhārana* [Bringing the mind to one pointedness]

Dhāraṇa is the condition in which the mind is confined to a single object. In this state, the student tries to keep the distractions away and points the mind towards the target. Internal yoga begins with *Dhāraṇa,* while the previous steps are preparation to control the mind.

(7) *Dhyāna* [Meditation]

Dhyāna or meditation is the condition in which the student succeeds in keeping his (her) uninterrupted attention on the object of concentration. This is a prolonged and deepened form of concentration. Distractions no longer come into the field of practitioner's consciousness. Self-awareness remains, though vaguely.

[8] *Samādhi* (Merging with Brahman)

Samādhi is the condition in which one loses self-awareness and there remains only the consciousness of the object of meditation. In this state, the mind sheds its notion of separate existence and merges with the object of concentration, the universal Self, called Brahman.

The adept established in a state of Yoga *samādhi* does not perceive sense impressions, becomes unaware of the progression of time and location in space.

TIME CONCEPT IN RELATION TO YOGIS

It is believed that time perception vanishes to the yogi while he is in a state of meditation and holding fast to its target. The essence of yoga is mental control. While the mind stays put at a point without wavering, there is no progression of perceptions and time is suspended in that state. *prāṇāyāma* requires suspension of mental activities and concentration of mind at a point. That is why time stops for a yogi while it flies for others. Time concept changes to one who is established in *prāṇāyāma* because it induces meditation and **the adept in yoga does not perceive progression of perceptions while in a state of meditation.**

You may be wondering whether the time concept changes from one person to another and one environment to another. Yes indeed, though not with every person, because most of us belong to the same flock and our perceptions are almost the same. If the two persons live in radically different environments and the environments have different frames of reference time concept changes.

Concepts of time depend on the state of development of mind. Animals have their own conception of time, which is different from ours. Animals and birds do not keep a calendar like us but they know instinctively when to mate, when to migrate or when to hibernate as winter approaches. These natural actions of animals and birds are guided solely by their instinctive minds.

ILLUSIONS CREATED BY MIND

The life spans of animals and birds are shorter than that of man, yet they think their short life spans are full life spans. To humans, twenty years life span of a monkey appears very short but to the monkey, it appears as full as hundred years appears to us because the mind can create illusions of making twenty years as full to one animal and partial to another. In other words, the mind can create the illusion of making hundred years appear as partial and twenty years as full. Every one of us is familiar with this phenomenon. When we are happy in a situation we like, time flies and we view it as contracting. On the other hand, when we are in an unhappy situation, time crawls and seems to be expanding. A story from the old Hindu scripture Srimad Bhagavatam illustrates this concept.

> Once, a great king took his daughter to the plane of the creator Brahma to know the name and identity of a worthy husband for his daughter. After arriving at Brahma's court, he waited for a few moments there and made his request to Brahma. To his surprise
>
> Brahma said "O king, when you go back to earth you will not see any of your people there-your relatives, friends, your subjects- and even your kingdom will not be there as they went into the womb of time. Though you arrived here only a few moments ago, these few moments here are equal to several thousands of years on earth and so all your people were gone. When you go back to earth it will be a new age and you will see Lord Krishna's brother, Balarāma who will be a worthy husband to your daughter." When the king came back to the earth after his few minutes in the world of Brahma, it was altogether New World he saw, new civilization, new faces and places. However, he found Balarāma and married his daughter to him. By this adventure, his daughter had to marry a person born in a subsequent age much younger to her. The point our interest here is not whether Balarāma agreed to marry a lady very much older to him but it is to emphasize the fact that time is not an absolute concept but relative.

This story should not be dismissed as silly even if it were fiction as it gave us a vital concept, which anticipated the now celebrated theory of relativity, and it simply dramatized the ancient Hindu concept of time's dependence on the frame of reference through which it is reckoned. As we now know from the theory of relativity, what is true in a particular frame of reference is not true in a different frame of reference.

These basic ideas of time can be translated into basic ideas of relativity postulated by Albert Einstein.

> A clock attached to a system moving with high velocity, appears to slow down to a stationary observer, compared to a stationary clock though both are of the same make. Again, a rod kept in the moving system appears to shrink to the stationary observer. However, if the observer also rides in the same moving system these changes do not appear to him. Just as speed shortens length, speed slows down time. These changes are not limited to rods and mechanical gadgets like clocks. Even physical, chemical, biological and mental processes will slow down if a person travels with very high velocity.
>
> Now let us consider the time concept as related to a passenger on a fast moving spaceship and his family on the earth. Suppose that in 2000 a space ship left earth to go to a distant planet in the solar system. Assume that it takes about 40 earth-years to complete the round trip, i.e. the space ship will return to earth in 2040 AD. One of the crew, who was 25 years old when the ship left the earth, leaves behind his wife aged 20 and a daughter aged one. When he returns home after a few hours (reckoned by his own watch) from his trip, he finds to his surprise, that his wife is no longer the young teenage beauty but middle aged woman of 60 with wrinkles on her face. His young baby daughter had grown up to be a mother. He himself is still a few hours past twenty-five. When he left the earth and traveled into space with tremendous velocity, he was no longer subject to the limitations of the time frame of earth.

When a yogi controls his mind and enters into the state of *samādhi*, he escapes the time frame of earthlings, having stopped the progression of mental perceptions. Thus, *prāṇāyāma* is a superb psychosomatic exercise.

The reader should have understood by now why the sage Patanjali defined Yoga as restraining of mental modes. If yoga is essential for radiant health and longevity, a placid mind is the sin-qua-non of yoga practice. It is not only in the Yoga system of natural healing and the *āyurveda* system of therapeutics that mind is given the prime of place, but also in Homeopathy which has a western origin. In Homeopathy, mental symptoms are given precedence over physical symptoms of a patient. Let alone the general mental dispositions of a patient like sanguine, phlegmatic etc but also mood changes like anxiety, fear, and dullness are used in prescribing a medicine. If two patients suffering with the same kind of ailment differ in mental symptoms they are administered different medicines. If that is the exalted position occupied by the mind, then, what is its status in the Allopathic system known as modern medicine? Almost nil.

One of the paradoxes of human life is that our mind is both our friend and our enemy. One great saint compared it to an intoxicated monkey stumbling on red-hot embers. It is our enemy until it is trained to behave well. Mind is extremely diplomatic, cunning and crooked. One of the mind's master strokes is it's artfulness in making the individual believe that he knows his mind perfectly well and cannot be led away by it; and at the same time to delude him totally. It is an arch-deceiver. It has the knack of making the unwary individual think herself the master while it makes a hopeless fool of her. Its deceptions are subtle. You have heard the saying "The devil can quote scriptures" for its purpose. Similarly, the mind can use a virtue to indulge in a vice. Mind has an inborn inclination to perversion. It can take the support of a perfectly good principle seemingly to justify even the most unscrupulous action. Its tricks can never be discovered unless the individual scrutinizes it dispassionately. Knowing this nature of mind if you force it to obey you, it rebels. Therefore, you should convince it carefully about the disadvantages of indulging in evil or worthless pursuits and gently persuade it to return to noble and worthwhile pursuits. As the senses constantly gather the material from outside world to feed the mind, a necessary adjunct to control the mind is withdrawal of the senses from their wanderings. Practice of meditation becomes possible only when you can restrain your senses at will, from their wanderings. Meditation requires no external adjuncts. Just get away from distractions, sit still and concentrate your mind on regulated breathing. As soon as concentration settles on breathing, wanderings of mind are curbed. With persistence in that mode, mind can be stilled.

4.Right Choices

Quite often in life, we are confronted with situations when we have to choose this or that. When we are at cross roads, we have to choose which way to go. Sometimes we do it unaware, sometimes we pause and think a little before choosing and sometimes we stop and deliberate on the course of action. Certain routine actions we do are really our choices put into practice and they become habitual. Whether these are habits or occasional choices, they have consequences on our lives. Different paths lead to different destinations though some of them may converge at times. Your present state of affairs is the result of the choices you made earlier and your future will depend on your present choices. Therefore, at every stage of life we have to make the right choices.

Some Hard Choices:

Choosing Between the Pleasant And The Good

Our actions are based on some criteria. We have freedom to choose a criterion but when once the choice is made there is less freedom to our actions. The nature of actions leading to a goal is the same though the actions may differ in detail. Thus, the nature of actions leading to bad goals is necessarily bad and vice versa. That is why we have to make good choices.

Why do people make wrong choices? Wrong choices are made sometimes because of the confusion in their goals and more often because of their preference of short-term results to long-term goals. This difficulty arises because of the distinction between the pleasant and the good. Pleasant things are more readily available than the good and they tempt; but pleasant things are not necessarily good nor do they always lead to the good. A beautiful story in Upaniṣads revolves round this idea of distinction between the pleasant and the good.

> Yama, the impartial god of Death grants three boons to a young celibate called Nachiketa, a spiritual aspirant and visitor to his abode, to recompense for his long wait of three days and nights at his door step while he was away. The first two boons were easily granted but the third one was not easy to manage. Nachiketa asked Yama to tell him what happens after a person dies. Does he exist or does he not exist? He wanted to know whether any entity called 'Soul' survives the death of body. Yama shirked answering this question, as it is futile to impart this knowledge to all and sundry, who were not well prepared to receive it. Therefore, he asks the young aspirant to release

him from this boon and choose another. As Nachiketa persisted with his question, Yama wanted to deflect his resolve, and lured him with sons, grandsons, wealth, a large kingdom and long life.

Says Yama: "Whatever objects are difficult to get in this world of the mortals, do thou ask for them according to your choice. These damsels, along with chariots and musical instruments – such as those not obtainable by ordinary men – be thou attended on by them whom I will bestow upon thee. Ask not anything about death."

Nachiketa replies: "O, God of Death, all these things you offer me are most transient and the whole span of life is but very short. I have no use for them. So, keep all these horses, music and dance for thyself. I choose that boon alone and nothing else." Thus the young aspirant persisted, as he was convinced of the utter meanness of sense pleasures and the fleeting nature of life. Even if it is extended beyond normal life span, that is again temporary. He thinks that it is impossible for anyone who has realized his mortal nature to ask for sense pleasures from immortal gods. He should ask for nothing but immortality.

Amazed at the steadfastness and pleased with the sincerity of his guest, Yama then reveals to his young guest the secret of soul in a famous discourse, which starts with his commendation of his young disciple for his wisdom in discriminating between "the pleasant" and "the good". "Both the good and the pleasant approach the man, the wise discriminate between the two, having examined them well. Yea, the wise prefer the good to the pleasant, but the fool chooses the pleasant, through avarice and attachment"[131]

The problem exemplified in this story is common to all of us. Quite often, we choose what is pleasant and not what is good. Is it because we don't know what is good for us? No, we know what is good. Situations in which we are unable to discriminate between the pleasant and the good are very rare. Does not the smoker know that smoking tobacco is injurious to his health? He knows, but he chooses deliberately the pleasure of smoking. Does not the eater of beefsteaks know that he runs the risk of heart disease by what he is eating? He knows, but he does it knowingly to satisfy his palate. Such choices indicate preference of sense pleasures, however temporary they may be, over more enduring values like good health and longevity. One may then ask how to know what is good and what is not. Is it not possible that 'the pleasant; is also 'the good?' It is not argued here that all pleasant things are to be eschewed always. The point emphasized here is that craving for sense pleasures even while the body is crumbling is not wise because one can neither satisfy his sense cravings nor attain the more enduring spiritual fulfillment, by hanging on to something that cannot support.

One crucial point emphasized by our ancient seers needs our careful attention. They did not prohibit the enjoyment of pleasures per say. What they say is they want to emphasize the transitory quality of sense pleasures in order to guide us to achieve pleasures that are more enduring. You may enjoy pleasures as they come but do not run after them when they go.

Choosing Between the Transitory And The Enduring

Wisdom of choosing between the transitory and the more enduring is not confined to spiritual matters; we see its value in mundane affairs. We sacrifice many comforts in the young age for education and for molding our personalities in preparation for the long-term goal of a good career. We sacrifice present consumption in order to save for the future. Every employee knows that he has to endure many temporary inconveniences of office going, pain of pin pricks from his boss and overtime work in order to keep his job and enjoy security of a permanent job. There are many such instances of our choice of the more enduring to the temporary.

Similarly, if you want to enjoy better health and aim for better score on longevity scale, you have to make many hard choices which involve giving up some temporary pleasures.

Choosing between Emotion and Prudence

A prudent person holds emotions in check. Life is not a bed of roses. Even if you think it is, remember roses do not come without thorns. If you are intimidated by the problems of life and resort to some suicidal act, obviously you are not a candidate for longevity. Have not many people jumped from the Golden Gate Bridge into the deep abyss below, torn by some problem of life! What longevity recipe can help such people? If you enter into duels with others rightly or wrongly, aren't you at risk of ending your life prematurely?

Similarly a person has to exercise due care in the daily conduct of life to avoid shortening of life and limbs. If you jaywalk on a busy road and if you are hit by a speeding automobile, what fountain of youth can help you? If you enjoy being an alcoholic, you have made your choice of temporary pleasure against longevity. If you enjoy the feelings of elation, by using unethical drugs you have revealed your preference to the fleeting pleasure of drug addiction over longevity.

If you develop a proper perspective of life, you can hold emotions in check while making your choice. Quite often in life, you have to make proper choices to prolong life. Right from the days of choosing your career, you have to make decisions, almost every day, that affect your life. The career you choose has something to do with your life, because of the various degrees of risk associated with various professions. If you choose the career of an infantry soldier, you are at higher risk than say a white-collar worker. The probability of death in an air crash is more to an airline pilot who is on travel circuit most of his time than to an artist who spends most of his time painting at home

CHOICE AND CHANCE

Chance and historical contingency give human life its élan and fascination. Can we always choose what we want? Obviously, we cannot. Sometimes we have little or no choice. When we have a choice situation, it may not be easy to choose. The choices we make may not be proper. They may not yield the desired results because we had to choose without proper and complete information, or we lacked the ability to judge under conditions of risk and uncertainty. Even when we feel satisfied that we made a good decision it may not yield the desired result due to intervention of chance events.

Nature and life may be simple to the maker who designed them and keeps them going but to the human mind, they continue to be mystery. In spite of the arrogance of man in proclaiming that he acquired control over Nature through science and technology, he is still a creature of chance. No man or woman can ever choose his parents and hence his genes. Who does not like to choose to be born to rich or famous parents? Has anyone done it? Right from the origination in mother's womb to the end in mother Earth's tomb, human being is haunted by chance. When the baby is in mother's womb, it is influenced by the habits and mind of mother. For example, the baby has no control over the birth defects caused by the mother's habits of smoking, alcoholism and drug addiction or by her mental attitudes. In the prenatal stage, the baby in womb, to a smaller or larger extent, shares all the influences to which the mother is exposed. At the time of birth, the human baby is utterly helpless, worse than an animal baby. It's future is completely determined by the behavior of it's parents and the environment into which it is thrown.

This can be illustrated by numerous examples but the life of warrior Karna♦ in Mahabharata epic is a glaring example of the fate of the human baby who is abandoned by it's mother. The place of birth has a lot to do with your comforts and progress in life but can you choose in which country you wish to be born? Even the time of birth has something to do with your life. If you are born the first in a royal family, you get the right of succession to the throne. Can you choose to be born the first in a royal family? How about gender? In some countries, only the male progeny shares inheritance. Can you therefore choose to be born a male? In most societies, males dominate females and ill-treat them. Can you therefore choose to avoid being born a female? A slew of life's events are determined by the time, place, sex of birth and parentage in which we have no choice. Even after birth, chance plays its role effectively in human lives.

Most of us are creatures of tradition and it takes a lot of courage to break the tradition and tread a different path. Sometimes we do not know which way is better. This is especially the case when we are sick and seek to get rid of sickness. There are no foolproof methods to get rid of diseases. In spite of all the propaganda in

---

♦Though Karṇa was the natural heir to the Paandava kingdom by dint of his being the eldest son of queen mother Kunti, and a valorous warrior endowed with qualities of charity and loyalty, he was denied the rights and privileges of royalty on the erroneous presumption that he belonged to a lower caste as he was raised by a suuta (a chariot driver), after he was abandoned by his unwed mother Kunti soon after his birth.

favor of modern medicine, it is a colossal failure when compared to other sciences. No medical doctor has ever succeeded in curing all the dangerous diseases. More often than not, they complicate the problems of diseases. That is the great tragedy of mortals. This argument is not to discourage you from seeing your doctor and emptying your pocket book on his table. Go ahead and take his advice, but remember you are wise enough to use your discretion to accept, modify or reject his advice. Medical doctors are neither gods nor their methods infallible. Choices are tough in these situations, but that is the name of this game and you have to minimize risk always if you cannot avoid it.

WHICH IS SAFER? AN AIRCRAFT OR A HOSPITAL!

Ask some critics of medical incompetence or negligence. There are situations in which we are forced to make decisions and yet cannot make foolproof decisions. Such situations occur when we are in need of medical care. We become victims of existing medical establishment due to its tight internal organization made impervious to external checks and the intense propaganda in favor of the infallibility of the medical personnel and their methods.

Tens of thousands of people in USA are killed by unnecessary surgical operations and medical errors. According to a report forgivingly titled "To err is human", by the National Academy of Sciences (released in November 1999), at least 44,000 people and possibly as many as 98,000 Americans die every year from medical mistakes made by physicians, nurses, pharmacists and other health care professionals. Even using the lower estimate, preventable medical errors in hospitals exceed deaths attributable to such feared threats as motor-vehicle wrecks, breast cancer, and AIDS.

That means admission to a hospital in America puts you at a greater risk than the risks of dreaded diseases like cancer, AIDS, and auto accidents combined.

This figure includes only two kinds of errors – those made by the physicians and errors in medications within hospitals – but not deaths due to infections picked up at the hospitals and deaths due to mistakes made by physicians outside hospitals. Picking up contagious disease in hospitals is real, and not mere fiction as portrayed by Robin Cook in his novel *Contagion*.

All these errors go unreported, as hospitals are not required to report their errors causing deaths. The federal health officials are unwilling to endorse a law requiring hospitals to report all such errors causing deaths or serious injury to patients. They seem to be protecting the industry by avoiding disclosure of information that could result in large damage claims. This attitude of suppression is not without dissent. At least one member of the NAS, Dr. Donald Berwick opposes this attitude. He counters *"Suppose an airline could crash an airplane and not tell anybody about it. That would be wrong. Likewise if people are seriously injured because of errors in medical care, don't they or their families have a right to know?"* Let us see how the number of people who die due to medical errors compares with the number of people who die in air crashes. If a passenger aircraft carries 268 people, it would take one air crash everyday[132] of the year to reach the number of deaths due to medical errors that go unreported. Would you keep quiet if there is an air crash every day?

According to this report, about 33.6 million people are admitted to hospitals each year in USA. Somewhere between 2.9% to 3.7% of these patients, suffer some 'adverse event' while in the hospital. Of these adverse events, about 8.8% to 13.6% are fatal. About 53% to 58% of hospital deaths are attributed to human mistakes or negligence. Based on these figures we can say that one out of every 500 people admitted to hospitals is killed due to medical negligence or errors. By comparison, the chance of dying in an airline crash is one in 8 million flights. The figure of 1 in 500 is based on information disclosed in patient records. "*Most errors and safety issues go undetected and unreported, both externally and within health care organizations. Silence surrounds this issue,*" says the report.

Therefore, you have hard choices to make when you are sick : first, to enter a hospital or not and then at whose mercy you have to put your life. Weigh the benefits and risks.

We have made elaborate study of how longevity is dependent on many factors like the food we eat, the way we use our body, the way we think, the choices we make for ourselves and of course the endowment with which we have come into existence. None of these occurrences is totally independent of chance.

If you follow the four-fold program given in this book, it will do you a lot of good. It extricates you from certain risks and pitfalls associated with the methods touted as biologically revolutionary. For one thing, all these are still in experimental stage. For the other, it is very doubtful whether these experiments conducted on small creatures and lower animals will benefit humans whose physiology is entirely different and who are endowed with minds with different modes, which affect their bodies. Therefore, it is a fair guess that the mechanical processes applied to worms and tiny animals cannot be routinely applied to human beings with success, unless their success is verified with human subjects. Even then you are still at the mercy of chance.

Why then should you waste your time waiting for something about which we are uncertain of achieving and uncertain about its consequences? If you are still euphoric about these so called revolutionary methods, nothing except perhaps your affordability, will stop you from trying them if and when they may be available later on, but it is to your advantage to follow meanwhile the four-fold program which has great potential of setting you on the path of lengthy life.

The Art of dying gracefully

Our caveat is still in operation. Prolonging life span by any means does not mean elimination of the eventual end of physical body. Death always haunts us. Therefore, it is essential to learn the art of dying gracefully, as much as the art of living long.

VARIETIES OF DEATH:

Death usually occurs in one of the following three ways.

- Death occurring as a natural phenomenon.
- Death occurring after a physically and morally depraved life style that misused the body.
- Sudden unexpected death due to disease or accident.

Death, from whatever point of view we look at it seems to be a break-up of the unity of the bodily organs.

It is a failure of the integrity of the body system in which various parts are rent asunder. But, we have to remember that such a break up may be of a healthy and normal type, or it may be of the nature of disease. In the first case, it may chiefly consist of the shedding of a worn out husk; to put it in the beautiful simile of ūpaniśads, like the casting off the slough of a snake on the anthill, or it is like the peeling off petals of a growing bud. Many old people seem to die in this way, the body being the scene of little or no disturbance or conflict, but simply withering away. Here too there is a certain dislocation, but no painful rending asunder. The center of life seems merely to retire to a more inward and subtle region as the outer body is peeled off as a sort of outworn shell while preserving the general unity of the various parts.

However, in the second type, death is undoubtedly very different. Instead of the one center simply withdrawing inward in the way indicated, non-cooperating and rebellious centers spring up to introduce serious conflict in the organism.

The culmination of the second case usually results from long abuse of body and living a morally depraved life, which inevitably leads to degenerative tendencies and ultimately to some distinct disease. Then breakdowns may lead to panic, hospitalization and losing control to doctors. A partial recovery may occur, but it is generally followed by a repetition of the whole series, without any improvement in the general health. Then of course worse breakdowns and panic, leading at last to violent drugs, intrusive procedures, tying down to machines and so forth, in the hope of prolonging existence for a few more hours are common scenes. Finally, death arrives, not graciously, but in the sense of a dismal defeat to everybody concerned. To the patient, it is a confused and embittered end, robbed of all decency and dignity.

Why should we behave in this feeble and foolish way leading lives which utterly unfit us for the peaceful end of life, and stricken with most incompetent panic and dismay when the very thing arrives which we have foreseen and for which we have had such ample time to prepare?

The readers may wonder whether there is preparation for death. They might have seen hundreds of books about the art of avoiding death but not about the art of living and dying and that is lamentable. Yes indeed, dying peacefully is an art to be cultivated. In fact, it should be the final chapter in any book on the art of living. Preparations are both physical and psychological. Physical preparations were outlined earlier in detail but psychological preparation needs to be elaborated a little more. No doubt, it is quite difficult to prepare for it but one can practice it as an art and get ready for it rather than remaining complacent thinking that you are going to live forever, courtesy some doctors or gerontologists and feeling bitterly disappointed and stunned when the long arm of death reaches you.

You have to program for eventual end as you would do for a tour. I am not talking of estate planning and other financial matters as plenty of professionals are available to advise you on them. None of them can prepare you psychologically because; the idea of death is such a taboo to many people that they shudder even to think of it.

The first step in this preparation is to get rid of this taboo and begin thinking about how to meet the inevitable end peacefully rather than being stunned, disheartened and confused in the last moments. The art of dying is an integral part of the art of living for; the end is determined on the way you live.

The first mode of death is normal and hence desirable, not the other two modes. You should endeavor to reach that kind of natural dissolution of the individual self, while constantly training the body with the psychosomatic practices advised earlier. You can simulate psychic death experiences with mock death now and then to get accustomed to the psychological atmosphere you will eventually have to face.

The great obstacle to the dissolution of individual self is, attachment to the feelings of 'my body', 'my kith and kin' and to 'my property'. If you overcome these feelings, death is not much different from sleep in which you are dispossessed of all of them temporarily. You have to endeavor to be loosened from these attachments mentally as a preparation.

> Think like this. "The idea of 'my property' is given excessive importance in modern cultures. Even otherwise, excessive possessiveness is not conducive to happiness. As advised by *Isavaasya* Upanishad, all that is there may be enjoyed without being attached to it. Kith and kin are just wayfarers who joined me like the faggots in a speeding stream and they will separate from me as casually as the faggots in the stream separate and go their way. So, Ill do my duty by them as dictated

by my moral and legal duties but why should I wail for them day and night?"

The hardest part in this exercise is to disown the body. Try this method.

> Ask yourself the question 'Who am I?' If I am this body, why am I referring to it almost every day as 'My body' as if there were something else that possesses this body? This consciousness, which makes me aware of body that I call mine, is present even when I sleep and dream. Therefore, some invariant permanent entity that is surely different from this perishable body must exist. I am that immutable entity. Let this body go peacefully when its time is over'. After all, when that animating principle, the self departs from the body, nobody, not the beloved wife and children, not the relatives and not the friends will allow this body to remain under my own roof. No sooner that consciousness departs the body than it will be thrown out by the very same people.

Such ratiocinating musings will help you to get rid of attachment not only to your body but also to the wayfarers whom you regard as your kith and kin. Frequent practice of these mental pictures will prepare you for giving up the body peacefully and allowing your individual Self to merge with the universal Self. Without learning the art of dying, the art of living is incomplete.

CHAPTER TWELVE

Conclusion

> To the eye of vulgar logic what is man? An omnivorous biped that wears breeches. To the eye of pure reason what is he? A soul, a spirit, a divine apparition.
>
> ---Thomas Carlyle.

We are now in a position to take stock of the human longevity situation. Before proceeding further, it is helpful to recapitulate our understanding of the concepts of "primary aging" and "secondary aging". Primary aging refers to the deterioration of structure and function of cells and tissues of the body by the mere passage of time, which in turn affects integrity of the body. Secondary aging refers to the decline in structure and function due to diseases and environmental assaults. Primary aging influences maximum life span whereas secondary aging influences the average life span.

Barring caloric restriction, scientists have not so far made any appreciable progress in extending human life span by reversing or decelerating aging. We have reached a dead end.

That is a reason why we should make some self-introspection about the essence and directions of our research. As observed at the beginning, gerontology and geriatrics have prehistoric and ancient roots. Evidence of efforts to prolong life and to remain youthful lifelong, can be found in ancient literature like the Vedas, Upaniṣads and other Sanskrit literature dating back to thousands of years. The trend has been continuing uninterrupted even in western thought. Some scientific ideas on this subject that have been touted as new inventions are not altogether new. Many contemporary professional gerontologists rarely allude, if at all they do, to these earlier efforts and pass them off as their new inventions. Some of them may really be unaware of their existence. Those who are unaware will do well to spend some time reading the older literature and convince themselves that many of their supposed inventions were known before they were born and they have not progressed much farther.

Professor Leonard Hayflick has brought to light some of these lapses with admirable candor. Himself a pioneering researcher, who helped to overthrow the dogma of perpetual cell replication, he honestly admits that the idea of a limit to cell replication was not really new and was found in the writings of Metchnikoff (winner of Nobel prize in 1908) who in turn gave credit to August Weismann. In a penitent tone, he says that, as a young researcher, he was not aware of this idea while he was conducting his landmark experiments that overthrew the Carrel dogma. Hayflick gives credit to Metchnikoff not only for discovering the idea of a limit to cell replication but also for finding the reason for encountering such limit. That was because of the caps at the end of chromosome ends now known as telomeres.

Another example he gives is, about the role of a single gene being responsible for aging. We really don't know even now whether a single gene is responsible for aging but he scoffs at the claim that it is a recent discovery. He tells us that H.S. Jennings first mooted the idea 68 years ago, in his 1939 essay written for a collection published by Arno Press under the title "Problems of Ageing."

I may now be permitted to give an example in which I have some personal experience. When I was reading about the role of caloric restriction in modern gerontology and given to understand that it was Clive McCay who first discovered about the usefulness of caloric restriction as a means to promote longevity, I remembered that I first heard about it from my grandmother who died in 1966 at the age of 90. She told me that she had been practicing abstemiousness in eating ever since she was 20 years old. Hayflick tells us that one Luigi Cornaro promoted the idea centuries before McCay experimentally verified the efficacy of caloric restriction.

Lest I should be misunderstood, I hasten to say that these examples are given neither to decry the gerontology research nor to deny credit to those dedicated scientists who experimentally verified the idea. They are given only to express my impression that we need to study our historical and ancient ideas also, give them the credit they deserve and make our efforts to advance from where they left.

Empirical studies on model animals have shed some light on how average life span and maximum life span behave under different life styles. These studies help us by isolating the conditions under which average aging and maximum aging can be slowed down by human intervention.

Maximum age is influenced perhaps by the factors that affect the structure and function of cells and tissues. Such changes may be caused by the genetic disposition of the body, and the environmental assaults that damage the DNA. The inherent ability of the animals to repair such damages may counteract such changes. Varying abilities of species to repair that damage may be regarded as one of the reasons for the wide disparity in the maximum life spans of various species.

In small animals some genotypes were found that could slow primary aging but in humans no such genotypes have been found so far. Whatever gains have been made were made possible by influencing the secondary aging caused by diseases and exogenous assaults on the phenotypes. Research has not made much progress in extending longevity by slowing down primary aging.

After several decades of research on aging, we have not been able either to understand correctly the mechanics of aging or to untangle the complex web of diseases and aging. However, the large body of published research shows that aging is a highly complex and multifaceted phenomenon. The rate of aging is highly variable among different species, in individual members within species, in organs within organisms, in tissues within organs and in cells within tissues. Because of this heterogeneous variability, we cannot isolate a single cause of aging. The erratic behavior of aging has led biologists to abandon their earlier notion of a genetic program as the cause of aging and to favor the idea that aging is a complex phenomenon involving many random variables.

The descriptive data gathered in recent decades does not suggest the existence of any universal markers of aging, Nevertheless, some general principles have emerged:

- Aging reveals itself in protected environments.

- Aging is the consequence of gradual failure of homeostatic mechanism in the organism.
- In stark contrast to the development that takes place in well-coordinated and programmed steps, aging is erratic and
- No genetic program determines the lifespan of an organism.

Why has aging research not gone farther than the general ideas like these? Has science arrived at its limit or have we reached a dead end in this particular branch of science? It is easier to deal with second question first. If we reach a dead end in a route, we can retrace our steps and take another route. Let us hope that is where we are now and try to change direction.

Changing direction in science requires courage and determination to give up dogmas and methods, which have outlived their usefulness. Cartesian reduction as a method of scientific analysis and Darwinism as an explanatory theory, to which we are bound, have outlived usefulness and we are in need of reevaluating them.

The contradictions in aging research

Biology is strongly wedded to two principles both of which are against the concept of life as an independent and autonomous entity. One is the reduction method, which claims that the behavior of the whole can be predicted by studying the parts of it and the other is the Darwinian theory of evolution with its implication that organisms must keep dying in order to facilitate evolution. Both of these concepts have turned out to be the conundrums of molecular biologists that have been trumpeting that their research would find a way to retard or reverse aging and prolong human life by hundreds of years. Reductionism cannot explain how the various parts of human body are integrated to function as a highly organized man or woman. Without that integrating principle, the body parts are utterly useless. Darwinism claims that individuals must perish in order to keep the lineage evolving and that death is an evolutionary necessity. How then, can the gerontologists who work mostly with molecular biology reconcile their efforts to prevent aging and death with these theories?

The Puzzle

The evolutionary theory has developed some arguments against the existence of genes that might have specifically evolved to cause aging and to determine the lifespan of an organism. Darwinian theory limits the role of genes to the extent of assuring a certain minimum duration of life during which the organism completes its reproductive functions. Post reproductive period is considered irrelevant, by the Darwinians, and hence Nature is supposed to ignore the organism. Then the puzzle arises. We see many people living long after they fulfilled their reproductive functions. If the theory of Natural Selection dispensing its favors for some time and then refusing to favor were correct, why should people continue to live long after reproductive period ceased? This is not an unobservable change like evolutionary change; it is change that everyone has been observing. This anomaly should be explained if we want to persist with this theory. Without answering this question satisfactorily, we cannot expect to solve this puzzle and develop a sound theory of aging.

Furious debates have been going on in the community of biologists, gerontologists, demographers and medical professionals about the feasibility of extending maximum life span. Some argue that there is no limit to maximum life span based on their laboratory experiments on tiny model animals and some based on mathematical models; but none has been able to produce a coherent theory as they find themselves caught in a whirlpool of contradictions brought about by their pet theories.

The difficulties experienced by biologists can be partly attributed to two theories that have outlived their usefulness and posed unanswerable questions. Reductionism cannot explain how the different parts of human body are integrated to attain the status of human being and the theory of mechanical automation never succeeded in producing life.

Such problems are however not unique to longevity research, many more can be found in the history of science but science progressed whenever we overcame dogmas and gave up straight-jacket thinking.

A simple answer to this vexing problem could have been to admit that the concept of natural selection's so called desire to keep organisms alive only up to reproductive period may be erroneous. However, professional biologists generally do not risk making such an admission. So, they beat about the bush and invent new words and phrases to make the idea acceptable. The present confusion in aging research seems to have arisen more or less as a legacy of these two theories.

If we have not succeeded in reversing aging or at least in stopping it, shouldn't we, at least, be able to determine the solvability or otherwise of the problem of life extension? We don't have unanimity even on the solvability question.

An unbiased reviewer cannot fail to recognize that like the earthlings bound by gravity to earth, biologists exploring extension of lifespan have been bound to the two laws whose sway has a limited domain. If they can escape from this domain, perhaps they will see a different frontier towards which they can move.

If we cannot give up Descartes' authority in scientific thinking, we should take the essence of his thinking rather than cling to a particular procedure outlined in his 'Method'. When the investigators reached the limit of the usefulness of this method why should they hesitate to pick up the essence of his philosophy? Consider what he said about the thinking person.

> I am not this assemblage of members called human body;I am not a rarefied and penetrating air spread throughout all these members (2nd meditation, pg 26).

'*Cogito ergo sum*' (I think, therefore I am) is the essence of his philosophy. Without investigating this 'I', that thinks, we cannot answer the questions whether we can live on indefinitely for several centuries or whether we can reverse aging and enable cringing ninety-year-olds turn into twenty-year-olds jumping with joy.

Darwinism came as an answer to Church bigotry

It is not a wonder scientists of those times felt relieved when Darwin imported the theory of mechanical causation into biology as they were fed up with the excesses of Roman Catholic Church which persecuted great thinkers and scientists of the time and tried every means available to them to suppress truth and perpetuate its mundane power. People could not openly express their dissent for fear of reprisals. Dissenters included in their ranks, the eminent scientist Isaac Newton, and the celebrated artist Leonardo Da Vinci, and the famed novelist Victor Hugo. They were in need of proffering an explanation other than a personal God to the phenomena of the universe, to break the power of the church. Darwinian explanation of proximate causation suited that need of the time quite well.

Darwin's theory cast in the physicalist mold avoids any reference to the role of spirit, the prime mover of that mechanism. Though Darwin never explicitly denied spirit and knew it to be the self-evolving Ultimate Cause, he cleverly maneuvered to project the proximate mechanism of change as the sole cause of evolution and called it "natural selection." Even though Darwin's friend and defender T. H. Huxley neither approved the doctrine of natural selection nor the materialism involved in it, many biologists who were looking for a theoretical foundation for their empirical work jumped on his bandwagon and promoted the materialist paradigm of science. Gradually an atmosphere developed in which students of biology were given the impression that to appear smart they should condemn religion and extol Darwinism. While the harm this distortion has wrought to science is substantial, its degrading effect on the modern society is enormous. The impression that religion and Darwinism are mutually exclusive is false, as it will be presently shown.

Reconciliation of Darwinism with Religion

Though Darwin brought in mechanical causation into biology, he left room for the Ultimate Causation by making a distinction between created forms and those produced by secondary laws. He believed that *"animals descended from at most only four or five progenitors, and plants from an equal or lesser number."* Again, he modified it to say, "*as far as it is at present known the germinal vesicle is the same, so that all organisms start from one common origin.*" He made oblique references to some superhuman agency regarding the origin of initial creation but confessed his inability to delve deeper than that into metaphysical problems. He took the abundance of nature and variation as a *fait accompli.*

When once it is admitted that the initial creation was made by an omnipotent agency as claimed by the First Cause argument, there should be no controversy about the proliferation of creation automatically by mechanical steps as formulated by the post-Darwinian synthetic theory of organic evolution. The dispute can be resolved by conceiving that nature existed with some forms of life and they were endowed with self-sufficiency and potency so that they could interact with each other and reproduce themselves into myriad forms of life.

Darwinians do not like to be confronted by questions concerning the origin of life, but if they wish to stick to their mechanical theory of life, they have the obligation either to account for the origin of life or to demonstrate their ability to create life independently and to sustain it.

Let alone duplicating the manifold ramifications of Nature, we cannot even defend ourselves when she gets furious and wields her scepter mercilessly, with earthquakes, tidal waves, typhoons and tornados. All that we can do is to run away holding on to our dear lives. We have not been able to find a Hercules in science to slay the hydra of diseases. All that we have been doing is, we are either suppressing one disease while another is raising its head, or we are expressing our hopelessness. If we were omnipotent we could have stopped them all or made our bodies immune to all diseases! How then can we resolve the question of retention of life in our bodies without trying to know what it is and whence it came? Thus, we are driven back to reexamine our hardened position that everything in nature operates mechanically without the motivation, drive or aid of some spirit and that nothing is beyond human capability.

Intelligent Design

Thus, we are back at the old question, 'How are the celestial bodies behaving so nicely without going helter-skelter?' Newton wondered at "the hand that hurled them into their orbits;" but Laplace tried to explain away everything with mathematics. When I asked my friend who is a physics teacher, whether he can explain me how the universe so ably explained by Laplace, came there first, he dismissed me with a flourish "You have to learn a lot more physics to understand." My daughter, when she was in primary school, asked me once, "Daddy, our teacher said that we have many problems because there are too many people on earth. Why don't we make another earth and put it over there so some can go and live there?" I said "I don't know dear, I'll ask my physics friend". I could never answer my daughter's question because physicists better informed than my teacher friend, told me that they couldn't help me in this adventure.

When I was wondering how to settle the tussle between poor old Newton who wondered at 'the hand that hurled the celestial bodies into their orbits' and the mechanical scientists, who refuse to see that hand, I remembered Bertrand Russell who was both a scientist and a philosopher. He showed the compromise by saying "What is the harm in allowing you to assume that God placed the celestial spheres in their positions as long as you allow me to perform my astronomical calculations without any hindrance?"

We attribute the celestial phenomenon to a superhuman agency (may hereafter be called God for brevity) because; even with all our pooled intellectual and material resources, we cannot accomplish such a task. When Newton's Principia Mathematica, failed to account for the stability of the solar system he proposed that "an intelligent and powerful being" was propping up the mathematical mobile. So, what is the harm in assuming that the life forms of the organic world fend for themselves faithfully following some secondary laws given them by the same "intelligent and powerful being?"

To use analogies easily understood in the present world, God can be imagined in the position of the head of a state or the managing director of a corporation. The head of state lays down the policy and delegates powers to the subordinates who follow the rules mechanically and run the state. Even the routine mechanical operations of a modern factory follow the blueprints designed by engineers and the rulebook laid down by the managing director. We don't deny his presence, simply because he is not visible in the routine operations.

Admitting Russell's compromise, we can view the so-called 'natural selection' as a secondary mechanism that aids the main thrust of intrinsic evolutionary drive. We can view it as God's method of propagating species. When we are unable to account for the origin of the abundance of nature and unwilling to accept the idea of a superhuman agency, we have invented the phrase 'natural selection' as a euphemism for 'God'. So, where is the contradiction between religion and evolution? Evolution may not fit into the religious schemes of some people but Vedanta, the great path breaker in philosophical thought, has no enmity with evolution, as will be explained presently.

This sequence of thoughts takes us back to the old Paley's watchmaker argument; the precise ticking away of the watch cannot happen without somebody winding the mainspring which when left to itself will unwind automatically. Darwinians exulted when Darwin said that this watchmaker could be replaced with the patchwork principle of circumstantial elimination of weaker elements called "selection." It is worth revisiting Paley's arguments to reexamine whether we are justified in rejecting him outright.

Revisiting Paley

It is generally agreed that in order to establish the validity of a theory, one has to identify all the possible alternate propositions and disprove them. The essence of Paley's theory may be stated in two statements:

- Organisms are well designed with a definite purpose.
- Purposeful design implies a designer.

Paley saw correlation between form and function as exhibited in the anatomical structures of organisms.

Paley cited the following competing arguments and refuted them.

Competing argument 1:

It is admitted that there is good design but that does not imply that it is designed for the current purpose. Suppose that form appears first on its own and function follows later as unanticipated but fortuitous fit.

Example: Fish skin is used by cabinetmakers to smoothen the mahogany wood.

Rebuttal:

Can we therefore say without exposing ourselves to ridicule that the skin of fish is made rough in order to be used like emery paper to smoothen our wooden cabinets? In trivial and isolated cases, we may allow such fortuitous fit but in general, nothing intricate and complex could be designed for one purpose and used fortuitously for another.

Take the case of eyes. Can anybody assert without inviting ridicule that eyes of organisms were formed aimlessly and the organisms later found that they could be used for seeing? Therefore, there cannot be form without an anticipated function. (Modern design principles support this argument)

Competing argument 2

Good design exists and implies production for its current purpose; but adaptations are built naturally, by slow evolution toward desired ends, not by a divine fiat. To this category belonged Lamarck theory that entails evolution by "use" and dissolution by "disuse."

Rebuttal:

Paley provided some powerful refutations to counter this argument. The case of elongated neck of zebra as Lamarck's example of evolution by 'Use' is well known. Paley countered this with a theoretical example referring to the case of elephant. He asks if the elephant's short neck implies great advantage for a long nose, what can a proto elephant do with one tenth of the usual length of trunk. He argues:

If it is suggested that trunk may have been produced in the long course of generations, by the constant endeavor of the elephant, to thrust out its nose, how could the animal subsist in the meantime before the elongation of snout was completed? What was to become of the individual whilst the species was slowly perfecting?

Regarding the argument of disappearance of organs by disuse, he countered by saying that centuries of disuse did not cause organs to disappear, or even to diminish; as in the case of mammae of men and the foreskins the races who have been following the custom of circumcision.

Professor Stephen Jay Gould, paleontologist and one of the

proponents of punctuated equilibrium theory, says that he accepts these two arguments but not Paley's conclusion as the latter failed to exhaustively disprove all the alternatives having failed to visualize a third alternative. There is a third alternative, which according to him, identifies natural evolution as a source of good design, thereby sinking Paley's central conviction that adaptation must imply creation by divine fiat. However, the argument that mechanical evolution is a source of good design is like placing the cart before the horse for it reverses the true nature of causation. Existence of a good design is the antecedent from which evolution takes place and not the other way round. Design always precedes the structure. The rise of great structures and advent of complicated machines was always preceded by purposeful planning and. design. Who can claim that the great cathedrals of Europe and the numerous architectural wonders of America came into being without prior design by engineers? Can we find a great architectural wonder that arose automatically from rubble without the skills of the architect and without the toil of workers? Has anyone seen great art that composed itself from a color palette without the imagination and skill of an artist? Has anyone read a literary classic that evolved itself out of alphabets, without the guiding intelligence of the author?

Can we say that an airplane assembled itself first by random arrangement and man found it useful for flying? I shudder to think that the airplane, which has enabled me to rise up high in the sky, has no design to provide it lift, thrust and aerodynamic stability. Can we deny honestly that evolution was made possible by the prevalence of abundance of species, variations among them and the instinct of self-preservation of animals? Nature was endowed with these attributes and left to fend for itself. Without the marvelous design in nature and its endowment, would it be possible for the mechanism suggested by Darwin to operate?

Indian Point of View

Some of these difficulties have arisen because of the lack of broader perspective about the origins of the idea of evolution and prevailing misconceptions about the idea of God and unfamiliarity with alternate ideas. Even an eminent evolutionary biologist like Ernst Mayr, seemed to be unfamiliar with the Indian thought in this respect, when he observed that "*Prior to Darwin, most philosophers were sure that species are constant.*" That notion is not true in the case of Indian philosophers, nay even Indian populace. In India, our mothers and fathers taught us in our boyhood days that nature has always been in a state of flux. We didn't have to wait to know about the changing nature until we were introduced to Darwinian theory in our college classes. The idea of evolution percolated from the philosophers down to every strata of Indian society. So much so, even ordinary folk in India talk about life originating as marine creatures, coming to shore as amphibians, and then the first mammal appearing as a boar. A mixture of lion and human forms represented transition to human form but soon it became extinct giving rise to a dwarfish human form. Then the valorous human form came with full stature and ripened into an ideal human being, valorous, truthful and embodiment of righteousness. As humans learned guile and created disorder, the next form outwitted guile with guile. Then came the rational human being. However, all strata of people there have a strong conviction that a superior agent has been at the substratum of all these evolutionary transformations and that the essence of all these forms is the same primordial God-stuff. Children there learn these things through stories and songs.

Sāmkhya Theory

Even the Sāmkhya theory of evolution and involution that sprang up on the Indian soil several centuries before Darwin, tried to undermine the foundations of supernatural religion, by substituting evolution for creation. However, it could not build up a cohesive system without introduction of a principle that activates the inactive nature of *Prakriti* (Nature), which itself is inert, dull and unproductive. Therefore, they introduced *Purusha*, in whose proximity *Prakriti* (Nature) evolves influenced by the former.

Sāmkhya follows the *Satkāryavada*' according to which cause and effect are the undeveloped and developed states of the same substance. The effect is contained in the cause; something else is required to liberate it from the causal state. Evolution means unfolding of what was already there in seed form. This is one variant of *Pariṇama vāda (*The Theory of Transformation*).* When the conditions are ripe for dissolution, the phenomena will fold and revert to the cause. This is a dualistic point of view of cosmogony.

Later thinkers rejected this idea on two grounds. First, just as clay cannot transform itself into pot without a potter, nature being inert cannot transform itself into cosmos without some driving force. Secondly, if a seed evolves into a tree, the seed must perish for a tree to come up. As the base ceases to exist in this process, no sound system can be built up on baseless theory.

Buddhism is based on what is known as *asatkārya vāda* or *sūnya* vāda, which propounds that universe, came out of void. śamkara dismissed this as too puerile to be worthy of serious consideration, as something cannot come out of nothing.

These and some more similar theories were condemned by non-dualistic Vedanta, which considers that all that exists is really Brahman (which is characterized by Existence, Consciousness and Bliss) and nothing else. Pure non-dualists contend that Brahman itself appears as phenomenal world due to ignorance, just as a rope is mistaken as snake in darkness. Independent world idea is illusory, as it has no existence apart from Brahman. This is known as '*Vivarta vāda*'.

Physically, man is the evolutionary product of this primordial Brahman through gradations of space, air, fire, water, earth, vegetation and food. Spiritually, he is Brahman itself in its limited form. Brahman is the substratum of all that we perceive and conceive.

Brahman, whose character is pure consciousness, is the first principle. From Brahman, everything else emerged. Life emerges in matter, not because of matter. Organic evolution with vegetation and other forms of life emerged in inorganic matter. From the abundance of variegated flora and fauna further evolutionary changes proceeded. All along, Self (Brahman) has been the motive force and substratum of all change. Therefore, all that is there is Brahman. This theory is not faith. Advaita Vedanta does not ask you to take it on trust. It is established on rational grounds. Any one who claims to be a rationalist cannot but be impressed by the precision and rigor of Advaita logic that establishes the non difference between Absolute Brahman and the phenomenal universe on one hand and the non difference between Individual soul and Brahman on the other. These theories emphasize, the literal meaning of evolution – unfolding of something that has already been there in dormant state. The banyan seed involves the banyan tree. No amount of jugglery or environmental influence can produce a banyan tree from a mustard seed.

Some scientists, who insist on empirical method as the only method of acquiring knowledge, are not satisfied with this explanation of cosmogony without realizing that empirical method explains only part of our experience. They remain silent about the rest.

BENEVOLENT GOD

The second misconception is the notion of benevolent God. The idea of benevolent God is like the idea of a benevolent judge, who sees no distinction between the criminal and his victim. It is a misconception, prone to lead to the fallacious argument that because we don't find a just and perfect world, therefore, there is no God. This difficulty arises only for those people who think that God is an embodied entity in the image of man, that God must necessarily be benevolent towards human species and that existence of God implies the existence of a perfect and happy world of human beings. These are all baseless suppositions.

To those, who are familiar with the Vedanta view, the idea of a benevolent God and the problems associated with it do not arise at all because God here means the Supreme Soul called "Absolute Brahman" which encompasses the thinking subject as well as the rest of the phenomenal world.

This difficulty vanishes when once the concept of God is reoriented. If at all a personal God exits, his work is seen not in perfections or imperfections in nature, not in his benevolence but in the operation of an invisible hand called 'ṛtam', that enforces a just order, handing out rewards as well as punishments to people according as they deserve, based on their past conduct.

ORTHOGENESIS

In the light of these theories, it is worth our while to reconsider the arguments of Otis Whitman, the dedicated American biologist who founded the Woods-Hole laboratory. Whitman proposed that evolution unfolds along pathways set by genetic and developmental programs of organisms.

He argued that natural selection can either accept or reject the variations that arise naturally among differing organisms in an inter breeding population. If the genetic and embryological systems prescribe a definite direction to this variability then natural selection cannot deflect the course of evolutionary change because it is too weak to do so. Natural selection could be a subsidiary force. As Darwin was aware of the power of this argument to destroy his theory, he countered it by saying that variation is random and has no inherent direction. Randomness of variation was an absolute prerequisite to Darwin's theory as otherwise he could not introduce an outside mechanism like natural selection to give direction to the random raw material. Whitman argued that internally directed variation is self sufficient to set an evolutionary trend.

Anyway, are organisms helpless like billiard balls buffeted hither and thither by the pool cues of natural selection? If we accept this argument, we are led to rigid determinism. On the fertile and tolerant soil of India, various profound philosophical systems arose. Even though they proposed the Theory of Karma, which enjoins everyone to reap the fruits of one's actions, not even one of them admitted the idea of fatalistic surrender. They argued free will could modify the impact of karma. To use Francis Galton's metaphor, we are like polyhedrons, not spherical billiard balls. A polyhedron can resist a push and settle down on one of its many faces. That means, evolutionary movement is the resultant vector determined by the internal and external forces, of which the internal force vector is the dominant. Placing good design before natural evolution, as it ought to be, may set one rethinking about the only objection raised by professor Gould to Paley's theory.

Many scientists somehow take the position of a man who denies the existence of a manufacturer of the beautiful car sitting in his garage because it is there ready for his use. Understanding the mechanism of your car and your ability to drive it do not imply that you created it, or the car just appeared there in your garage automatically. So also, mere understanding of some laws of nature does not entitle us to claim that we have created this complex universe or that we can duplicate it.

Science, which confines itself to correlating phenomena, is not in a position to pass judgments about the underlying reality unless the scientist goes further and admits hypotheses about reality, which were intuitively formulated and tested by earlier thinkers.

The antagonism between science and philosophy has arisen mostly because some scientists and philosophers overreach their respective domains and they try to view the vista in the other domain with the instruments meant for their own particular domain. In the middle ages, the Christian Church did irreparable harm to the cause of philosophy by exerting itself in an area in which it was not competent to pass judgments. Some scientists tend to commit the same mistake of applying science to ideas in which empirical method of science is either irrelevant or insufficient.

Albert Einstein defined science as the

> Endeavor to bring together, by means of systematic thought, the perceptible phenomena into a cohesive whole. It is an attempt at the posterior reconstruction of existence by the process of conceptualization.

The phrase 'posterior reconstruction' clearly debars any claims to our ability to create the universal phenomena. Recognition of the mechanism suggested by Darwin as a proximate cause of speciation need not preclude the Ultimate Cause. Then, there is no real contradiction between religion and Darwinism as far as the latter is applied to sub human species and if it stops there.

In our enthusiasm to celebrate Darwin, we cannot deny everything that appears mysterious to our fallible intellects. How can we deny the Supreme Self, from whom all of us have emanated? If I deny the supreme Self, it amounts to denying myself because the entity behind that 'I' is the Supreme Self. As a Cartesian, I cannot deny myself, so I cannot deny the Supreme Self that projects itself as nature wherein all the contenders exist as symbols of that all-pervading Self. This imperishable Self, variously called Brahman or ātman is not adventitious; it was there before we became conscious of our existence and it will be there after we cease to feel our existence, and ever afterwards. It does not depend on anybody's recognition.

Operability of Darwinism

Now consider whether the idea of survival of the fittest implicit in the mechanism of 'natural selection' is operable in the realm of human affairs that are now governed by the principles of democracy, rule of law and natural justice. In most of the communities, the tendencies of the stronger to annihilate the weaker are checked by the state and the weaker are protected by the force of law. We are no longer living in a primitive state as conceived by Darwin where the weaker are devoured by the stronger. The recent history of mankind shows that this doctrine cannot automatically come into operation in the realm of human affairs as it may in the animal world.

Post Reproductive Period

Neo-Darwinians, even as they dismiss teleology in evolution, attribute a purpose to the amorphous entity called "natural selection." It is the all-important purpose of regulating the duration of man's life span. They say that evolution is interested (notice the word 'interested') in protecting the individual only up to the time of cessation of reproduction. This amounts to personifying natural selection and attributing motives to it, thus bringing in through backdoor teleology that was challenged by them at the front door. This conclusion is particularly intriguing, as it is not sustained by practical experience. Most of the people live long after ceasing to reproduce.

. It is common knowledge that adults who are either advanced in age or not interested in sex at all, have produced great works of art, literature, science and philosophy, all of which have great utilitarian value to the younger generations. Do we consider them all a waste? Do we discard all those books in our libraries and dismantle the laboratories simply because they were established by the older generations? Do we ask our youngsters to reinvent the wheel or encourage them to absorb the accumulated knowledge?

Another paradox needs to be addressed in this context. Women, in general, attain menopause around the age forty while men, even in their seventies, are fertile. If evolution protects individuals as long as they are reproductive, men should outlive women, but the facts are otherwise. In a monogamous society, a male is stopped from reproduction around the age of fifty, not because he becomes infertile but because his spouse attains menopause and then he lives on for about forty to fifty more years, acquiring experiences, doing useful work and improving his faculties in various ways. Should we deprecate all this effort as wasteful while admiring his earlier effort at reproduction as fruitful?

Can we deny that the older generations perform the useful function of passing on their knowledge and wisdom to the younger generations and help them to evolve faster into more powerful and better human beings? Else, what is the difference between animals and human beings?

The rapid proliferation of scientific culture in many parts of the world during the last century shows that there is not any particular ethnic group with special scientific endowment, which survived the struggle for existence weeding out the non-scientific groups. Such a change, if it were Darwinian, would not have taken place in a short time. Rapid spread of scientific culture has been made possible by the fund of knowledge accumulated by seniors and their organizational efforts. Such changes are Lamarckian rather than Darwinian[1]. Who is ready to discard the accumulated scientific wealth because numerous older scientists were involved in building it up? Does anybody throw away the inheritance because older people accumulated it? Therefore, post-reproductive life cannot be regarded as useless.

The reason that has forced the natural scientist to adopt such a line of argument is the need for a mechanism that transmits characteristics from one generation to the next. Scientists trained only in the positivistic school of thinking, naturally, see no alternative but to rely on reproductive process through sexual union, as the conduit for transmission. However, this line of reasoning has several shortcomings.

First, it implies that only characteristics developed before the cessation of a short reproductive period can be transmitted. Positivistic thinking cannot explain what happens to the character-

---

[1] Lamarckian changes are directed changes taking place in an observable span of time while the changes conceived by Darwin are random and beyond the pale of comparative observations as they are supposed to take place over millions of years.

istics developed over long post reproductive period.

Secondly, we have to answer clearly, whether teleology is admitted or not in evolutionary process. If admitted, why should we presume that Nature has a bias towards supporting the individual only through the reproductive period?

Thirdly, how to reconcile the contradiction between the posited interest of Nature in keeping the individual alive through the fertile period and the experience of individual living long after cessation of reproductive functions?

Finally, this approach fails to explain how mental characteristics are transmitted. These difficulties leave us with a lingering suspicion that exclusive dependence on positivistic approach may not be adequate to solve this problem.

What then is the compelling reason behind the argument that the post-reproductive period of animals is not of interest to Nature? It is because of the usual understanding that inheritance of characteristics is ensured by reproduction. This argument loses steam when it is extended to humans who are at a higher level than animals in the evolutionary scale. This line of reasoning presumes that the be all and end all of human life is reproduction and all other great endeavors of mankind are useless.

It is because of this knotty problem of inheritance, Lamarckism lost favor, but even Darwinism cannot explain what would happen to the characteristics one develops in the long post reproductive period. Later developments in molecular biology, notably the so called central dogma of biology which affirmed that the information contained in the proteins cannot be transmitted back to the nucleic acids, dealt a death blow to Lamarckism because according to the prevailing knowledge, the characteristics of phenotype cannot be transmitted back to the genotype. Unless the acquired characteristics are transferable to the genotype, inheritance cannot take place. However, such reverse transmission was declared impossible by the authority of the central dogma.

However, subsequent researches buckled the rigid central dogma. One of them is Barbara McClintock's finding that transmission is possible from RNA to DNA. The other is the recent research finding that Genome cannot directly translate instructions into actions of the body but its actions are mediated by what is known as epigenome. Epigenome is a collection of chemical markers and switches that lie along the length of the double helix and they can influence the gene expression by turning them on and off. The point critical to the central dogma is that these chemical markers can themselves be influenced by as simple a maneuver as diet regulation. These two researches have opened up the possibility of raising the ghost of Lamarck from his grave.

Differences between Animal and Human

This enigma has arisen because of the assumption implied in Darwin's theory that the processes of organic evolution apply to human beings also as mechanically as they apply to subhuman species.

Nature has developed the mechanisms of built in equilibrium, first thermostatic and next homeostatic, in mammals at different stages. This phenomenon is immensely important for two different reasons. First, the built in thermostatic system enabled mammals to survive the changing temperatures on the earth. The second is of particular significance to humankind as the automatic equilibrating system of homeostasis uses only a portion of the brain leaving the other part for functions surpassing the wonders of homeostasis itself.

Referring to the difference between animal and man neurologist, Grey Walter made a very illuminating comment.

> Thus the mechanisms of the brain reveal a deep physiological division between man and ape ... If the title of soul must be given to the higher functions in question, it must be admitted that the other animals have only a glimmer of the light that so shines before men ... The nearest creature to us, the chimpanzee, cannot retain an image long enough to reflect on it, however clever it may be in learning the tricks or getting food that is placed beyond its natural reach. ...The activity of the animal brain is not checked to allow time for the choice of one among several possible responses, but only for the one reflex or conditioned response to emerge. The monkey's brain is in thrall to its senses. ... The brain of lion, tiger, rhinoceros and other animals lack the mechanism of imagination, or we would not be here to discuss the matter. They cannot envisage changes in their environment, so they

> have never sought to alter it in all their efforts to retain lordship of their habitat.[133]

Therefore, as contended by eminent thinkers like the biologist Julian Huxley, paleontologist Pierre Teilhard de Chardin and eminent Eastern philosophers, the kind of organic evolution postulated by Darwin and Wallace should be limited to pre-human stages of evolution. Man with his brain and its neural system is entirely on a different pedestal. With the emergence of humans on the scene, the process of organic evolution yielded place to the psychological and social evolution.

If by evolution, we mean progression from the lower to the higher, evolutionary process does not stop at mere reproduction. The evolutionary changes in human kind have been psychological and social. The changes the world has witnessed in the recent past – from stone age to industrial revolution, then to scientific revolution and now on the threshold of biological revolution – are all dependent on brain. Evolution proceeds from the sentient to the conscious, conscious to the rational and rational to the supra rational.

If there is no preconceived design of the universe, well, there cannot be a design either to favor those in fertile period. We encounter this pitfall, by assuming that the purpose of life is only reproduction. We cannot accuse Nature of capriciousness to accommodate our theories.

INHERITANCE

There is not much dispute about inheritance of physical features from parents and forefathers. The challenging question is how the mental qualities are transmitted from one generation to the other. Scientists have now come round to the view that each animal and man is born with a fund of experience but the point of controversy is about the manner in which this experience is transmitted. Steeped in genetic theory, they view this experience as belonging to the body and it is transmitted by the physical mechanism of meiosis. The hereditary theory takes for granted without any proof, the questionable proposition that mental characteristics can be stored and involved in matter. If it were true, children of the same parents should have identical mental characteristics, but such inheritance is not borne out by experience.

Secondly, the experiences acquired in the long post-reproductive period would be incapable of being transmitted, as they have to perish along with the body in which they are supposed to reside. Isn't it inconsistent to assume that actions of human beings in the first part of their lives have their consequences while their actions in the second part simply perish without consequences? Actions done regardless of whether they were done before or after the reproductive phase ought to have their consequences if Nature is not capricious, but genetic theory cannot account for the actions of human beings in almost half of their lives. This conundrum cannot be solved unless we realize that empiricism is not the only source of knowledge and that epistemology admits other sources of knowledge like inference and verbal testimony of selfless scholar-saints who had intuitively experienced the verities of Ultimate Reality. Why, even the principle of selection was inferred by Darwin and not empirically established.

Limit to Darwinian Evolution.

Most of these difficulties have arisen because of our habit of viewing man as the corporeal body, the exterior and nothing more than that. As Darwin's organic evolution limits itself to the exterior, it cannot adequately explain the human phenomenon. The noted paleontologist Chardin says:

> 'The time has come to realize that an interpretation of the universe - even a positivist one - remains unsatisfying unless it covers the interior as well as the exterior of things; mind as well as matter. *The true physics is that which, will one day, achieve the inclusion of man in his wholeness in a coherent picture of the world.'* (Italics by the original author)

Up to the end of nineteenth century, science never bothered to look at the world other than from without. Physics and astronomy have earlier concerned themselves with the external phenomena of nature. The emergence of quantum mechanics has forced many great physicists to modify their attitude, to look into the interior and to recognize the unavoidable link between the object and the observer's consciousness.

The British physicist Sir James Jeans found that the final picture of the universe emerging from 20th century physical science will be one in which the notion of matter would lose significance, "mind reigning supreme and alone."

While the 20th century physics was thus rapidly turning away from thoroughgoing materialism, contemporary biologists with the exception of a few progressive thinkers were hesitating to stray from positivism and the beaten track of Cartesian reductionism.

In a symposium held at the university of Chicago on commemorating the centenary of Darwin's "Origin of species," late Sir Julian Huxley, the noted biologist gave a new orientation to the evolutionary process. He said:

> 'Man's evolution is not biological but psychosocial; it operates by the mechanism of cultural tradition, which involves the cumulative reproduction and self-variation of mental activities and their products. Accordingly, major steps in human phase of evolution are achieved by breakthroughs to new dominant patterns of mental organization of knowledge, ideas and beliefs ideological instead of physiological organization[134]

Revealing the trend of evolution, he went on to say:

> 'It shows us mind enthroned above matter, quantity subordinate to quality.'[135]

Psychosocial Evolution

Let us see now what is meant by psychosocial evolution. From the level of living cell up to man, biological evolution was motivated by numerical increase, organic satisfactions and organic survival. With the appearance of man, these motivations became secondary, the primary motivation being fulfillment of the ends of life, aptly defined as the fourfold ends of life (*Chaturvidha Purusharthas*) by the Hindu philosophers. Evolution, at the stage of man is no longer Darwinian. It becomes conscious, deliberate and teleological (goal oriented) unlike the blind and accidental processes at the pre-human stages. This change can be attributed to the fully developed cerebral system in man by virtue of which evolutionary process itself undergoes a revolutionary change. With the aid of brain, man can invent organs or substitute organs he may need, more efficiently than the slow evolutionary process. Modern biotechnology bears testimony to this aspect. What was organic evolution then becomes psychosocial evolution now.

The most important change that manifests at the stage of man is the **appearance of ego** and emergence of new capacities like imagination, reasoning, judgment and will power. These capacities made post-Darwinian evolution possible by conferring two specific powers on humans, first the **power to analyze and process raw experience** (raw data) into knowledge and applying that knowledge to manipulate nature and the second the **ability to discriminate between the Self and non-self.** The first power manifest as Cartesian reduction method helped in the growth of scientific knowledge. However, the second power has remained unexplored.

Man is unique in many ways. The field of awareness of all other organisms is confined largely to external environment and to a small extent to their own bodies. Man alone has the simultaneous awareness of being "Self," the subject of experience and of cognizing the "non-self," the object of experience. The spiritual value of awareness grows in richness and variety, as we move up the evolutionary ladder. The self-awareness which neurology considers as the source of dominance of man over nature is recognized by Vedanta philosophy as the basis on which the further evolutionary destiny of man depends. The mystery of the universe stirs in man as the mystery of the Self. Vedanta thought regards the human evolution as a process of progressively greater manifestation of spirit.

Resistance to Change:

In spite of the claim of scientists that science is based purely on rational foundation, a few irrational factors can be identified even in scientific research:

- Perceptual Bias: The perceptions used in scientific research vary from subject to subject. Because of the selectivity of perceptions and the underlying system of subjective values, conclusions cannot be extricated from subjective coloring. Same evidence may lead different scientists to different conclusions depending on their predispositions.

- Influence of paradigm: Practice of science at any one time is a function of the existing paradigm. Paradigm is a particular view of the world and its study, in which certain assumptions, rules, methods and instruments are fashioned by the leaders and shared by the rest of the scientific community. Rather than splaying out their efforts in sundry directions, followers focus their research efforts on specific areas and try to make slight extensions with known methods consistent with existing paradigm. The problem with paradigms is that when once they are entrenched, they cannot be easily replaced. That becomes a hindrance to admit novel ideas. Thus, problems and ideas that do not fall within the parameters of existing paradigm are dismissed as metaphysical and not worthy of consideration. Paradigms have a high degree of inertia because of the human consideration of self-preservation. Most of the academicians are reluctant to embrace novel ideas that stray far from the established principles because their reputations and careers depend on their credibility and respect of their peers. They are unlikely to welcome new ideas that might undermine their own work and positions.
- Reality Beyond Instruments: Ideas such as the indestructibility of embodied soul and its reincarnation that indicate a reality beyond their microscopes and telescopes, rattle their self assurance of mastery over nature and they feel threatened in their high positions. Therefore, it is natural that they develop resistance to such ideas. To gain insights into the phenomenon of reincarnation, the skeptic has to admit that he may be wrong in denying the existence of a power greater than his body-mind-intellect apparatus. Such admission is an anathema to the average scientist. Therefore, he ends up re-

jecting it altogether.

Need for Religion

The conflict between science and dogma bound religion of medieval Europe led to the eclipse of the latter. Unfortunately, the avalanche of fascination for the physical comforts flowing from modern science and technology has not only eclipsed the dogma bound religion, but also has taken in its sweep all religion shaking the belief of people in the importance of religion in their lives. However, in the midst of power, plenty and pleasure provided by science and technology, modern humans are feeling inwardly impoverished. They are tense and melancholy haunted by doubts and insecurity. They feel a void in their lives. Due to lack of inner satisfaction, they are smitten by ennui arising from the limitations of sense bound *Weltenshaung* of modern era. They have become a burden to themselves.

This malady cannot be cured either by creating more amenities using physical sciences or by manipulation of social structure; it can only be cured by cultivation of a balanced view of life and its purpose provided by the science of religion. While medieval Catholicism underrated sciences, modern civilization has overrated physical sciences and technology. There is an urgent need for correcting this imbalance.

Just as Western science has provided a framework, uniform concepts and nomenclature for the practice of positive sciences, Vedanta system has provided the world the necessary framework, concepts, nomenclature and methods for practicing the science of religion. Easterners have studied and absorbed sciences and technology from the West, it is now the turn of the Westerners to study and absorb the principles and methods of Vedanta in order to consummate the union of science and religion, and to integrate the West and the East.

Life cannot be comprehended in its fullness by logical reason alone. Self-consciousness is not the ultimate category of the universe. There is something transcending self-consciousness, variously called by the names "super consciousness," "revelation" or "intuition."

If we do not acquire the strength of spirituality from within and depend for our happiness on external sources alone surrendering our destiny to social engineering techniques like molecular biology and genetics, before long, human societies would be turned into animal farms.

Some of the professional western scientists reject religion, as their view of religion is confined to Christianity and its account of genesis. Vedanta view of Nature is that it is the outer form (*para rupa*) of the all-inclusive Brahman, while the spirit or consciousness is its inner form (*pratyak rupa*). Science and religion are not enemies as misunderstood by many modern people but they are allies complementing each other – speaking in two different languages, complementing each other by telling us a part of the same story.

Common ground of Science and Hindu Religion

Any ill-conceived suggestions prompting people to eschew religion in favor of science may be receptive to the adherents of those religions, which collided with science, but not to the Hindu religion, that has a common ground with science.

The methods of investigation followed in sciences are largely the same as those in Hindu religion, collection of facts, classification and subjecting them to logical analysis to discover the laws underlying them for application of that knowledge for alleviation of human suffering, for enrichment of human life and for its spiritual growth.

Whenever the ancient sages of India undertook this kind of scientific study of inner life of human beings, a galaxy of followers and subsequent sages verified and ratified the insights of their predecessors and thus, they left us a rich legacy of scientific practice in the field of religion.

As we said earlier, Advaita system doesn't ask you to take anything on trust. It assures you that if you maintain a passionate craving to know the Ultimate Reality and work for it, you can experience the Absolute as surely you can experience the pleasures and pains. Its instrument is the reasoning faculty of man, considered precious by scientists.

Because of its strict adherence to rationality and experiential base, Hindu spirituality has stood the test of time. Because of these qualities, it is not only hospitable to science but also proud of its achievements.

When science eclipsed dogma bound religion in the West, it created a spiritual vacuum there. The decline of religion in the wake of scientific advance has created an uneasy feeling in the minds of many thinkers that something of immense value has unnecessarily been sacrificed because of the misbehavior of a few shortsighted people enjoying power. Stung by the stigma that was attached to everything that was religious, some thinkers began reassessing the meaning and scope of religion in the calmer atmosphere prevailing after the power of Catholic Church declined.

To this task of bridging the gulf between science and religion, several thinkers, especially Indian philosopher-saints made invaluable contributions. They tried to reconstruct the mental life of modern man based on a unified view of his mundane existence and the true conception of religious life, by drawing distinction between reason-based religion on one hand and superstition and obscurantism on the other.

Discursive thinking, however extended cannot lead us to apprehension of Reality. It is difficult to know how far our knowledge of the external world that science investigates is objective. (This is the dilemma faced by the quantum physicists while investigating the subatomic particles).

Reason can carry us to the threshold of the Promised Land but cannot let us in. Reason is necessary but not sufficient. Insight or intuition is also required, as the sufficient condition.

The Great Transition

In stark contrast to their counterparts in the nineteenth century, humility was discernable in the approach of some of the great scientists of the 20th century, who began realizing that the existence of a "within" can no longer be evaded. Physicists of the first quarter of this century faced with the challenge of the revolutionary discoveries of the theories of relativity and quantum mechanics, turned into bold philosophical thinkers by venturing to apply critical reason not only to the observed data of the physical world but **also to the observer**.

Starting with thinkers like Bertrand Russell, physicists like Arthur Stanley Eddington, James Jeans, Max Planck, Einstein, Erwin Schrodinger, Neils Bohr, Heisenberg and other creators of 20th century physics; the trend has persisted through the last six to seven decades.

Some views expressed by a number of eminent 20th century scientists on the importance of consciousness in science are worth noting here.

Science professionals are wont to dismiss any idea beyond their reach as 'mysticism' using that word in a pejorative sense, but look what the great philosopher-mathematician Russell says about mysticism in his 'Mysticism and Logic'.

> The greatest men who have been philosophers have felt the need of both of science and mysticism.

In his book "The Nature of the Physical World", the British astronomer Sir Arthur Stanley Eddington observed

> Those who in search for truth start from consciousness as a seat of self knowledge with interests and responsibilities not confined to the material plane, are just as much facing the hard facts of experience as those who start from consciousness as a device for reading the indications of spectroscopes and micrometers.

Sir James Jeans, the British physicist, famous for his work in stellar dynamics observes in his book 'Physics and Philosophy':

> For the captives in the cave, the shadows constitute the whole world of appearance - the phenomenal world - while the world of reality lies forever beyond their ken.

On the usefulness of mystic thought in science, the Berkeley professor of physics, Fritzof Capra, writes vividly: Referring to the basic unity of the universe as viewed by the eastern mysticism and modern physics, Capra says:

> The basic oneness of the universe is not only the central characteristic of the mystical experience but is also one of the most important revelations of modern physics.[136]

Regarding consciousness, he went on to say:

> In modern physics the question of consciousness has arisen in connection with the observation of atomic phenomena. Quantum theory has made it clear that these phenomena can only be understood as links in a chain of processes, the end of which lies in the consciousness of the human observer.[137]

Eugene Wagner, another physicist expressed similar view when he observed:

> It was not possible to formulate the laws (of quantum theory) in a fully consistent way without reference to consciousness.

Consciousness is the very essence of Self as heat is of fire. While the content of experience changes, Self does not. It is present even in deep sleep when there are no objects to be known. The conditioned Self (ego) dies nightly when we sleep while the unconditioned Self stands witness to our existence. We cannot trace the steps through which it crept into awareness of itself.

The legendary mathematical physicist Von Neumann showed us how the validity of quantum mechanics implied the distinctness of consciousness of observer from all other aspects like the observer's body and brain.

Winds of Change

However, some biological scientists also have been recognizing, of late, the deficiencies of mechanical reductionism. New discoveries in biology, especially genetics have enabled the surviving vitalists to feel the winds of change.

When Watson and Crick proposed in 1962 a double helix model with strings and beads to represent genetic systems, it was readily welcomed and hailed as a great scientific discovery largely because its interpretation conformed to the prevailing canonical view of scientific procedure. - The strings of the helix were chromosomes and the beads genes, with the implication that evolution proceeds slowly based on the genetic variation (mutation) by minute changes in the beads (substitution of one nucleic acid base for another yielding a different amino acid). - This explanation fits with the official view that genetic systems could be broken down to minute parts and studied. It seemed to support the view that completed organisms could be explained ultimately as the translated products of an inherited program coded by a simple mechanism of four bases arranged in triplets. In this mechanism, DNA makes RNA and RNA makes proteins in a one-way flow of information. This unidirectional sequential flow of information is what Crick called 'the central dogma' of modern biology.

This rigid dogma, which has been ruling the biological world for a long time, has however buckled with the discovery of jumping genes (transposable elements) by Barbara McClintock.

The deficiencies of Darwinism are however understandable because those were the times when science was just emerging and fighting with entrenched Christian dogma. Up to the end of nineteenth century, science was wrapped up in final judgments. There was an air of finality in its pronouncements. The theory of evolution arrived at a time when the climate was dominated by the materialistic determinism of physics and led to the equating of man and animal to each other and both to a machine. However, attitudes began to change, with the advent of 20th century.

Molecular biologists generally work in reductionism mode trying to establish linear chains of cause and effect working with the simplest organisms. Their explanations are basic, being at the molecular level. They tend to concentrate on the repeatable properties of large groups and deliberately avoid the individual properties of any creature under study. Whatever might have been its benefits in the past, this method has reached its limit. A time has come when we have to study the complex interactions of individual creatures within their genetical systems.

Biological determinism tells us that death is a Darwinian imperative and that individual's final destiny is bodily death though the species may survive. Votaries of strict Cartesian reductionism ignore individuals and deal with groups of entities, but their quest is the immortality of the individual, whom they have rejected on principle. If, on one hand, we profess absolute faith in Darwinism, and on the other, reject the individual on principle, there should be no further struggle for individual longevity or immortality as the case may be. If immortality is a concept applicable only to the species and not the individual and if the individual is sacrificed on the altar of Darwinism and reductionism, how can we aspire for individual immortality?

Self Integrates

It is high time we realize that, when physicists who were once the hard core of physicalism, are now suggesting that we need to take our consciousness into consideration even to understand inanimate atomic phenomena, we can ill afford to ignore giving due place to consciousness in our human models. We cannot make much headway by doggedly sticking on to the reduction dogma in dealing with human models.

Take the case of human being. Is there any doubt that human body is the greatest machine that we have ever seen? This incredible machine has been in existence millions of years before Cartesian reductionism began to assert in the halls of science. Therefore, we cannot say that it is the outcome of experiments based on reductionism.

What is really the human being? No sensible person ever claimed that he (she) is an assemblage of hands, feet, eyes and ears. On the other hand, everyone claims these limbs as his (her) possessions when he (she) says "my head," "my eyes" etc. All these organs are controlled by the mind, which itself is controlled by the Self, that stands as a witness even during sleep. Think for yourself, who is the "I" when you say "my body" quite often. To whom does the possessive pronoun, "my" refer?

The human being is an inseparable whole, a grand hierarchical system in which many subsystems, including brain system, are connected by feedback loops, all of which are controlled by Self. We feel that consciousness as "ego". Remove that Self, all the sub systems and their parts fall asunder. No amount of laboratory juggling with the fallen parts can make them useful without that Self.

Some biologists hope that they can somehow enable the body to work, by manipulating its parts without the participation of the presiding Self; but they will realize sooner or latter that they hoped in vain too long.

It is time we realize that pure reductionism based science can no longer cater to the need of problems associated with animate entities, without an integrating principle. The Self integrates the various parts and functions of the body.

The Missing link

Scientists, as they claim to be lovers of Truth, should at least consider what other selfless intellectuals say about the missing link in what appears to them as unfathomable mystery.

Selfless saints and brilliant philosophers both in the East and in the West investigated this problem long ago. Indian philosophers of various schools and hues – orthodox and heterodox, theistic and atheistic – who investigated this problem arrived, surprisingly, at the same result. Their investigations were totally independent as they were separated by centuries of time and hundreds of miles of space, in an age when modern communication facilities were unknown. Their investigations were unbiased as they were pursuing Truth untainted by desires for comfortable positions, prizes, and publicity. Their independent investigations led them almost to the same inevitable conclusion that physical death is not the end of the individual. The soul migrates into another body. Although many of them arrived at this result aided by intuition, they supported their conclusions by irrefutable logic, and left no scope to the critics to dub them as mere faith.

Referring to the idea of transmigration of souls, even Thomas Huxley, the foremost champion of Darwinian evolution said none but hasty thinkers would reject it on the ground of inherent absurdity. Like the doctrine of evolution itself, the doctrine of transmigration has its roots in the realm of reality, for transmigration is the vehicle that carries on the post human-evolution with the potential of transforming ordinary consciousness into super consciousness. The German philosopher Schopenhauer quoted earlier voiced similar opinion.

TRANSMIGRATION AS A VEHICLE OF POST HUMAN EVOLUTION

What is it that survives death if not the body? The great thinkers of the East propounded and argued that the real man is not the body that perishes but the Soul that persists. When we raise the question of future life, we are concerned with the individual human soul. Soul is the name of the composite nature which one knows as oneself and which functions as one person though it passes from one life to another. It must have existed before it began to animate the body of this life and will exist after it ceased to animate it.

The characteristics of human life are determined jointly by those, which the Self had before it began to animate the organism and by the nature of the organism, it animates. Then the question arises: How does a certain soul get associated with a certain organism when the latter is purely a product of biological causation? It is plausible to think that a pre-existent Self becomes associated with a certain organism at the time when the latter is conceived. This kind of adaptive affinity is well known in nature, as in chemical affinity and the selective affinity of spermatozoa to the ovum of the same species. Mind stuff and organisms attract each other in the same way. The reincarnating ego is attracted to parents from whom it can inherit a particular set of qualities. The natural body derived according to the laws of heredity is taken over by the soul. The psyche appropriates the body necessary for the realization of its goal.

Thus, while the physical heredity is derived from the parents, social heredity (tradition) is acquired from the family, society and religion. The actions and thoughts of an individual in the previous lives are carried forward by the invisible subtle body (*sukṣma śarīra*) and they exclusively determine the psychological heredity. We cannot judge the reality of a thing by its visibility, because of the limitations of our sense organs and the distortions caused by them. (Visible spectrum is very narrow with a wavelength of just 0.00003 cm). Indeed, many things – atoms, electrons and other elementary particles – cannot be seen in this narrow range with our naked eyes. If this theory cannot be empirically proved it cannot be disproved either and at its worst, it stands on a footing higher than Darwin's theory of selection. To the curious, the following account found in the Upaniṣads, gives some idea about the mechanics of transmigration.

> When a person dies, his soul is said to carry with it the sum total of all its actions and thoughts as tendencies (*vasanās*). The resultant of all its works and thoughts determines and directs its path. If the accumulated actions and thoughts demand a new body consequently, the soul will then be steered to go to parents who can provide it with a suitable

new body. Thus, it acquires a new body sharing the bodily characteristics of the parents but carrying forward its own fund of mental impressions accumulated in previous lives. Soul is thus embodied.

Science cannot disprove this proposition nor can it provide a more rational explanation for the basis of many human problems like seemingly unjust inequalities and undeserving sufferings of good people, which would remain inexplicable otherwise.

Though independent, soul expresses itself through a body and mind, it is neither body nor mind, but it controls both. Soul is the entity that consolidates the individual cognitions produced by the organs of the body and produces an integral experience. Taittiriya upaniṣad says that both words and mind return unable to define it.

The theory of transmigration of soul, which cannot be proved or disproved by empirical methods, can be inferentially proved. It fills up the gaping voids in the mindless materialism, synthesizes the fragments of the reductionism into a whole, and endows it with life. It assures us that human efforts, at no part of human's life, (even those after the post reproductive period) are wasted. It accounts for every action and thought, even unto the last breath. It satisfies the physical law that each action has its inevitable consequences, even in the psychic realm.

How to know the invisible Soul?

If so, the question arises as to how we can detect and recognize this soul. Soul is not an object to be cognized by our senses. How then can it be detected? It can be intuited. Upaniṣads give us the closest mental glimpse of Ultimate Reality when they tell us:

> 'That, from which these beings are born, that by which they are sustained and that into which, on ceasing to live, they are absorbed, know that as Brahman.'[138]

All the same, sincere seekers can intuit it.

Even in science, there are invisible entities. The fundamental indefinable entities in physics are length, mass and time, but by common agreement among scientists, they chose certain metrics as the basic measuring units for their categories whence science was built up. Similarly, in philosophy we start with certain indefinable entities like 'I' and 'Thou' conceptual. Further enquiry emanates from them. When properly understood, the "I" conceptual refers, not to the body but to the consciousness behind the "I" concept. Soul is not a thing that can be cognized with senses, it can only be intuited. Sincere seekers can find the way to intuit it.

Intuition is integral insight different from sense observation and reasoning. Rousseau called it, "sovereign intelligence, which sees in the twinkle of an eye the truth of all things in contrast to vain knowledge." Intuitive knowledge dawns only upon those who made some earlier efforts to attain certain goal.

In intuitive vision, one perceives clearly the whole solution, though it may be complicated to express in words. *Taittiriya Upanishad* says that speech and mind return unable to describe such a vision.

Subjectivity in Science

Subjective element had played a dominant role in acquisition of valuable insights throughout the history of knowledge. The Upanishads, which have been hailed all over the world as the unparalleled documents of human knowledge that demonstrate the verities of life, have been the results of the subjective insights revealed to the forest dwelling seers who recorded them. Great mathematicians like Karl Frederick Gauss and Bernhard Reimann admitted that some of their great results occurred to them intuitively in a flash without their being aware of the exact path through which they were lead to them. The great Indian mathematical prodigy Srinivasa Ramanujam went a step further and claimed that some goddess was dictating to him the theorems he was effortlessly producing. Though subjective element has had an honored place in the history of knowledge, it has been relegated to the background in the recent past due to a number of reasons like the emergence of Utilitarian philosophy and the success of science in providing physical amenities to people. People who were already under the influence of hedonism were misled to believe that the mechanistic model presented by science was the whole truth.

As indicated earlier, this situation is in need of change. Change does not imply giving up the logical rigor and objectivity of the empirical-theoretical models, but it involves admittance of the subjective elements and their demonstrated ability to create anomalous situations in our experiences which we hitherto considered to be purely objective. The much-quoted anomalies occurred in modern physics in the context of resolving the dispute between particles and waves as the true nature of light and again in the context of determining simultaneously the position and velocity of an electron. The former dispute was compromised by the principle of complimentarity while the latter problem was given up as indeterminable due to the interference of subjective element. This experience suggests that we can view the objective and subjective sides as complimentary.

Lest the opponents of subjectivism may charge that introduction of subjective elements in scientific models may involve introduction of religion they abhor, we hasten to add that it could be done without aiming at any ontological implications. They should realize that objective science has much more potential than they imagine, if only light is allowed to enter into our self-woven cocoons. It is for us to choose whether we want to set limits to science by halting at the portal of personality or allow it to climb to loftier heights by entering the portal.

Quantification of Consciousness

We are not completely in the woods as regards quantification of human traits. Examples are not lacking even now, about how the human consciousness can be studied quantitatively using already known methods, to know how it affects objects outside its domain. I think the earliest efforts to quantify human behavior were made by, Von Neuman in his great work 'The Theory of games and Economic Behavior', produced in collaboration with Oscar Morgenstern.

Can the behavior of external objects be changed without a medium of contact? This crucial question determines the fate of the candidacy of consciousness as an element of scientific models. If the answer is yes, there is no reason why we should exclude it. If biologists are wondering how to quantify consciousness to be introduced into experimental research models, the work done by the PEAR group in Princeton University offers some encouragement and provides some leads.

With the introduction of Consciousness into the human experience model, the lingering controversy between the physicalist tradition and the vitalist tradition can subside and the random walk methods of experimental biologists can be given some fruitful direction viewing Soul as the invariant substratum of human experience. Biological models that include Chaitanya (activity and consciousness) as one of the invariants, may perhaps one day fortify the psychosomatic models based on ancient Yoga system described earlier. In an earlier chapter, we have given an account of experiments conducted in Princeton University seeking to find the influence of mind on matter and showed that strong human volition can affect matter.

In this backdrop of science, we have to view the current longevity research in its proper perspective. All this research is concerned with the body, but the real person is the entity that keeps the body alive. Efforts to keep the body alive by repairing it and replacing body parts ignoring the force that animates it, is like trying to run a worn out automobile by replacing parts and repairing it, without putting fuel in it. Even as repeated repairs cannot prevent automobiles ending up in junkyards, any number of interventions cannot prevent the bodies going to the graveyards. That is the nature of all composite things. Nothing in the world is permanent and it is foolish to ask anything to last forever. It is even more foolish not to make good use of it when we have it.

With this caveat in place, one can follow the methods adumbrated earlier to keep one's body and mind healthy and together for as long as possible with the aim of realizing the final goal of emancipation from our thralldom to matter. We must realize that while it is possible to achieve greater degree of health and longer life span, the end of physical body is unavoidable.

The paradox of immortality is this. Any desire including craving for immortality propels your soul to take rebirth in order that you may try to satisfy your desires. Taking rebirth violates the idea of immortality as it involves death of the present physical body. Cessation of all desires prevents the soul from seeking perishable bodies thus leading to immortality. Whosoever realizes that he is not the corporeal body but the soul attains immortality.

> I decline to accept the end of man... I believe that man will endure; he will prevail. He is immortal, not because he alone among creatures has an inexhaustible voice but because he has a **soul**, a **spirit** capable of compassion and sacrifice and endurance."
>
> ----William Faulkner

(-From the Speech of Acceptance of Nobel Prize for literature -10th December 1950.)

---(0)---

APPENDIX

Table A3-1: Top ten Oldest Persons ever (in economically developed countries) (Source: Wikipedia).

#	Name	Age	Born	Died	Notes
1	Jeanne Calment (F)	122 years 164 days	21 February 1875	4 August 1997	Oldest documented and fully validated French super-centenarian ever
2	Shigechiyo Izumi (M)	120? Years 237 days	29 June 1865?	21 February 1986	Oldest Japanese male super-centenarian (disputed)
3	Sarah Knauss (F)	119 years 97 days	24 Sept. 1880	30 December 1999	Oldest in U.S. history, oldest recognized in world 1998–99
4	Lucy Hannah (F)	117 years 248 days	16 July 1875	21 March 1993	Oldest African American
5	Marie-Louise Meilleur (F)	117 years 230 days	29 August 1880	16 April 1998	Oldest Canadian, oldest recognized in world 1997 – 98
6	María Capovilla (F)	116 years 347 days	14 Sept. 1889	27 August 2006	Oldest in Ecuador ever, oldest living person 2004 –06
7	Tane Ikai (F)	116 years 175 days	18 January 1879	12 July 1995	Oldest Japanese woman
8	Elizabeth Bolden (F)	116 years 118 days	15 August 1890	11 December 2006	Oldest living U S person at time of her death, Tennesseean
9	Carrie White (F)	116? Years 88 days	18 November 1874?	14 February 1991	Formerly oldest in U.S. history, fourth-oldest American ever recorded (disputed)
10	Kamato Hongo (F)	116? Years 45 days	16 September 1887?	31 October 2003	Third-oldest Japanese person (disputed)

Source: http://en.wikipedia.org/wiki/Image:JCalment_001.jpg.

Table A3- 2: Top ten Oldest living men (in economically developed countries)

	Name	Age as of 14 September 2007	Born	Residence
1	Tomoji Tanabe	111 years, 361 days	September 18, 1895	Oldest man in the world (since 24 January 2007). - Lives in Miyazaki, Japan
2	Sukesaburo Nakanishi	111 years, 183 days	March 15, 1896	Lives in Japan
3	Henry Allingham	111 years, 100 days	June 6, 1896	Lives in England, UK. World War I Veteran
4	George Francis	111 years, 100 days	June 6, 1896	Lives in California, U.S.
5	Walter Breuning	110 years, 358 days	September 21, 1896	Lives in Montana, U.S.
6	Augusto Moreira de Oliveira	110 years, 343 days	October 6, 1896	Lives in Portugal
7	Walter Seward	110 years, 336 days	October 13, 1896	Lives in New Jersey, U.S.
8	Nicholas Kao Se Tseien	110 years, 242 days	January 15, 1897	Lives in Hong Kong, China
9	Aarne Arvonen	110 years, 41 days	August 4, 1897	Lives in Uusimaa, Finland. Finnish Civil War Veteran
10	Louis de Cazenave	109 years, 333 days	October 16, 1897	Lives in France. World War I Veteran

[Source: Wickipedia].

Table A3-3: A sample of long-lived people of India

Life spans of some recent Yogis of India

Name(Sex) and other description	Born	Died	Duration of life	Notes
(1) Trlinga Swamy*	19 DEC 1607	26 Dec1887	280 years-8 days.	Died in Varanasi-See details below.
(2) Sorakayala swami (M)	1700	9 Aug 1902	202 yrs	
(3) Sadasiva Brahmendra swami (M)--Itenerant scholar- monk	1560	1750	190	Died in South India
(4) Adoni Tikkalakshmamma (F)	1815	16 May 1993	178	
(5) Machilipatnam Dontulamma (F)	1807	9 Feb 1932	125	
(6) Neelkanta Satchidananda Ghanendra (M)	1788	30 Nov 1907	119	
(7)Kaivaram Amara Narayana	1730	2 June 1840	110	
(8) Gunturu Nalla Mastan (M)	1685	23 May 1895	110	
(9) Hazarat Ismail Baba (M)	1825	22 July 1934	109	
(10) Narayana Teertha (M) Sanskrit scholar & saint	1580	1680	100	Died in South India
(11) Yogi Vemana (M)- A very popular people's poet	1550	1650	100	Lived & died in Andhra Pradesh

Source of Tabular data: B.Ramaraju, National research professor, Hyderabad "*Andhra Yogis, Part-1*"

Note*: Trilinga swami was a contemporary of stalwarts like Swami Dayananda Saraswati, founder of the Arya Samaj, Ramakrishna parama hamsa, Vivekananda, and Lahiri mahasaya, guru of Yogananda, founder of *Self Realization Fellowship* of Los Angeles. His biography by Umacharan Mukhopadhyaya is available in Bengali.

Source of image: *http://www.spiritual-note.com/Trailinga_Swami*, and some other similar web sites.

1607-1887

Trilinga Swami

Born 19 Dec 1607 * Died 26 Dec 1887

Table A9- 1: determination of frame size

Height	Wrist measurement	Frame Class
Short	(a) <5 1/4"	Small
	(b) 5 1/4 to 5 3/4"	Medium
	(c) > 5 3/4"	Large
Medium	(a) < 5 1/2"	Small
	5 1/2 to 6"	Medium
	(c) > 6"	Large
Large	(a) < 5 3/4"	Small
	(b) 5 3/4 to 6 1/2"	Medium
	(c) > 6 1/2"	Large

Table A9- 2: Typical Vital Statistics based on Medium bone structure

Vital Statistics corresponding to medium frame and different height groups (in inches)

	Heights			
Body Part	Under 5'-4"	5'-4" to 5'-7"	5'-7" to 5'-9"	Over 5'-9"
Upper Arm	9	9 1/2	10	10 1/2
Waist	22 to 23	>23 to 25	>25 to 27	>27 to 29
Thighs	19	19.5	20	20.5
Calf	12.5	12.5	13	13.5
Ankle	7 and 3/4	8	8 and 1/4	8.5

Notes:

1.To get ideal measurements for large frame add 1", for small frame subtract 1"

2.To get ideal measurement for large frame add ½", for small frame subtract½"

3.To get ideal measurement for large frame add ¼", for small frame subtract ¼"

Table A9- 3: Ideal Weights of Women and Men

Ideal weight ranges (lbs) based on Heights (ft-in)						
Height	Women			Men		
	S F	M F	L F	S F	M F	L F
4-10	89-97	94-106	102-118	--	--	--
4-11	92-100	97-109	105-121	--	--	--
5-00	95-103	100-112	108-124	--	--	--
5-01	98-106	103-115	111-127	--	--	--
5-02	101-109	106-119	114-131	112-120	118-129	126-141
5-03	104-112	109-123	118-135	115-123	121-133	129-144
5-04	107-116	113-128	122-139	118-126	124-136	132-148
5-05	111-120	117-134	126-143	121-129	127-139	135-152
5-06	115-124	121-136	130-147	124-133	130-143	138-156
5-07	119-128	125-140	134-151	128-137	134-147	142-161
5-08	123-133	129-144	138-156	132-141	138-152	147-166
5-09	127-137	133-148	142-161	136-145	142-156	151-170
5-10	131-141	137-152	146-166	140-150	146-160	155-174
5-11	135-145	141-156	150-171	144-154	150-165	159-179
6-00	139-149	145-160	154-176	148-158	154-170	164-184
6-01	--	--	--	152-162	158-175	168-188
6-02	--	--	--	156-167	162-180	173-193
6-03	--	--	--	160-171	167-185	179-199
6-04	--	--	--	164-175	172-190	182-202
SF= Small frame, MF= Medium frame, LF= Large frame						

Table A9- 4: Body-Mass Indices

Obese — Over weight — Normal — Under weight

Weight

Height (FT-In)	100	105	110	115	120	125	130	135	140	145	150	155	160	165	170	175	180	185	190	195	200	205
5-0	20	21	21	22	23	24	25	26	27	28	29	30	31	32	33	34	35	36	37	38	39	40
5-1	19	20	21	22	23	24	25	26	26	27	28	29	30	31	32	33	34	35	36	37	38	39
5-2	18	19	20	21	22	23	24	25	26	27	27	28	29	30	31	32	33	34	35	36	37	37
5-3	18	19	19	20	21	22	23	24	25	26	27	27	28	29	30	31	32	33	34	35	35	36
5-4	17	18	19	20	21	21	22	23	24	25	26	27	27	28	29	30	31	32	33	33	34	35
5-5	17	18	18	19	20	21	22	22	23	24	25	26	27	27	28	29	30	31	32	32	33	34
5-6	16	17	18	19	19	20	21	22	23	23	24	25	26	27	27	28	29	30	31	31	32	33
5-7	16	17	17	18	19	20	20	21	22	23	23	24	25	26	27	27	28	29	30	31	31	32
5-8	15	16	17	17	18	19	20	21	21	22	23	24	24	25	26	27	27	28	29	30	30	31
5-9	15	16	16	17	18	18	19	20	21	21	22	23	24	24	25	26	27	27	28	29	30	30
5-10	14	15	16	17	17	18	19	19	20	21	22	22	23	24	24	25	26	27	27	28	29	29
5-11	14	15	15	16	17	17	18	19	20	20	21	22	22	23	24	24	25	26	26	27	28	29
6-1	14	15	15	16	16	17	18	18	19	20	20	21	22	22	23	24	24	25	26	26	27	28
6--1	13	14	15	15	16	16	17	18	18	19	20	20	21	22	22	23	24	24	25	26	26	27
6--2	13	14	14	15	15	16	17	17	18	19	19	20	21	21	22	22	23	24	24	25	26	26
6--3	13	13	14	14	15	16	16	17	17	18	19	19	20	21	21	22	22	23	24	24	25	26
6--4	12	13	14	14	15	15	16	16	17	18	18	19	19	20	21	21	22	23	23	24	24	25

Table A9- 5: Recommended K.Caloric Intake

Recommended Caloric Intake based on height						
Height	Calorie ranges for Women			Calorie ranges for Men		
	SF Low	MF Mean	LF High	SF Low	MF Mean	LF High
4-10	1335-1455	1410-1590	1530-1770	--	--	--
4-11	1380-1500	1455-1635	1575-1815	--	--	--
5-0	1425-1545	1500-1680	1620-1860	--	--	--
5-1	1470-1590	1545-1725	1665-1905	--	--	--
5-2	1515-1635	1590-1785	1710-1965	1680-1800	1770-1935	1890-2115
5-3	1560-1580	1635-1845	1770-1890	1725-1845	1815-1995	1935-2160
5-4	1605-1740	1695-1920	1830-2085	1770-1890	1860-2040	1980-2220
5-5	1665-1800	1755-2010	1890-2145	1815-1935	1905-2085	2025-2280
5-6	1725-1860	1815-2040	1950-2205	1860-1995	1950-2145	2070-2340
5-7	1785-1920	1875-2100	2010-2265	1920-2055	2010-2205	2130-2415
5-8	1845-1975	1935-2160	2070-2340	1980-2115	2070-2280	2205-2490
5-9	1905-2055	1995-2220	2130-2415	2040-2175	2130-2340	2265-2550
5-10	1965-2115	2050-2280	2190-2490	2100-2250	2190-2400	2325-2610
5-11	2025-2175	2115-2340	2250-2565	2160-2310	2250-2475	2385-2685
6-0	2085-2235	2175-2400	2310-2640	2220-2370	2310-2550	2460-2760
6-1	--	--	--	2280-2430	2370-2625	2460-2760
6-2	--	--	--	2340-2505	2430-2700	2595-2910
6-3	--	--	--	2400-2565	2505-2775	2685-2985
6-4	--	--	--	2460-2625	2580-2850	2730-3060

Note: The values given are in Kilo Calories. Kilo calorie=1000 calories.

In the nutrition context, the word "Calories" is used by many people synonymously with Kilocalories. However, the reader should understand that the word "Calories," in this context is used to mean. K.Calories, which is equal to one thousand calories.

Table A9- 6: Protein & Calorie values of various foods. - Part1

Food Category & Substance	Limits of daily use (gm)		Serving size		weight in serving size (gm)		Energy content	Calorie to protein ratio
Daily allowances	Min	Max	Kitchen measures	weight (gm)	Total protein	Usable protein	Calories	
Grains, Cereals & bread								
Wheat, bulgar	50	100	1/2 cup	100	11	7	361	51
Wheat, whole grain	50	150	1 cup	200	20	12	722	60
Wheat, germ	6	18	1 tb sp	6	2.7	2	21	11
Wheat, bran	6	18	1 tb sp	6	1.6	0.9	21	23
Rice, brown	50	200	1 cup	180	15	11	712	64
---Polished	50	200	1 cup	200	12	8	570	71
Barley	50	200	1/2cup	90	9	6	357	59
Cereals, Rye	20	80	1 cup	80	12	7	313	45
---Rolled Oats	20	80	1 cup	80	14	9	284	32
Bread-rye, regular	21	75	1 slice	25	2.25	1.1	63	57
Bread-whole wheat	25	75	1 slice	25	2.25	1.1	64	58
Legumes: Dried beans and lentils								
Soy beans	50	75	1/3 to 1/2 cup	50	17	10	200	20
Mung beans	50	75	1/3 to 1/2 cup	50	12	7	175	25
Broadbeans	50	75	1/3 to 1/2 cup	50	13	6	170	28
Black beans	50	75	1/3 to 1/2 cup	50	12	5	150	30
Cowpeas	50	100	1/3 to 1/2 cup	50	12	5	165	33
Kidney beans	50	75	1/3 to 1/2 cup	50	12	5	195	39
Garbanzos	50	75	1/3 to 1/2 cup	50	11	5	186	37
Lima beans	50	100	1/3 to 1/2 cup	50	10	5	94	19
Lentils	50	100	1/2 to 1 cup	100	26	9	423	47
(Flours - Values given refer to per cup size, not serving size)								
Soybean flour (non fat)			1 cup of flour	140	65	40	440	11
Whole wheat flour			1 cup of flour	120	16	10	400	40
Peanut flour			1 cup of flour	100	48	21	693	33
Oatmeal			1 cup of flour	80	10	7	154	22
Cornmeal			1 cup of flour	120	10	5	400	80
Rice flour			1 cup of flour	120	8	4	168	42

Protein& Calorie values (part 2)

Food Category & Substance	Limits of daily use (gm)		Serving size		weight in serving size (gm)		Energy content	Calorie to protein ratio
Daily allowances	Min	Max	Kitchen measures	weight (gm)	Total protein	Usable protein	Calories	
Nuts & seeds								
Pignolia nuts	10	28	2 tbsp	28	9	5	180	36
Pumpkin seeds	10	20	2 tbsp	28	8	5	160	32
Sunflower seeds	10	30	3 tbsp	20	7	4	160	40
Peanuts	50	100	1/2 cup	100	26	11	629	57
Peanut butter	14	28	2 tbsp	28	8	3	153	51
Cashews	50	100	1/2 cup	100	17	10	598	60
Sesame	20	40	3 tbsp	40	5	3	172	57
Pistachio nuts	20	30	2 tbsp	28	5	3	180	60
Brazil nuts	20	30	2 tbsp	28	4	2	196	88
Dairy Products								
Whole Milk	120	500	1 cup	244	9	7	151	22
Skim milk	120	500	1 cup	244	9	7	85	12
Condensed milk	60	250	1/2 cup	122	9	7	84	12
Butter milk	120	500	1 cup	244	9	7	100	14
Yogurt	120	240	1 cup	244	9	7	156	22
Ice cream	100	200	1 cup	200	10	8	180	23
Cheese, American	25	100	1" cube	25	8	6	93	16
---Cheddar	30	120	1" cube	30	8	6	121	20
---Cottage (low fat)	60	240	1 cup	60	5	6	68	11
---Monterry	30	120	1" cube	30	5	4	41	10
---Mozzrella	30	120	1" cube	30	5	4	76	19
---Swiss	30	120	1" cube	30	8	6	111	19
Eggs								
--Whole	50	50	1 med	50	7	6	82	14
--Yolk	29	29	1 med	29	5	4	100	25
--White	15	15	1 med	15	2	0.5	7	14

Protein & Calorie values of foods (part.3)

Food Category & Substance	Limits of daily use (gm)		Serving size		weight in serving size (gm)		Energy content	Calorie to protein ratio
Daily allowances	Min	Max	Kitchen measures	weight (gm)	Total protein	Usable protein	Calories	
Seafood (Ave. size 3 1/2 oz)								
Clams	70	70	3 1/2 oz	70	9	7	52	7
Cod	100	100	3 1/2 oz	100	18	14	75	5
crab (steamed)	50	100	3 1/2 oz	100	17	14	88	6
Halibut	100	100	3 1/2 oz	100	20	9	93	10
Herring	50	100	1 herring	50	9	7	84	12
Lobster	50	100	1 cup	145	28	22	91	4
mackeral	100	100	3 1/2 oz	100	19	15	184	12
Red fish	100	100	3 1/2 oz	100	18	14	83	5
Oysters, med size	50	100	2 selects	25	2.1	18	16	2
Salmon-Atlantic	100	100	3 1/2 oz	100	23	15	209	14
----Canned	100	100	6 oz	185	41	33	363	11
Sardines, Paciic	50	100	3 1/2 oz	100	19	15	157	10
Shrimp, raw	50	100	3 1/2 oz	100	18	14	85	6
Sole	100	100	3 1/2 oz	100	17	14	75	5
Sword fish	100	100	3 1/2 oz	100	19	15	112	7
Tuna-canned	50	100	1/2 cup	115	32	26	137	5
Meat & poultry products								
Chicken, breast	100	100	3 1/2oz	100	23	15	110	7
Lamb, lean	100	100	3 1/2oz	100	20	13	257	10
Pork, lean &fat	100	100	3 1/2oz	100	29	19	275	18
Turkey, roasred	100	100	3 1/2oz	100	24	17	159	9
Hamburger	1/4 lb	1/4 lb	3 1/2oz	100	26	17	255	15

Note: {The calorific value given in the above three tables are kilo calories.

Each kilocalorie = 1000 calories.

GLOSSARY

Acquired Characters: Characteristics of an organism's phenotype that result from the influence of environment rather than inheritance.

Base Pair: A pair of hydrogen-bonded nitrogenous bases that connect the two strands of DNA double helix.

Central dogma: The assertion that the information contained in proteins cannot be translated back to nucleic acids.

Determinism: The theory that the outcome of any process is strictly predetermined by definite causes and natural laws. (Therefore, predictable at least in theory).

DNA: Deoxyribonucleic acid, the molecule that transmits genetic information.

Emergence: The theory the occurrence of characters at the higher level of integration, which could not have been predicted from knowledge of lower level components.

Entropy: The constant tendency of nature to sink into disorder. Expenditure of energy is required to overcome entropy and establish order. In biological systems, homeostasis is the force that opposes entropy to keep the organism alive and functioning. Eventually entropy gains upper hand and destroys the organism.

Fitness: The relative ability of an organism to survive and to reproduce.

Gamete: A germ cell (egg or sperm) carrying half of the full set of the organism's full set of chromosomes, especially a mature germ cell capable of participating in fertilization.

Gene: A sequence of base pairs in a DNA molecule that contains information for the construction of one protein molecule, or in some cases for construction of transfer RNA or ribosomal RNA.

Genetic Code: The code by which the genetic information contained in the base-pair sequence of DNA is translated into amino acids.

Genetic Recombination: Artificial reshuffling of an organism's genes to produce a new variety.

Gene Expression: The process by which information stored in DNA is converted into protein. In this process information is first transcribed into a single stranded molecule called RNA. The 4-letter nucleotide is then translated from RNA to amino-acid language of proteins. Gene expression is tightly controlled and it plays a major role in determining the amount of each type of protein a cell contains.

Genome: The primary repository of genetic information of an organism--generally refers to the DNA molecule and its nucleotide sequence. The bacterial genome is a single DNA molecule whereas the human genome consists of DNA molecules from each of the chromosomes.

Genotype: The totality of genes (genetic information) carried by an individual.

Homeostasis: Tendency to maintain normal internal stability in an organism by coordinated responses of the organ systems that automatically compensate for environmental changes.

Life: Biologists define 'Life' as a set of chemical reactions organized in such a way that this set can reproduce itself. This cryptic definition can mislead readers into believing that we can create life by organizing some chemicals. No one has ever-organized chemicals to produce life. All that the definition implies is that we can **recognize** life by its ability to reproduce itself.

Life Table: A tabular statement of data consisting of the number of births and deaths per unit of a population at different ages. It is used by demographers to compute birth rates, death rates and life expectation at any given age.

Life Expectation (also called 'Life Expectancy'): The number of years an individual of a given age may expect on the average to live, as projected in the statistical tables. Thus, the life expectancy of a person at birth becomes his/her total life span, while the life expectancy at a particular age after birth indicates the expected number of years remaining for that person.

Meiosis: The process of sexual reproduction wherein the developing germ cells divide twice consecutively resulting in pairing and segregation of homologous chromosomes.

Mitosis: The process by which (including its chromosome set) divides into two daughter cells.

Metabolism: A collective term for all the chemical reactions occurring in a body. For example, the chemical reactions that occur in the body during the production, use and breakdown of sugars is known as sugar metabolism.

Molecule: A group of atoms tightly held together by what are known as chemical bonds. The arrangement of atoms in a given molecule is quite specific and this arrangement gives each molecule specific chemical and physical properties.

Molecular Biology: The study of living organisms at the molecular level. Unlike physicists and chemists, molecular biologists do not go into the study of atomic structure and electronic configuration etc.

Natural Selection: A phrase used by Darwin to indicate a process of automatic retention of characteristics favorable to the reproductive success of organisms while eliminating others by the process known 'as survival of the fittest' for reproduction.

Phenotype: The totality of the observable characteristics of an individual resulting from the interaction of the genotype with the environment.

Phylogeny: The pathway of descent from ancestors.

Physicalism: An emphasis (or belief in) on certain principles dominant in physical sciences, such as determinism and reductionism etc.

Reductionism: A theory propounded by Rene Descartes which states that all phenomena and laws relating to complex phenomena (living organisms included) can be explained only by reducing them to their smallest components and that the higher levels of organization of the phenomena can be explained through a knowledge of these components.

Reproduction: Defined as the duplication of information content of the cell followed by segregation of this information into two newly formed daughter cells.

Saltationism: The belief that evolutionary change is the result of the sudden origin of a new kind of individual who becomes the progenitor of a new lineage.

Species: A group of reproductively isolated populations, which share some common characteristics and can interbreed with one another.

Stasis: The maintenance of a constant phenotype by an evolutionary lineage through geological time.

Telomere: Gene-free region of DNA at the end of chromosomes made up of thousands of DNA bases and associated proteins.

Telomerase: An enzyme that rebuilds the nucleotide sequences of telomeres. This is a double-edged sword. It's over activity leads to cancer while its deficiency is suspected to be behind the phenomenon of telomere shortening leading to cell senescence.

Teleology: The actual or seeming existence of goal-oriented processes in Nature and their study.

Transposon: Genes that move from one chromosome to another.

Yeast: Yeast is a one-celled organism containing a true nucleus and mitochondria. Many biochemical properties are similar to those of higher organisms. They are supposed to be related closer to mammals than to bacteria.

Vitalism: The belief that living organisms have a vital force that cannot be found in inert matter.

Zygote: A fertilized egg; the individual that results from the union of two gametes and their nuclei.

NOTES

Chapter 2-Quest for Immortality

[1] Whitney, *Jl. of American Oriental Society*, iii, 292, (quoted by S. Radhakrishnan in *Indian Philosophy*, vol.1,pg.83)

[2] *Rg Veda* viii, 48.3

[3] Sacred Books of the East, Vedic Hymns, part 1

[4] Walford, Roy L., *Maximum Life Span*, W. W. Norton & Co., New York

[5] Swift, Jonathan, *Gulliver's travels*, Ch7, "Voyage to Luggnagg"

[6] dehinosmin yadha dehe, kaumaram yauvanam jarā tadha dehantaraprāpihi, dhīrastatra namuhyati. ---(Bhagavdgita, 2-12)

[7] vāsāmsi jīrnani yadha vihāya, navāni gruhnāti naro parāni
tadha śarirāni vihāya jīrnā, nyanyāni samyati navāni dehi. ---(Bhagavdgita, 2-22)

[8] A fallacy of composition is an idea that is true if individual units are considered but false when the same individual units are composed into a single unit.

Chapter 3-How to think about aging?

[9] Paramahamsa Yogananda, '*An autobiography of a Yogi*', Self Realization Fellowship, San Rafel Ave., Los Angeles, CA

[10] J. R Wilmoth., et. al, "Increase of maximum life span in Sweden, 1861-1999,". *Science*, September 2000, vol. 289, PP 2366-2368.

[11] The formula used for this computation is: (death rate sans accidents) ^ x =0.5, where x represents median age at death and 0.5 represents the median survival. For example age specific accidental death rates as per actuarial data compiled in (19xx) were:

99 year-olds----15/1000 = 0.015 Survival rate = 1- 0.015=0.985

69 year olds----1/1000 = 0.001 ---ditto--- = 1- 0.001=0.999

39 year-olds----.39/1000 = 0.00039 ---ditto--- = 1- 0.039=0.99961.

[12] Arius; *Indica* XXIIC

[13] James W. Vaupel., et al. "Bio-demographic trajectories of Longevity", *Science*, vol., 280, 8th May 1998, pp855-860.

Chapter 4-The Mystery of Aging

[14] If two thin flat objects A and B with surface areas of 1 sq. cm and 2 sq. cm respectively, are left out in a winter day, B will lose heat 2 times faster than A because B has two times the surface area of A.

This argument can be extended to two objects of the same shape but of different sizes. Now let two cubes each of side 1 cm and 2 cm respectively take the place of flats in the above example. A's surface area is 6 sq.cm, and B's surface area is 24 sq.cm. Since B's linear dimension is twice that of A, it follows that B's surface area is 4 times that of A. Therefore, we expect B to lose heat 4 times faster than A, but it doesn't happen so when A & B are not flats but solids.

Let us consider volumes also now. A's volume is 1x1x1=1Cu.Cm., whereas B's volume is 2x2x2=8 Cu.Cm (=8 times that of A). Consider now surface to volume ratios. A's ratio is 6/1=**6** and B's ratio is 24/ 8 = **3**. As the heat losses are proportional to their respective surface/ volume ratios, A loses heat twice faster than B. by A. (If heat production or metabolic rate in animals follows the same rule, smaller bodied animals are supposed to lose metabolic heat faster than bigger bodied animals).

[15] B.L. Strehler, J.D. Ebert, H.B. Glass, and N.W. Shock, ed. *The Biology of Aging*. American Institute of Biological Sciences, Washington D.C., 1960, pp14-29.

[16] Om, bhadram karnebhihi śrunuyāma devā
bhadram paśeyma ak□abhiryajatarā
sthirairangai tuuvāmaha tanubhir
vyasomhema devahitam yadāyuh
(---Invocation at the beginning of *Mundaka upaniṣat*)

Chapter 6 - Battles to beat aging

[17] W. French. Anderson, "Gene Therapy", *Scientific American*. Sept. 1995, pp124-128.
[18] M.R. Klass., *Mech. Aging De v*,.22, 279-286 (1983).
[19] D. B. Friedman. and T.E Johnson., *Genetics* 118, 75(1988).
[20] *New York Times*, January 13, 1998.
[21] *Nature*, February 17, 2000, quoted in 'News- of- the- week' section of Science, February 18, 2000, vol.287, pg. 1181.
[22] Feature article report, *New York Times*, May 4, 2000.
[23] L Hayflick, and P Moorhead (1961), *Exp.Cell.Res*., 25, pp 585-621.
[24] L Hayflick (1965), "The limited 'in vitro' life time of human diploid cell strains", *Exp. Cell. Res*.37:pp 614-636.
[25]. G.M. Martin, C.A. Sprague, and C.J. Epstein., *Lab Invest*. 23, p86 (1970)
[26] J.G. Rheinwold, and M. Green., C*ell* 6, p331(1975).
[27] As per previous note
[28] S J. Goldstein, *Invest Dermatology*., 73, p19(1979).
[29] T.H. Norwood, H Hoehn, D Salk., and G. M. Martin, Ibid, 73, p92(1979).

[30] E E. Medrano, et al., *Mol. Biol. Cell* 5, 497 (1994).
[31] S Goldstein, Exp.Cell *Res.* 83, 297 (1974).
[32] A. M. Olovnikov, (1971), *Doklady Biochem.* (English Trans.), 201, 394-397
[33]Richard C Allsop, et al, *Proc. Natl. Academy of Sciences, USA*, vol. 89, pp10114-10118, Nov 1992.
[34]Robert P Lanza, et al, "Extension of cell life-span and telomere length in Animals cloned from senescent somatic cells." *Science*, vol.288, 28 April, 2000, pp 665-668.

[35] Gustavo Barja and Asuncion Herrero. "Oxidative Damage to mitochondrial DNA is inversely related to maximum life span," The *FASEB Journal*, vol.14, February, 2000, pp312-318.

[36] Ron Hart, and Richard Setlow, "Correlation between DNA Excision Repair and Life span in a number of Mammalian Species," *Proc. Natl l Acad. Sc. , USA*, 71:2169-2173 (1974).

[37] Simon Melov, et al., "Extension of Life-Span with super oxide Dismutase/Catalase mimetics." *Science* (2000), vol.289, pp1567-1569.

[38] Rubner M., "Das problem der Lebensdauer and seine Beziehungen sum Wachustumund Emahung, Oldenburg, Munich", quoted by Van Voorhies in "Metabolism and Life span", *Experimental gerontology* 36 (2001) 55-64.
[39] Pearl R., "*The Rate of Living*" (1928), Knopf, New York.
[40] D Harman., 1956, ":Ageing: A theory based on free radical and radiation chemistry".

[41] K.B Beckman, B.N Ames,1998. "The free radical theory of ageing matures.", *Physiol.Rev*.78, 547-581.

[42] Gordon Rattray Taylor, "*The Biological Time Bomb*", p172, The world Publishing Co, Cleveland, OH.
[43] Report, *Scientific American,* summer 2000, vol.11, No.2, pp66-67
[44] C Moreschi, "The Connection between nutrition and tumor promotion" *Z.Immunitatsforsch.* 2:651-675
[45] P Rous. The influence of diet on transplanted and spontaneous mouse tumors, J. *Exp. Med.* 20: 433-451.
[46] C. M. McCay., M.F. Crowell, and L. A. Maynard, "The effect of retarded growth upon the length of the life span and upon the ultimate body size.", J. Nutrition, 10: 63 -79.
[47] A Tannenbaum, 1940, "The initiation and growth of tumors.- Introduction. 1.Effects of under-feeding", *Am J. Cancer*, (38): 335-350
[48]------1942a, "The genesis and growth of tumors.2. Effects of Caloric Restriction per se", *Cancer Res.* 2:460-467
[49]------1942b, " The genesis and growth of tumors.3 Effects of a high fat diet", *Cancer Res.*2: 468-475.
[50]------1945a. "The dependence of tumor formation on the degree of caloric restriction", *Cancer Res.*5 (11): 609-615.
[51]------1945b. "The dependence of tumor formation on the composition of on the calorie restricted diet as well as on the degree of restriction," *Cancer Res.* 5(11): 616-625

[52] D. G Jose, J.S. Welch, and R. L. Doherty, 1969, " Humoral and cellular immune responses to streptococci, influenza and other antigens in Australian aboriginal school children", Austr. Pediatr. J. 5:209-218.

[53] R.A Good, D.G. Jose, W. C. Cooper, G. Fernandez, T. Kramer, and E.J. Yunis, 1977. "Influence of Nutrition on antibody production and cellular immune responses in man, rats, mice, and guinea pigs". *In Suskind*, Ed. Malnutrition and Immune response. Raven, New York.

[54] S.A West. and G Fernandes, 1980, "Nutritional modulation of immune responses. "Fedn.Proc" 9:3089-3104

[55] Ktesis, Ancient India.. trs: J..W. McCrindle, pgs. 18,25,61.

[56] William Mair, Patrick Goymer, Scott D. Pletcher, and Linda partridge, "Demography of dietary restriction and death in drosophila,", *Science* vol301, 19 Sept. 2003

[57] Su-Ju Lin et al., *Science* vol.289, 22 Sept.2000, pp2126-2128

[58] "Compact Dexterous Robotic Hand," U.S. patent No.6, 244, 644 (12th June 2001.)

[59] Larry Thompson, Washington post columnist, Feature article published in *Houston Chronicle* dated March 7,1988.

[60] Taylor, Gordon Rattray, *The Biological Time Bomb*,p120, The world Publishing Co., Cleveland, OH.,

[61] M.Thallmair et al., *Nature, Neurosc.* 1, 124 (1998)

[62] O. Raineteau, et al., *Proc. Natl. Acad.Sci, U.S.A.* 98,6929-6934(2001)

[63] F.Oberpenning, J. Meng. J.J. Yoo, and A.Atala. *Nature Biotechnology*. 17,149-155(1999).

[64] L.Griffith, B.Wu, M.J.Cima, B.Chaignaud, and J.P.Vacanti, *Annals of New York Acad. of Sci.* 831,382(1997)

[65] S.C.Strom, *Sem. Liver Dis.* 19, 39(1999).

[66] E.A. Ryan et al.,*Diabetes* 50,710-719(2001).

[67] Advanced Cell Technology and Aastrom Biosciences are two of such companies.

Chapter 7 – Appraisal of anti-aging theories

[68] Kael Linhard, Ronald A. Depinho et. al, "Inhibition of Experimental Liver Cirrhosis in Mice by Telomerase gene delivery", *Science* vol,287 18 February 2000, pp 1253-1262.

[69] Robin Cook, *Chromosome 6*, G. P. Putnam & Sons, New York.1997.

Chapter 8 - Consequences of long-living-societies

[70] Mary Barlove, "The food crisis: The shortages may put the 'Have nots' against the 'haves' " *Wall Street Jl.*, Oct 2, 1974, pg20.

[71] Frances Moore Lappe,, *Diet for a Small Planet*, (New York: Ballantine Books 1982) , pg. 69.

[72] The President's Science Advisory Committee, "The World Food Problem", A Report, Vol. II, May 1967

[73] As per note 71.

[74] Boyce Resenberger, "World Food Crisis: Basic Ways of Life Face Upheaval from Chronic Shortages", *New York Times*, Nov 5,1974, Pg14.

[75] As per note 71.
[76] Robin Hur, and David Fields, "Are High-Fat Diets Killing Our Forests?" *Vegetarian Times*, Feb '94.
[77] Ibid.
[78] *Acres, USA*, Missouri, vol. 15, No.6, quoted by John Robbins in "*A diet for a new America*".
[79] Brune Williams, State Conservationist, Soil Conservation Service. Des Moines, Iowa, "Testimony before Senate Committee on Agriculture and Forestry," July 6, 1976.
[80] King Seth, "Iowa Rain and Wind Deplete Farm Lands", *New York Times*, December 5, 1976, pg.61.
[81] Curtis Harnack, "In Plymouth County, Iowa, The Rich Top Soils Going Fast, Alas!" *New York Times*, July 11, 1980.
[82] Ibid.
[83] Robin Hur and David Fields, as per note 76.
[84] *Lappe, as per note 71.*
[85] Altshul, quoted by Lappe, as per note 71.
[86] John Robbins, A *Diet for a new America*, pg.372.
[87] David Pimental,, " Energy and land constraints in food protein production", *Science*, Nov 21, 1975, pg. 754.
[88] H.A Jaseorwski., "Intensive systems of animal production", *Proceedings of the III world conference on Animal production,* ed. R.L.Reid, Sydney, Sydney university press. 1975, pg. 384
[89] Jackie Robbins, "Environmental Protection Technology Service", Cincinnati, USPEA, Office of the research and development, Environmental Research Information Center, Feb 1978, Pg. 9.
[90] Report, *News week,* Nov 8, 1971, pg. 85
[91] Georg Borgstrum, cited by Lappe in '*Diet for a small planet,*' 1975 Ed, pg. 22
[92] Ibid.
[93] G.M Grossman, and Krueger, A.B., *U.S.Free Trade Agreement*, ed. by P.Garber, M.I.T.Press, Cambridge, MA, 1993, Pp. 165-177—Cited By Kenneth Arrow et al. *Science* magazine, vol. 268,pg. 520.
[94] Shaik, N, and Bandopadhy S. "Economic Growth and Environmental Quality: "Time Series And Cross Country Evidence", Background Paper for *The World Development Report*, *World Bank*, Washington, D.C., 1992.
[95] Lucas Reijnders and Sam Soret, "Quantification of the environmental impact of different dietary protein choices." *American Jl of clinical Nnutrition* 2003; 78 (suppl). 664s - 8s.

Chapter 9 -The Role of diet in longevity Planning

[96] The National Research Council, "*Recommended dietary allowances*", 9th ed., Washington, D. C
[97] News item, '*The Blade*', 19th Oct. 2003

[98] Mary Barlove, "The food crisis: The shortages may put the 'Have nots' against the 'haves " *Wall Street Jl.*, Oct 2, 1974, pg20.

[99] Frances Moore Lappe, "*Diet for a small planet*, " Ballantine Books, New York, 1982, pg. 69.

[100] President's Science Advisory Committee, "The World Food Problem", Report, Vol. II, May 1967.

[101] Lappe, as per note 99.

[102] Boyce Resenberger, "World Food Crisis: Basic Ways of Life Face Upheaval from Chronic Shortages", *New York Times*, Nov 5,1974, Pg14.

[103] *Lappe, as per note 99.*

[104] Council on foods and nutrition, "Diet and Coronary Heart Disease", *Journal of American Medical Association*, 222 (13) 1972, pg. 1647.

[105] Board of scientific consultants of The American Heart Foundation, "Position statement on diet and coronary Heart Disease", *Preventive Medicine 1:253-286,1972.*

[106] National Research Council, "*Diet and Health Implications for reducing Chronic Disease Risk*", (National Academy Press, Washington, D.C). (1989)

[107] Walter C. Willet, American Cancer Society, American Institute for cancer research, quoted by *Newsweek*, November 30,1998.

[108] Joan Sabate, *Am Jl of clinical nutrition,* 2003; 78 (suppl); 502s-7s

[109] Barnard Neal D. et al, *Preventive Medicine* 24, 646- 655,1995.

[110] Irving Fisher, "The Influence of flesh eating on endurance", *Yale medical Journal*, 13(5): 205-221,1907

[111] Ibid

[112] Ioteyko, J., et al, Enquete scientifique sur les vegetariens de Bruxelles, Henri Lamertin, Brussels, pg.50.

[113] Schouteden, A., "*Ann de Soc. Des science Med at Nat. de Bruxelles (Belgium)*", quoted by John Robbins in 'A diet for a new America', pg.158

[114] Feature report, *Energy Times*, October 1996.

[115] Richard Passwater, *Life Extension*, 1995 vol. 1, No.6.

[116] James Privitera, M.D., *Journal of longevity research* 1995, vol.1, No.6.

Chapter 10 - Prospects of extending longevity

[117] S.J. Olshansky, B.A Carnes., and C Cassel., *Science* 250,634 (1990).

[118] S.J.Olshansky, Bruce A, Carnes, Aline Desesquelles, *Science* vol. 291. Feb 2001 pp1481-82.

[119] B L. Strehler and A S. Mildvan (1960), "General theory of mortality and aging." *Science* 132: 14-19.

[120] Bruce A Carnes, S. Jay Olshansky, and Grahn, Douglas (2004), Biological evidence for limits to the duration of life, Biogerontology 4(1): 31-45.

Chapter 11 – Intimations of immortality

[121] J O. Holloszy and W.H. Kohrt. "Exercise." In: Mason E.J., ed. *Handbook of physiology: Aging*, (New York, NY: Oxford University Press) 1995).

[122] The seven dhatus mentioned in the *ayurveda* are chile (rasa), blood {*rakta*), flesh (*mamsa*), fat, (*meda*). Bones (*asthi*), morrow (*majja*), and semen (*sukra*)

[123] The three humors or doshas are called *sleshma, vata* and *pitta*

[124] "*Atamate vatak*", Rg Veda VII.87.2

[125] New England Journal of Medicine, June 1985

[126] New England Journal of Medicine, Fall 1991

[127] *Isavasya Upanishad*, 1-1, Cp: Patanjali yoga sutras: "Samatvam Yoga Uchyate

[128] Yogschittavritti nirodham

[129] One of the best books available in English language on the topic is "*Light on Yoga*" by B.K.S. Iyengar, J.J. Evans and D. Abrams, 2005.

[130] Pracchardana vidhāranābhyām vā prānasya, ---(Yoga Sutras of Patanjali)

[131] Sreyaśca preyaśca manushyametah | Tausamparītya vivinakti dhīrah
Sreyohi dhīrobhi preyaso vrunute | Preyo mando yogakśemat vrunute||

--Katha Upanishat 2-2

[132] Total number of deaths due to medical errors/deaths due to a single air crash = 98000 / 268 = 366 approx.

Chapter 12 - Conclusion

[133] W. Grey Walter, "*The living Brain*', pg2

[134] Julian, Huxley, "The Evolutionary Vision – The Convocation address," pp. 249-261, in "*Evolution after Darwin*," Vol III, Issues in Evolution, The University of Chicago Centennial Discussions, University of Chicago Press, Chicago, 1960.

[135] Ibid, vol III,pg 262.

[136] Fritzof Capra, "*The Tao of physics*". P25.

[137] Ibid.

[138] yatova imāni bhootāni jāyante, yena jātāni jīvanti, |
yatprayantyabhisamviśanti, tad vijñansasva, tad brahmeti.

INDEX

Made in the USA
Monee, IL
11 November 2026

68342111R00243